Smart Materials in Mechanical Engineering

Smart Materials in Mechanical Engineering: Innovation and Applications serves as a comprehensive educational resource that bridges traditional mechanical engineering principles with contemporary smart technology applications. This essential guide explores how artificial intelligence, automation, and advanced materials are revolutionizing mechanical engineering across the manufacturing, transportation, energy, and healthcare industries.

Readers gain access to a complete framework for transforming mechanical engineering through smart materials, IoT integration, and AI-driven systems. Content builds progressively from fundamental concepts to advanced topics, including machine learning algorithms, digital twins, and adaptive systems. Each chapter features learning objectives, worked examples, and end-of-chapter problems to support both theoretical understanding and practical application. Detailed case studies examine real-world applications, such as autonomous vehicles, smart prosthetics, and self-healing materials, while addressing critical challenges including cost analysis, security considerations, and ethical implications. Coverage highlights future trends in sustainable mechanical innovation and explores emerging applications across key industry sectors.

This book is designed for undergraduate and graduate students in mechanical, industrial, and materials engineering programs, as well as engineering professionals pursuing continuing education in smart technologies. The structured content supports both classroom instruction and self-directed learning, making it ideal for integrating smart technologies into traditional mechanical engineering practice.
Comprehensive practice problems, extended case studies, and practical applications appear throughout. A complete solutions manual is available for qualified textbook adoption to support effective course delivery and student engagement.

Smart Materials in Mechanical Engineering

Innovation and Applications

Riyadh A. Al-Samarai and Yarub Al-Douri

CRC Press
Taylor & Francis Group
Boca Raton London New York

CRC Press is an imprint of the
Taylor & Francis Group, an **informa** business

First edition published 2026
by CRC Press
2385 NW Executive Center Drive, Suite 320, Boca Raton, FL 33431

and by CRC Press
4 Park Square, Milton Park, Abingdon, Oxon, OX14 4RN

CRC Press is an imprint of Taylor & Francis Group, LLC

ISBN: 978-1-041-27035-5 (hbk)
ISBN: 978-1-041-28273-0 (pbk)
ISBN: 978-1-003-75855-6 (ebk)

DOI: 10.1201/9781003758556

Typeset in Times
by Newgen Publishing UK

Access the Instructor and Student Resources: www.routledge.com/9781041270355

Contents

Preface

The integration of new materials, automation, and artificial intelligence into traditional mechanical systems has radically transformed mechanical engineering in recent decades. The design, analysis, and operation of mechanical systems have been radically transformed by the advent of smart technologies, giving rise to a new, interdisciplinary field called smart mechanical engineering. This book, *Smart Materials in Mechanical Engineering: Innovation and Applications*, aims to provide a comprehensive reference that bridges contemporary smart systems with classical mechanical concepts. This effort stems from the growing need for energy-efficient, self-optimizing, adaptive mechanical systems capable of responding to changing environmental conditions. By integrating sensors, actuators, artificial intelligence, and data-driven control into traditional mechanical systems, smart mechanical technologies are revolutionizing sectors such as manufacturing, transportation, energy, and healthcare. These developments support the broader goals of sustainability and digital transformation across various engineering fields, while improving system performance and reliability.

This book is designed to provide readers with a theoretical understanding of the rapidly evolving field of smart mechanical systems, along with practical insights. It begins with fundamental ideas that define the relationship between smart control, materials science, and mechanical design. Subsequent chapters discuss the role of smart materials, including magnetorheological fluids, piezoelectric composites, and shape memory alloys, in enabling adaptive mechanical behavior. To illustrate the shift from static to dynamic engineering paradigms, the book details the integration of digital twins, machine learning algorithms, and Internet of Things (IoT) technologies within mechanical systems. The applications section provides an overview of current uses of smart mechanical systems in a range of fields. Examples include autonomous vehicles with real-time feedback control, sensor-monitored precision manufacturing, biomechanically adaptable smart prosthetics, and resource-efficient smart energy systems. Each example highlights how information technology and mechanical design can combine to improve the utility and efficiency of mechanical systems. This book also discusses the challenges associated with these developments from a research and development perspective. Engineering discussions continue to focus on issues such as cost-effectiveness, material constraints, data security, and system compatibility. The ethical and social implications of autonomous mechanical systems must also be considered in future design and deployment approaches as artificial intelligence advances. Policymakers, computer scientists, materials researchers, and mechanical engineers must work together to address these complex issues. This book addresses contemporary technology, examining emerging trends that will influence the future development of intelligent mechanical engineering. Artificial intelligence, robotics, and nanotechnology are likely to combine to create systems with unprecedented precision and flexibility. The future of mechanical innovation lies in the creation of self-healing materials, self-learning devices, and hybrid cyber–physical systems. Furthermore, sustainability-focused engineering will foster the development of intelligent systems that improve durability and performance while reducing their negative environmental impact.

Students, researchers, and professionals working in applied physics, materials science, and mechanical and industrial engineering are among the diverse audiences targeted by this work. Practitioners seeking to innovate new intelligent products and processes or integrate intelligent technologies into traditional mechanical systems will also find this book useful. The book seeks to stimulate creativity and critical thinking in addressing contemporary mechanical engineering challenges by combining academic rigor with practical experience.

This book, *Smart Materials in Mechanical Engineering: Innovation and Applications,* highlights the transformation of mechanical systems into intelligent, interconnected, and sustainable entities. The book summarizes how engineering has shifted from an exclusive focus on mechanical design to a comprehensive, data-driven, and resilient field. This contribution aims to serve as a reference for the next generation of academics and engineers committed to influencing smart mechanical technologies.

This section of the book is divided into 10 in-depth chapters, each addressing a different but interconnected aspect of smart materials and their technological uses. Chapter 1 introduces advanced materials and their role in Industry 4.0, emphasizing the effective use of the Preferred Selection Index (PSI) approach to identifying the best material choices for contemporary engineering systems. Chapter 2 examines the mechanical and physical properties of advanced materials, focusing on how these essential attributes impact performance, reliability, and resilience in industrial environments. Chapter 3 addresses the processing and manufacturing techniques needed to create advanced and useful materials with organized microstructures and enhanced properties. Chapter 4 introduces the fundamentals of nanomaterials, providing an understanding of nanoscale phenomena and the influence of surface, size, and shape effects on mechanical and physical behavior. Chapter 5 compares the synthesis processes, fundamental properties, and potential technological uses of metallic, ceramic, and polymeric nanomaterials.

Chapter 6 explores the industrial applications of nanomaterials, highlighting their impact on a range of industries, including manufacturing, energy, and healthcare. Chapter 7 addresses the principles of composite materials, including design techniques, structural bonding, and properties necessary to achieve customized multifunctional performance. Focusing on their advanced roles in optoelectronic, catalytic, and energy harvesting applications, Chapter 8 explores perovskite composites: structure, modification, and advanced properties. Chapter 9 examines the applications of self-healing polymer nanocomposites in automotive and aerospace, delving into smart material systems that improve durability, efficiency, and safety under harsh operating conditions. Finally, Chapter 10 discusses the electronics and energy storage applications of self-healing polymer nanocomposites, summarizing current developments and potential paths forward in developing smart materials for advanced electronic devices and sustainable energy systems.

Riyadh A. Al-Samarai
University of Samarra, Iraq
Yarub Al-Douri
Al-Bayan University, Iraq
Piri Reis University, Turkey
University of Malaya, Malaysia

Authors

Riyadh A. Al-Samarai is the Assistant Dean for Scientific Affairs, Faculty of Engineering, at the University of Samarra, Iraq. Prof. Al-Samari has more than three decades of academic and professional experience, including 32 years of research activity, 23 years of university teaching, and 16 years in academic leadership and administrative positions. Prof. Al-Samarai played a pivotal role in establishing the Bachelor's degree programs in the College of Engineering at Samarra University, as well as in launching the MSc program in Electromechanical Engineering at the same institution. In recognition of his outstanding scientific and academic contributions, he was awarded the Iraqi Scientific Excellence Award in 2024 and was honored by the Iraqi Minister of Higher Education and Scientific Research for authoring the Best Engineering Book among Iraqi Universities for the year 2024. His honors and awards also include winning the Creativity Award for Scientists in Malaysia, the Iraqi Forum for Intellectuals and Academics (IFIA) Award, and the Best Research Paper Award at the Conference of Asian Universities in Malaysia. He is the managing editor of the *Samarra University of Engineering and Science Journal*, with research interests including advanced manufacturing, applied mechanics, materials engineering, and electromechanical systems.

Yarub Al-Douri is a Professor at Al-Bayan University, Iraq; Piri Reis University, Turkey; and University of Malaya, Malaysia. He has 31 years of research experience, 25 years of university teaching experience, 12 years of administrative experience and more than 860 publications. Dr. Al-Douri has supervised many postgraduate students, reviewed for 34 peer-reviewed journals, and organized and chaired international conferences. He is a Fellow of the European Academy of Sciences, the Islamic World Academy of Sciences, and the Arab Thought Forum, as well as an IETI Distinguished Fellow. His awards include the NCPD International Prize (2025), the Khalifa International Award for Date Palm and Agricultural Innovation (2024), the ICTP-Arab Fund Associateships (2024–2026), and the IAAM Scientist Award (2022). Dr. Al-Douri is known for his contributions to nanotechnology and renewable energy, with research interests spanning renewable energy, nanotechnology, nanoelectronics, nanomaterials, modeling and simulation, semiconductors, and optical studies.

1 Introduction to Advanced Materials and Their Role in Industry 4.0

1.1 INTRODUCTION

Materials science has always been fundamental to the advancement of technology. The pace of industrial development has been determined by the discovery and use of new materials, from the Bronze Age to the Silicon Age. However, the need for stronger, lighter, smarter, and more sustainable materials has increased significantly in the 21st century. From this need has sprung the concept of advanced materials engineered materials designed to exhibit specific mechanical, physical, and chemical properties that surpass those of conventional metals, ceramics, or polymers [1–4]. Nanomaterials, composites, high-entropy alloys (HEAs), functional biomaterials, and smart materials that respond to external stimuli are just a few examples of the many categories that fall under the umbrella of advanced materials. The design concept that connects these disparate categories is that they are engineered at multiple scales from the atomic to the macro level to meet specific industrial and societal needs, rather than relying solely on natural properties.

Advanced materials have several key properties that make them essential to contemporary engineering: High strength-to-weight ratios, toughness, and corrosion resistance are examples of superior mechanical performance. Thermal and chemical stability are examples of resistance to oxidation, corrosion, and harsh environments. Versatility is the ability to combine structural and functional roles (such as load-bearing composite materials with self-sensing capabilities). Scalability represents the potential for large-scale industrial production using modern manufacturing technologies such as powder metallurgy and additive manufacturing [2,5]. For example, carbon fiber-reinforced polymers (CFRPs) enable fuel-efficient cars and aircraft, as well as being stronger than steel. Similarly, high-energy-efficient (HEA) composites exhibit exceptional mechanical stability and high-temperature stability, which were previously thought to be mutually exclusive. Although Industry 4.0 focuses on automation and the integration of digital technologies, as shown in Figure 1.1, it tends to prioritize machine efficiency over human engagement. One of the most significant drawbacks of Industry 4.0 is the potential dehumanization of workers. Robots perform tasks such as assembly in highly automated environments, such as the automotive industry, but this can reduce employment opportunities for humans. Industry 5.0 addresses this limitation by prioritizing human–machine collaboration.

DOI: 10.1201/9781003758556-1

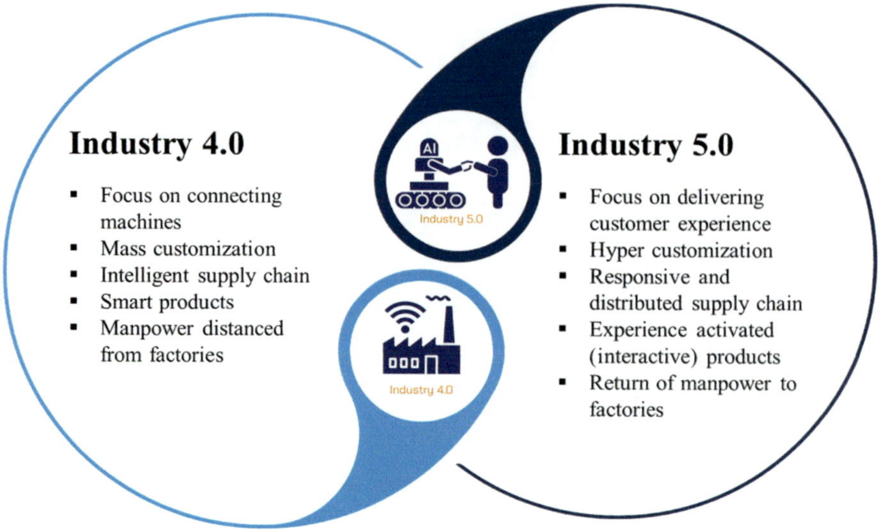

FIGURE 1.1 Industry 4.0 and Industry 5.0 highlights [From Ref. 1] [Open access].

Smart manufacturing, cyber–physical systems, and the integration of digital technology are hallmarks of the Fourth Industrial Revolution, also known as Industry 4.0. Advanced materials are the hidden enablers of this change, but digital technologies such as automation, robotics, and artificial intelligence (AI) are receiving the most attention. The full potential of Industry 4.0 cannot be realized without materials capable of operating in multiple ways, operating in harsh environments, and resisting higher stresses [6]. Examples of this synergy include nanomaterials that enable the manufacture of sensors, wearable electronics, and miniaturized medical devices; smart vehicles used to monitor the integrity of bridge infrastructure, aircraft wings, and industrial equipment; HEA and ultra-high-temperature materials that support next-generation turbines and space propulsion systems; and lightweight materials used in energy-efficient drones and autonomous vehicles [1].

Digitizing materials research is another aspect of Industry 4.0. Historically, the process of developing new materials involved a laborious cycle of incremental improvement and trial and error. This paradigm is currently being reshaped by advances in AI, machine learning (ML), and materials informatics. Predictive modeling, computer simulations, and large datasets are accelerating the production of new materials, allowing for "designed materials" rather than "random materials."

This computational approach has significantly reduced the time it takes from concept to implementation. To eliminate costly and time-consuming testing campaigns, ML systems can, for example, predict the mechanical behavior of high-performance energy composites (HEAs) based on composition and processing factors. Similarly, before building actual prototypes, engineers can test performance under simulated settings using digital twins, which are virtual representations of materials and components [1–4].

Beyond technical performance, innovative materials are critical because they have a direct impact on economic competitiveness, sustainability, and human well-being. Advanced materials are essential for wind turbine blades, fuel cells, solar cells, and hydrogen storage devices in the energy sector. Biocompatible metal alloys and nanocoatings are revolutionizing tissue engineering and implants in the biomedical field. Lightweight composites and high-temperature alloys in the aerospace and defense industries make vehicles faster, safer, and more efficient.

Moreover, the pressing issue of sustainability requires the use of modern materials. They contribute to the global transition to a resource-efficient, low-carbon economy by facilitating recycling, durability, and energy efficiency. In the future, collaboration between Industry 4.0 technologies and advanced materials is expected to intensify. This future is shaped by three key trends: Integration of AI and materials design—using predictive models to accelerate discovery; development of multifunctional materials—combining sensory, electrical, and structural properties into a single material system; and designing materials with a low environmental impact and the ability to be recycled is the third circular economy technology. New materials are transforming the way engineers innovate, produce, and apply products across various industries. These materials represent a cornerstone of Industry 4.0 because they offer exceptional performance and facilitate the implementation of smart, digital, and sustainable technologies. In addition to highlighting the revolutionary potential of advanced materials, this introduction paves the way for in-depth studies of high-HEAs, composites, nanomaterials, and their multiple engineering applications [2–4].

1.2 DEFINITION AND CLASSIFICATION OF ADVANCED MATERIALS

The term "advanced materials" refers to a new class of engineering materials that surpass the performance of conventional metals, ceramics, and polymers. Unlike conventional materials, which are widely used in their natural or molded forms with minor modifications, advanced materials are specifically designed and engineered at multiple levels, from the atomic level to the macrostructure, to achieve specific mechanical, physical, chemical, or functional properties. The growing demands of contemporary engineering fields, such as aerospace, automotive, energy, electronics, and biomedical applications, where conventional materials can no longer meet the criteria of lightweight design, high efficiency, and multifunctionality, are driving their development. Therefore, a working definition is: Advanced materials are engineering systems that enable innovative solutions in Industry 4.0 and beyond by delivering superior mechanical, thermal, electrical, or functional performance through controlled microstructure, novel composition, or intelligent response.

1.2.1 Classification of Advanced Materials

Advanced materials fall into several main categories, despite the breadth and continuous development of this field: [Verified]

1. *Nanomaterials:* Nanoparticles, nanowires, nanotubes, nanofilms, and nanocomposites are just a few examples. These materials exhibit exceptional strength, reactivity, and electrical properties by exploiting size effects at the nanoscale. Examples include titanium dioxide nanoparticles used in photocatalysis and carbon nanotubes, which possess exceptional tensile strength [2].
2. *Composite Materials:* Composites are made by combining two or more different phases to produce materials with complementary properties. Due to their excellent strength-to-weight ratio and corrosion resistance, fiber-reinforced polymers, metal matrix composites (MMCs), and ceramic matrix composites (CMCs) are widely used in the automotive and aerospace sectors.
3. *HEA:* This pioneering group of metallic materials, known as HEA, is composed of five or more elemental elements in approximately equal proportions. Unlike conventional alloys made of one or two elemental elements, high-entropy alloys exhibit special properties, such as lattice distortion, slow diffusion, and high compositional entropy, resulting in exceptional strength, toughness, and stability at high temperatures.
4. *Smart and Useful Materials:* Stress, temperature, magnetic fields, and electric fields are examples of external stimuli that these materials interact with. Examples include self-healing polymers, piezoelectric ceramics, and shape memory alloys (such as NiTi). Their versatility is essential for use in biomedical devices, robotics, sensors, and actuators. Table 1.1 shows intelligent and practical materials and their uses in Industry 4.0.
5. *Sustainable Materials and Biomaterials:* Advanced biomaterials used in tissue engineering and implants include biocompatible metals such as titanium alloys, naturally degradable magnesium alloys, and nanocoatings. Furthermore, within the framework of green engineering, sustainable materials with low environmental impact and recyclability are gaining increasing importance. Nanomaterials, composites, advanced bioenergy compounds, smart materials, and newly developed sustainable biomaterials all fall under the umbrella term advanced materials. Each category makes a distinct contribution to meeting the demands of Industry 4.0, where lightweight, high-performance, and versatile materials are essential. Their classification is not rigid; rather, it is changing as a result of ongoing research producing hybrid and versatile materials, making it difficult to distinguish between traditional classifications. In the context of Industry 5.0, Figure 1.2 illustrates the interplay between environmental sustainability, human-centered technologies, and additive manufacturing. As an integrative platform that connects and enhances various components, Industry 5.0 holds a crucial position. It highlights how human capabilities, environmental concern, and innovative industrial technologies can be integrated. For example, using additive manufacturing to design and print aircraft components with enhanced AI can increase flight efficiency by reducing weight to improve performance. Custom implants and prosthetics are examples of personalized medical solutions that can enhance patient outcomes in the healthcare sector. Local additive manufacturing and on-demand spare parts can improve supply chain efficiency by eliminating the need for bulk inventories and ensuring the availability of critical parts. The use of recycled materials and renewable

TABLE 1.1
Smart and Useful Materials and Their Applications in Industry 4.0

Material Type	Equation	Variables	Industry 4.0 Application
Piezoelectric	$D = d * \sigma$	D: electric displacement (C/m^2); d: piezoelectric coefficient (C/N); σ: stress (Pa)	Smart sensors, actuators, energy harvesting
Shape memory alloy (SMA)	$\varepsilon = \alpha\ \Delta T$	ε: strain; α: thermal expansion coefficient; ΔT: temperature change	Robotic grippers, adaptive components
Thermoelectric	$V = S\ \Delta T$	V: voltage (V); S: Seebeck coefficient (V/K); ΔT: temperature difference	Waste heat energy conversion, smart power systems
Magnetorheological (MR)	$\eta(H) = \eta_0 + \alpha H^n$	η(H): viscosity under magnetic field; η_0: base viscosity; H: magnetic field intensity	Adaptive dampers, brakes, robotic systems
Electrochromic	$\Delta T = f(V)$	ΔT: transmittance change, V: applied voltage	Smart windows, adaptive lighting control

Source: From Ref. 1 [Open access].

FIGURE 1.2 Relationship between AM and Industry 5.0 [From Ref. 1] [Open access].

energy sources, along with improved printing procedures and resource utilization, can lead to sustainable production. Designers can now build complex, organic shapes and internal systems that were previously impossible due to the limitations of traditional production. Groundbreaking developments are expected to revolutionize manufacturing, personalize products, and usher in a new era of efficient and sustainable production as additive manufacturing and Industry 5.0 continue to evolve [2–4].

Efficiency, specificity, and scalability are some of the factors that determine the advantages and challenges of any technology. Advances in synthesis techniques, which enable the development of new materials, drugs, and technologies, continue to stimulate innovation. The accuracy and utility of synthesis in solving contemporary scientific problems are expected to increase with the continued integration of diverse technologies. The production of nanomaterials and nanostructures is a fundamental component of nanoscience and nanotechnology. Novel physical properties and applications of nanoparticles will only be possible if nanomaterials of appropriate size, shape, crystal structure, and chemical composition are available. Advances in science and engineering have driven the development of nanoparticles to achieve unique properties that distinguish them from bulk materials [1]. These synthetic techniques cause the dispersion of liquid particles and nanomaterials. Depending on the intended use, physical, chemical, biological, and hybrid methods are usually chosen to produce nanomaterials [2,3]. These techniques fall under the category of conventional methodologies and are widely used.

1.3 KEY PROPERTIES AND ADVANTAGES OVER CONVENTIONAL MATERIALS

Unlike conventional metals, ceramics, and polymers, advanced materials boast enhanced properties that overcome the drawbacks of conventional materials in contemporary engineering applications. To enable stronger, lighter, and more versatile solutions than ever before, these materials are designed to deliver exceptional performance in mechanical, thermal, chemical, and functional areas. The demands of the Fourth Industrial Revolution and smart manufacturing are closely linked to their development, where high efficiency, reliability, and multifunctionality are critical. Several recycling strategies (i.e., reuse, reduce, recycle, refurbish, remanufacture, etc.) should be considered to address circular thinking regarding the challenging end-of-life potential of composite materials (Figure 1.3).

1.3.1 Mechanical Properties

Advanced materials often exhibit significantly better mechanical properties. These properties include exceptional fatigue resistance, enhanced toughness, greater stiffness, and a higher strength-to-weight ratio [2–4]. For example, carbon fiber-reinforced composites are ideal for automotive and aerospace applications due to their combination of low density and high tensile strength. Similarly, even at high

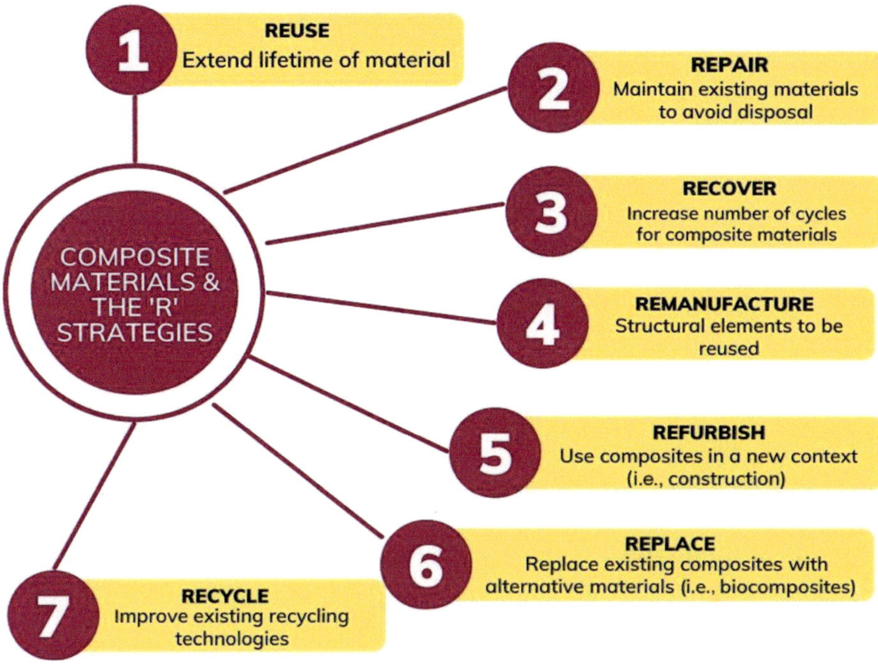

FIGURE 1.3 Applying "R" technologies in composite material waste management [From Ref. 5] [Open access].

temperatures, high-entropy alloys (HEAs) exhibit exceptional strength and resilience, outperforming conventional alloys under challenging conditions.

1.3.2 Chemical and Thermodynamic Stability

Many advanced materials are designed to withstand harsh chemical and thermal environments. For example, CMCs are suitable for turbine engines and high-temperature industrial processes because they retain their mechanical integrity at temperatures where metals would otherwise be damaged. Furthermore, materials such as nanocoatings and corrosion-resistant alloys protect structures from chemical damage, increasing their lifespan and reducing maintenance costs.

1.3.3 Smart and Functional Properties

The functional response of many modern materials is one of their defining characteristics. Shape-memory alloys, piezoelectric ceramics, and magneto-rheological fluids are examples of smart materials that react dynamically to mechanical stress, temperature, and magnetic fields. This enables engineers to create adaptive systems that are not possible with conventional materials, such as vibration dampers, self-healing parts, and sensitive sensors.

1.3.4 Multipurpose and Lightweight Design

Advanced materials enable weight reduction without compromising structural integrity. This is critical in sectors where performance and energy efficiency are critical. Furthermore, many of these materials are multifunctional, combining properties such as heat control or electrical conductivity with structural strength. For example, nanocomposites can simultaneously enhance thermal conductivity and mechanical durability, providing integrated solutions for aerospace and electronics applications.

Quantitative data and reports are needed to study the environmental, social, and economic issues related to recycling, as well as to track the emergence of a new mindset, to facilitate the smooth monitoring and reporting of any action taken to promote the circular economy concept in relation to vehicles [3]. Key performance indicators (KPIs), end-of-waste (EWC) criteria for composite materials based on their composition, material flow analysis (MFA) from design to manufacturing and disposal, and other widely used scientific methods are the only ways to provide such data (Figure 1.4).

1.3.5 Economic and Environmental Benefits

By reducing material use, lowering energy consumption, and extending component life, the superior performance of advanced materials often translates into long-term

FIGURE 1.4 Developments in composite materials that drive the economy towards a circular economy [From Ref. 5] [Open access].

economic benefits, despite their potentially higher initial costs. Furthermore, the production of recyclable and sustainable advanced materials supports environmental goals by reducing industrial waste and aligning with the principles of the circular economy.

New materials offer substantial advantages over conventional materials due to their key properties, including mechanical strength, thermal and chemical stability, smart performance, lightweight, and multifunctional design. These qualities enhance the efficiency, sustainability, and adaptability needed for Industry 4.0 and beyond, while also opening new avenues for creative engineering solutions. By leveraging these superior qualities, engineers can create safer, more reliable, and environmentally conscious next-generation systems [2–4].

1.3.6 Nanostructured Materials

Because they are crafted with structural properties at the nanometer scale (1–100 nm), nanomaterials are among the most promising types of advanced materials. Materials often exhibit special properties at this extremely small scale that are fundamentally different from those seen in their typical bulk form. The remarkable behavior of these materials can be primarily attributed to their high surface-to-volume ratio, interface dominance, and quantum confinement effects, all of which work together to offer performance customization options unattainable with conventional materials. These materials can be found in thin films, coatings, nanoparticles, nanotubes, nanowires, and nanocomposites, among other forms. For example, nanoparticles are widely used in catalysis and biological applications due to their high surface reactivity. Nanotubes and nanowires are examples of one-dimensional (1D) structures that combine exceptional mechanical strength with electrical conductivity, making them suitable for use in nanoelectronics and sensors. On the other hand, nanocomposites enable the integration of lightweight matrices with nanoscale reinforcements to achieve superior strength, stiffness, and toughness, while thin nanoscale films and coatings are widely used to improve the stiffness, wear, and corrosion resistance of engineering components.

Nanomaterials offer exceptional properties in a variety of fields. Because of the strengthening of nanoscale grain boundaries, they often exhibit increased mechanical stiffness, strength, and corrosion resistance. They also exhibit semiconductor behavior and tunable conductivity due to the electron confinement phenomenon, a phenomenon of critical importance in modern electronic devices. Their controlled heat transfer properties and thermal stability enable them to be used in thermal management and energy storage. Unique optical interactions between nanostructures and light enhance absorption, reflection, and emission properties, essential for photonic devices, coatings, and sensors. These materials are highly effective catalysts for chemical reactions and environmental cleanup due to their large surface area, which significantly increases chemical reactivity [3].

Nanomaterials are used in many different industries. These materials are used in lightweight structural composites, corrosion-resistant coatings, and advanced cutting tools in engineering and manufacturing. They are essential in the manufacture of batteries, sensors, transistors, and flexible displays in electronics. Examples of biomedical applications include antimicrobial surfaces, tissue engineering structures, drug delivery systems, and contrast agents in imaging. Environmental technologies use nanostructures in air purification units, water purification systems, and catalytic converters to reduce pollution, while the energy sector uses them in fuel cells, photovoltaics, hydrogen storage, and supercapacitors. Figure 1.5 shows a variety of nanocarriers used in cosmetics. This section discusses the specific properties and functions of some nanocarriers or nanomaterials used in cosmetics delivery.

However, despite these tremendous prospects, some barriers remain. Widespread acceptance is limited by concerns about stability, recyclability, and safety, and the large-scale synthesis of nanomaterials with structured properties remains difficult and expensive. Furthermore, extensive research and the development of safe handling procedures are required due to concerns about their potential impacts on the environment and human health. Future studies will focus on developing cost-effective and sustainable production processes, improving characterization methods, and integrating these materials into next-generation technologies in the environmental,

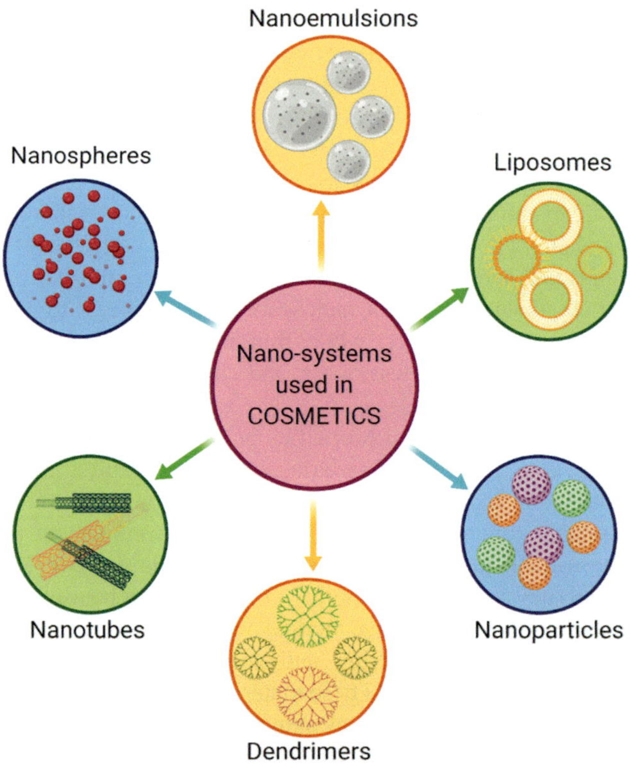

FIGURE 1.5 Cosmetics using a variety of nano-catalysts. The BioRender.com template was used to create the shape, which was exported according to the premium subscription requirements [From Ref. 6] [Open access].

biomedical, and industrial fields. Nanomaterials are expected to push the boundaries of contemporary science and engineering with further development.

1.3.7 Composites

A class of engineering materials, known as composites, is made by combining two or more different components to produce properties that are superior to those of the components alone. A composite typically consists of a dispersed phase, called the reinforcement, which imparts the material's strength, stiffness, or other functional properties, and a continuous phase, called the matrix, which binds and protects the structure. The advantage of special composites lies in their ability to combine the properties of different materials into a single system, resulting in lightweight structures with improved mechanical, thermal, and chemical performance [4].

Fibers, particles, or filaments can be used as reinforcements in composite materials, while the material itself can be polymeric, metallic, or ceramic. Due to their exceptional strength-to-weight ratio and versatility, fiber-reinforced composites such as those made of carbon, glass, or aramid fibers are the most common type. CMCs are valued for their thermal stability and ability to operate under extreme conditions, while MMCs are used when improved wear behavior and ductility at high temperatures are required. To customize the behavior of materials for highly specialized applications, hybrid composites combining multiple types of reinforcements or matrices are also produced [5]. The type, orientation, distribution, and volume fraction of reinforcement, as well as the bond strength at the matrix-reinforcement junction, significantly influence the properties of composite materials. Composite materials typically have lower density, better strength, stiffness, toughness, and fatigue resistance than conventional materials. They also offer advantages, including corrosion resistance, thermal insulation, and resilience to impact and harsh environmental conditions. Through careful design of structures at both the micro and macro scales, engineers can create composite materials with highly specialized performance to meet specific industrial needs.

Composite materials are used in virtually all areas of contemporary engineering. They are essential for producing durable, lightweight parts for the automotive and aerospace sectors, improving performance and reducing fuel consumption. Composite materials are used in civil engineering to extend the life of infrastructure through structural panels, bridge reinforcement, and repair solutions. They are also used in pipelines, pressure vessels, and wind turbine blades in the energy sector. Their versatility and adaptability are further demonstrated by their use in electrical appliances, sports equipment, defensive armor, marine structures, and biological implants [6]. Despite these advantages, composite materials suffer from some drawbacks, particularly with regard to large-scale, cost-effective manufacturing, ease of repair, and recyclability. While problems such as delamination and degradation under long-term service conditions remain concerns, complex manufacturing processes often require precision and sophisticated solutions. Furthermore, because composite materials are heterogeneous, recycling and disposal raise environmental concerns. Therefore, to improve performance while considering environmental impact, research is being conducted on the production of sustainable composite materials, improved interface engineering, and the study of nano- or bio-based reinforcements [7]. Overall, composite materials are a key component of contemporary materials engineering, offering a balance between high mechanical and functional properties and lightweight performance. Composite materials are expected to have a greater impact on the development of advanced technologies in the future, as industries continue to demand stronger, lighter, and more flexible materials.

1.4 THE ROLE OF ADVANCED MATERIALS IN INDUSTRY 4.0

By combining cyber–physical systems, digitalization, AI, and the Internet of Things (IoT), Industry 4.0 often referred to as the Fourth Industrial Revolution represents a paradigm shift in manufacturing and production systems. While digital technology is at the heart of this change, improved materials are equally important for enabling and sustaining Industry 4.0 breakthroughs. The full potential of automation, smart factories, and next-generation products will not be realized without the development of high-performance materials. Advanced materials, such as functional polymers, high-entropy materials, composites, nanoalloys, and smart coatings, provide the foundation for improved performance, durability, and adaptability. With their exceptional mechanical strength, thermal stability, corrosion resistance, and light weight all of which directly impact productivity and efficiency these materials are specifically designed to withstand harsh operating conditions. For example, nanocoatings increase the wear and corrosion resistance of precision tools, extending service life and reducing downtime, while carbon fiber-reinforced composites enable the aerospace and automotive industries to produce lighter, stronger components with lower energy consumption and emissions. Three important revolutions have shaped industry, each of which has significantly changed certain aspects of production and greatly benefited humanity and society [8]. Figure 1.6 lists the four distinct industrial revolutions documented in the literature.

Advanced materials also facilitate digital and smart systems within Industry 4.0. Shape memory alloys, self-healing polymers, and piezoelectric ceramics are examples of smart materials that offer instant response, flexibility, and integration into cyber–physical environments. These materials bridge the gap between digital control and physical structures by acting as sensors or active actuators in automated systems. Furthermore, functional materials used in electronics, energy storage, and data communications, such as graphene, conductive polymers, and solid-state electrolytes, support the infrastructure of connected devices. Applications go beyond structural and functional functions to include sustainability, as improved materials support circular economy models and greener production. Energy-efficient metals, recyclable polymers, and bio-based composites contribute to solving environmental problems while maintaining industrial competitiveness. However, the production of specialized powders, resins, and hybrid materials is critical for additive manufacturing, a key Industry 4.0 technology that enables complex engineering, mass customization, and rapid prototyping.

However, scaling up the manufacturing of advanced materials, maintaining cost-effectiveness, and addressing long-term reliability and environmental impacts remain challenging tasks. To accelerate development cycles and maximize performance, digital design, simulation, and predictive modeling must be integrated with materials innovation. Therefore, to fully leverage the potential offered by new materials in the

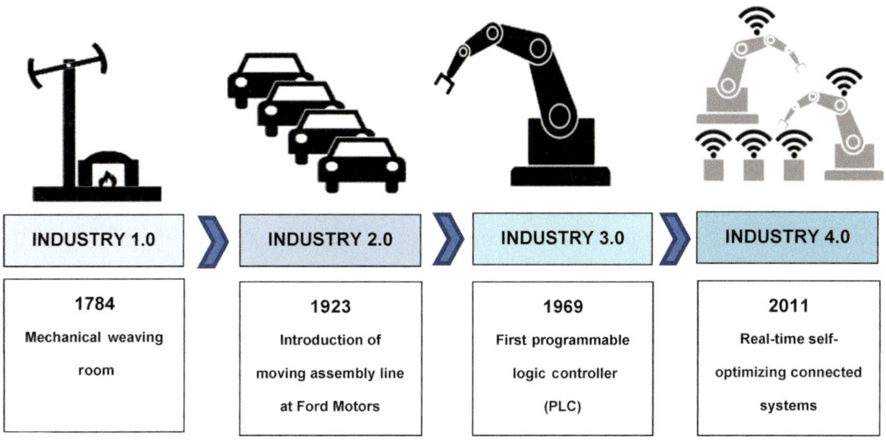

FIGURE 1.6 Revolutions in industry [From Ref. 7] [Open access].

FIGURE 1.7 Levels of the framework hierarchy [From Ref. 7] [Open access].

Industry 4.0 era, collaborative research involving materials scientists, engineers, and digital technology experts is essential.

Advanced materials actively facilitate technological developments in Industry 4.0, rather than being passive components. Their strategic role ensures that Industry 4.0 is not just a digital revolution, but also a materials revolution, reshaping the future of industrial innovation. They support the shift towards highly efficient, interconnected, and sustainable products [9]. In general, the framework hierarchy is divided into seven sections, each corresponding to a specific level. Generally, the hierarchy levels are designed using a horizontal perspective of how an organization initiates and completes its Industry 4.0 adoption [10–14]. Firmware can be organized hierarchically into several layers corresponding to different maturity levels [7]. Figure 1.7 illustrates the levels of the proposed firmware hierarchy.

To provide organizations preparing for the Industrial Revolution with a comprehensive view and to predict the pace and direction of Industry 4.0 integration, this study used trend analysis based on expert opinion to conduct a regression analysis. Figure 1.8 illustrates the S-curve evaluation.

The analysis highlighted the company's growth trajectory toward full integration of Industry 4.0 technology, as well as its current maturity level in this area based on the proposed operations management framework. Figures 1.9 and 1.10 illustrate the professional backgrounds and areas of competence of the participants. The majority of participants ranged from 11 to 20 years of experience. The study primarily considered the experience of the company's process development engineers.

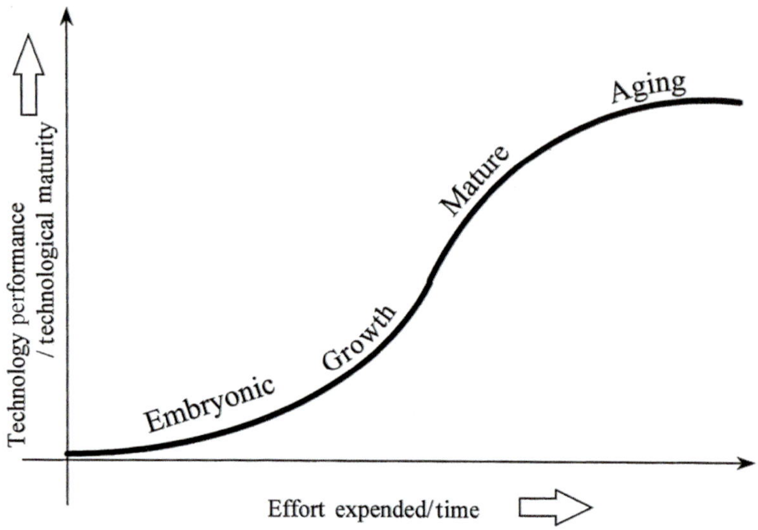

FIGURE 1.8 Technology growth S curve. [From Ref. 7] [Open access].

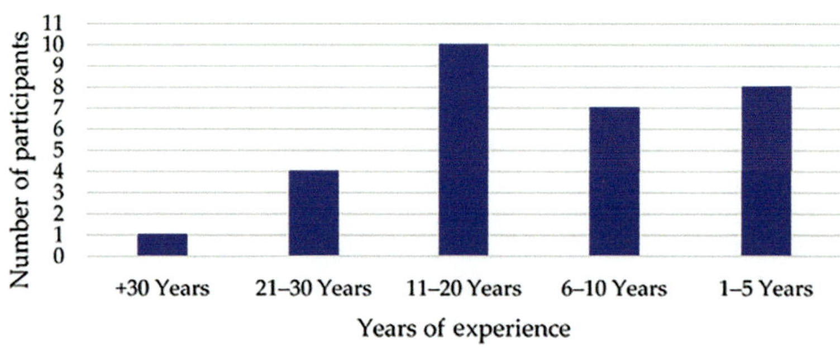

FIGURE 1.9 Work experience of participants. [From Ref. 7] [Open access].

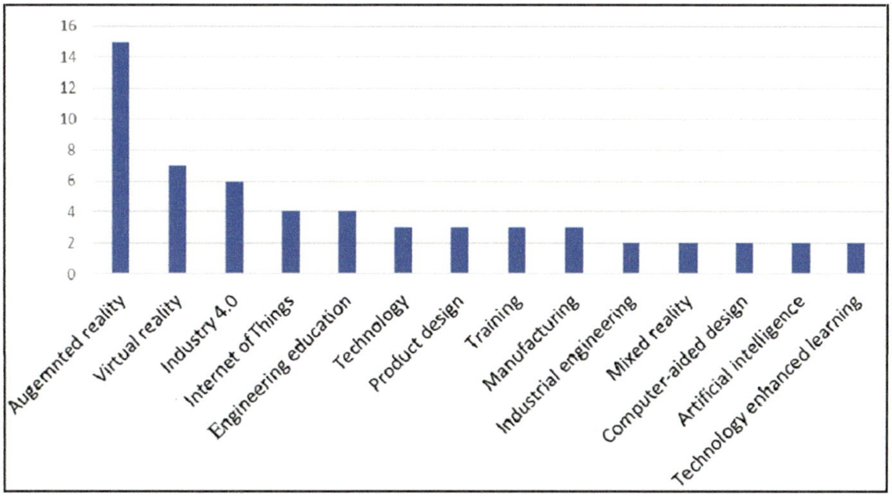

FIGURE 1.10 Expertise of the participants. [From Ref.7] [Open access].

1.5 DIGITAL TRANSFORMATION IN MATERIALS DEVELOPMENT

Modern materials science is evolving because of digital transformation, which is fundamentally changing how new materials are created, tested, and used. Materials development has traditionally relied heavily on long, iterative cycles of synthesis and testing, along with trial and error. While this approach has achieved many significant breakthroughs, it has often been expensive, time-consuming, and limited in its ability to study a wide range of potential material combinations. As a result of digital technology, the materials sector is undergoing a radical shift toward data-driven innovation, simulation-led discovery, and accelerated design pipelines in line with Industry 4.0 concepts. Computational modeling, ML, and high-throughput simulation are at the heart of this change, enabling researchers to predict the behavior and properties of materials prior to physical fabrication. Molecular dynamics, density functional theory, and finite element analysis are examples of advanced computational methods that provide powerful tools for simulating atomic interactions, mechanical performance, and environmental durability. When combined with AI and big data analytics, these technologies can uncover hidden patterns, accurately identify optimal formulations, and significantly reduce the time required to produce new materials for specific uses [15–19]. The idea of ensuring excellent quality of service (QoS) based on low latency accuracy is central to the design of the IoT architecture. High QoS supports user tasks, customer events, IT services, and latency-sensitive real-time applications in the ecosystem [20]. Figure 1.11 provides an overview of several smart technologies related to the IoT.

Digital twins, a key concept of Industry 4.0, are emerging in the materials industry. By building virtual copies of materials, structures, or processes, engineers can test multiple conditions, track performance in real time, and optimize properties without

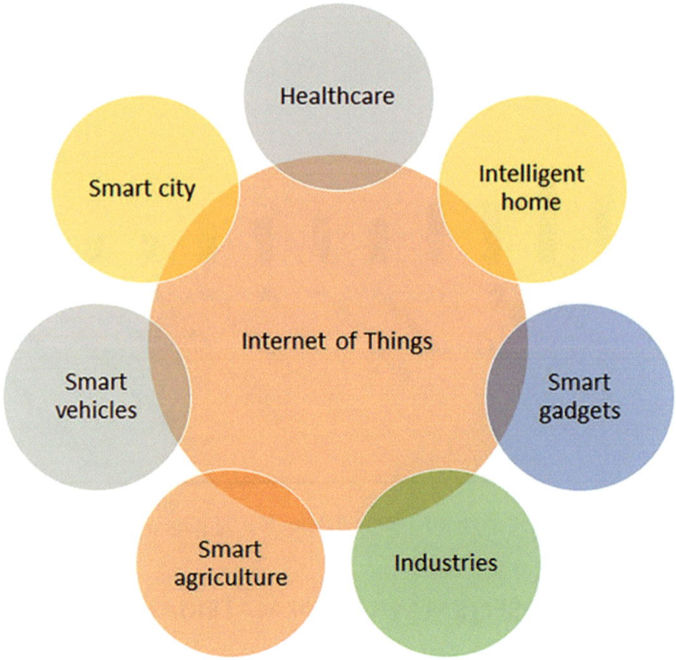

FIGURE 1.11 Several smart technologies related to the IoT [7] [Open access] [Confirmed].

spending large sums on practical experiments. In addition to increasing productivity, this approach facilitates predictive maintenance and adaptive design in industrial systems. Furthermore, the increasing use of cloud platforms and collaborative databases is fostering a more open and connected research environment, enabling the global sharing of experimental data, computational models, and design tools [21–25].

Another area where digital transformation and materials science converge is additive manufacturing, also known as 3D printing. The production of specialized raw materials that are amenable to digital optimization, such as powders, resins, and composites, is essential to the success of this technology. When combined with field observations, simulation-based design ensures quality is maintained and defects are minimized, enabling complex engineering and mass customization unattainable using traditional techniques. The following three main components of control should also be added: the organizational system, the technological system, and the human-operating system [6]. Socio-technical systems view the organization, in whole or in part, as a set of interconnected subsystems (see Figure 1.12).

The circular economy and sustainability are two additional benefits of the digital transformation in materials development. Data-driven design can reduce waste, enhance resource efficiency, and enable the production of recyclable and environmentally friendly materials. Digital solutions that facilitate life cycle analysis ensure that new materials are high-performance and also aligned with international sustainability goals [26–30].

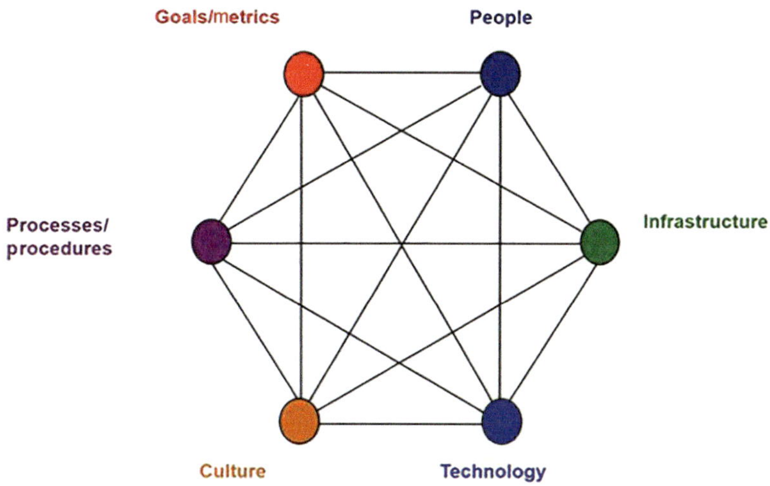

FIGURE 1.12 Socio-technical systems view the organization as a set of interconnected subsystems [8] [Open access] [Confirmed].

Despite these encouraging developments, challenges remain. Significant obstacles to progress include the need for large, high-quality datasets, digital platform compatibility, and data standardization. Furthermore, seamless collaboration between different disciplines is essential to combine computational predictions with experimental validation. As science becomes increasingly digitized, ethical issues related to data protection and ownership are becoming increasingly important. By combining advanced computing, AI, and experimental innovation, digital transformation is transforming materials development. It facilitates the development of next-generation materials that meet the demands of Industry 4.0 and beyond, accelerating discovery, lowering prices, and improving sustainability. As this transformation evolves, it can expand the scope of materials science and establish a faster, smarter, and more sustainable innovation path.

1.6 SOCIETAL, ECONOMIC, AND ENVIRONMENTAL IMPACTS

Beyond industrial productivity, the rapid growth of contemporary technologies, particularly in the context of Industry 4.0 and the production of advanced materials, has profound implications for society, the economy, and the environment. These impacts are closely linked, affecting not only how industries are managed but also how communities, economies, and ecosystems are maintained. Understanding these aspects is essential to ensure that technological developments lead to sustainable and inclusive growth. From a societal perspective, new materials and advanced technologies significantly improve people's quality of life [20,31–33]. Innovations in healthcare, such as biocompatible composite materials for implants or nanoparticles for the delivery of concentrated medications, are improving medical treatments and longevity. Wearable technology and energy-efficient homes are just two examples of how smart materials are being integrated into everyday life to change lifestyles and provide greater convenience and connectivity. The 5SEnSU model (Figure 1.13), which considers five sectors with specific roles, was used to evaluate a small number of variables for established and developing economies in the G19 countries, aiming to predict a sustainable economy that contributes to sustainability.

At the same time, the digital transformation of industry is creating new job prospects in high-tech industries, necessitating advanced skills in automation, materials engineering, and data science. However, this transformation also has its drawbacks, including job losses in more established industries, a digital divide between skilled and unskilled workers, and the need for continuous education and training to ensure equitable access to the new economy. Economically, the use of Industry 4.0 technologies and advanced materials enhances industrial competitiveness by enabling the production of innovative goods, reducing production costs, and increasing efficiency. Energy-efficient metals, durable nanocoatings, and lightweight composites extend product life while reducing operating costs. Research cycles are shortened, and new technologies are rapidly commercialized when digital tools are incorporated into materials development, opening the door to global market leadership and economic expansion. However, the high costs of advanced manufacturing technologies and the difficulty of scaling up new materials may make them inaccessible to small- and medium-sized enterprises (SMEs), potentially widening the economic gap between developed and developing countries. Digital manufacturing and advanced materials

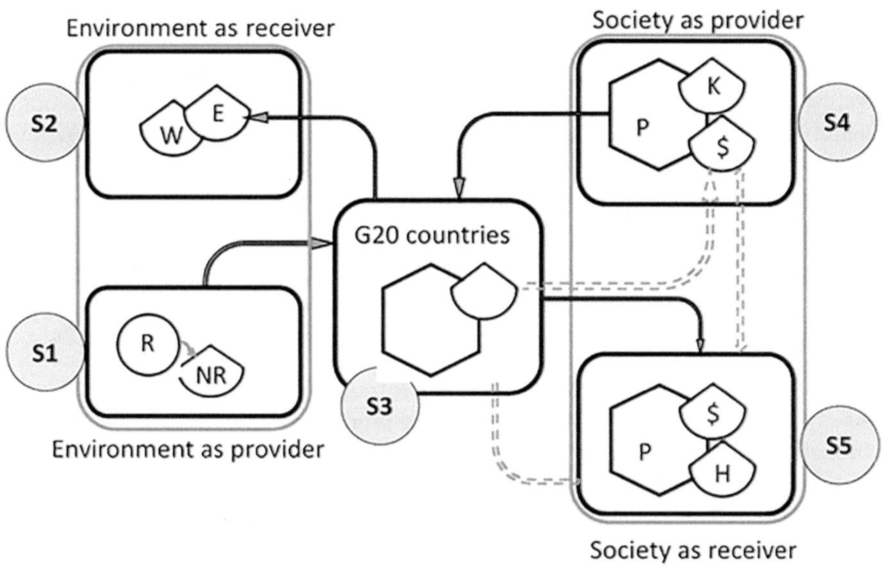

FIGURE 1.13 A diagram illustrating the 5SEnSU model. P stands for people; K for knowledge; H for happiness; E for emissions; R and NR for renewable and non-renewable resources; and W for waste and sanitation. The white arrows indicate monetary transactions, while the straight lines indicate material and energy transfers [9] [Open access].

present both potential and environmental challenges. Positive impacts include the development of sustainable, recyclable, biodegradable, and energy-efficient materials that promote circular economy models and reduce waste. While renewable energy technologies rely on high-performance materials in solar panels, wind turbines, and energy storage systems, lightweight vehicles, for example, reduce fuel consumption in transportation. Resource management and pollution control can be improved through digital transformation, which also enhances environmental monitoring through sensors and smart systems. However, the production of advanced materials often requires energy-intensive processes, scarce raw materials, and the generation of nanoscale waste, all of which can have long-term negative impacts on the environment and human health if not effectively addressed. Balancing innovation, sustainable practices, and rational resource use remains a challenge. Advanced materials and Industry 4.0 technologies have significant and diverse impacts on society, the economy, and the environment [35–39].

While these technologies raise questions about equity, cost, and environmental responsibility, they also offer revolutionary prospects for enhancing sustainability, increasing industrial productivity, and improving healthcare. Policymakers, businesses, and researchers must collaborate to develop frameworks that consider ethical considerations, support workforce development, and encourage sustainable innovation to maximize benefits and minimize risks. There is no way to build a comprehensive and balanced future except by following this comprehensive approach [2–4,40,41]. The foundations of this research question were identified in the updated literature. To model the term university social responsibility (USR) from the concept of corporate social responsibility (CSR), some notes are provided on the terminology and concepts used in the context of this research. Consequently, Figure 1.14 illustrates the conceptual framework of the research question, which incorporates the terms CSR and USR, addressing the literature review and the Sustainable Development Goals identified in the study.

1.7 INFORMATION TECHNOLOGY AND ELECTRONICS

History has shown that there have been many revolutions in human history. The revolution in information and communication technology (ICT) is the most recent. These advances have increased the need for this technology. Thus, compared to traditional resources, the importance and use of latter have increased, which has increased interest in new ICTs among both enterprises and individuals [32,33].

The world is currently experiencing a rapid increase in the quantity, quality, and availability of information tools needed to make the best decisions, given numerous changes and changes in the business environment. This eloquently illustrates the growing importance of information systems and their ability to meet the information needs of businesses. This highlights how important is to create these systems and improve their efficiency and effectiveness. The fact that these systems have come to rely on advanced technologies that have significantly reduced the cost of access and made it easier to access information is one of the major factors contributing to their success. Technological advancements supporting the telecommunications industry have brought about a significant transformation of the sector in a short period. It

FIGURE 1.14 USR's conceptual framework [10] [Open access].

currently serves as the basis for the so-called "knowledge economy," which depends on the fastest and least expensive ways to disseminate information. The acquisition of information and communications technology represents a new challenge for businesses as a result of the incredible development of sector and extent to which it has benefited all industries, especially in light of using satellites, mobile phones and Internet.

Information technology is undoubtedly a revolutionary and fundamental advancement that has had a significant impact on people's lives over the past century. ICTs are used in all walks of life and not a passing phenomenon or a fad. This policy does not exempt any domain, making it easier for the organization and its employees. This chapter discusses nanomaterials, their processing methods, and their potential uses. Even if nanomaterials have been consistently successful in all aspects of life, they still have many hurdles to overcome. Although nanomaterials have many applications, the impact of nano is limited by their unpredictable use and release into the environment, raising concerns about potential health risks. Most research and development in nanotechnology is still theoretical and critical research has not been conclusively linked to large-scale production and profitable commercial production. Safety and risk studies are also necessary to ensure public acceptability. There must be policies that prioritize standards, education, infrastructure, and workforce readiness. Commercialization of nanotechnology can be enhanced by establishing new regulations and a clear plan that will attract investors and make marketing easier [20,33,34].

The actions of government agencies must be coordinated. Energy-neutral materials (ENMs) have the potential to address current and future challenges, such as energy shortages, climate change, and industrial and medical needs, while improving living standards, maintaining industry standards, and sustaining competitiveness.. There is a need to create more organic methods and tools to perform metrological studies quickly and easily. Further research is also needed to identify ethical challenges, explore public opinion, and explore regulatory issues. Future scientists and engineers can be prepared by integrating nanoscience and nanotechnology into science and engineering curricula. Research findings and advances in the field of ENM permeate every aspect of our lives, including those related to energy, agriculture, defense, materials science, aerospace, environmental science and medicine. Nanotechnology is a rapidly developing field that is being studied by more and more people, several government agencies, venture capitalists, scientists, engineers, and doctors. Nanoporous materials, nanorobotics, nanodevices, and nanolithography are a few current areas of research, including pharmaceuticals, molecular manufacturing, nanopowders, nanocomputing, nanostructured catalysts, nanolayers, and nanobiology (e.g., disease prediction, prevention, and treatment). Early treatment and prevention of diseases are the focus of attention of clinical researchers and biomedical engineers around the world. Processes occurring at the nanoscale are important for modern technologies. Lithography, ion exchange, adsorption, drug discovery, catalysis, composites, and plastics are some examples of these technologies [35–37].

The current era of global transformation is driven by manufactured nanomaterials and nanotechnology. Especially in the 21st century, this pioneering discipline has attracted the attention of the general public, scientists, researchers, and business community. With a wide range of applications, nanotechnology has become an established scientific field that has achieved significant growth and progress. The process of moving from research and development to commercial applications remains complex and time-consuming, even as the nanomaterials field evolves and matures. A major source of difficulty is the need for robust and reproducible technology, which requires rigorous standardization, testing, and evaluation at each stage of process development. Significant research and investment by organizations and investors are needed to address these challenges and facilitate successful technology transfer and commercialization. This chapter provides an overview of the major developments, applications, and challenges associated with nanotechnology and manufactured nanomaterials. It discusses the usefulness of these technologies and offers ideas for potential future research directions. To fully realize the potential of nanotechnology and ensure its successful integration into other sectors, it will be necessary to address these challenges and explore new research directions as the field evolves.

1.8 FUTURE PROSPECTS AND CHALLENGES

In addition to unprecedented prospects, the future of advanced materials and Industry 4.0 technologies presents a number of challenging problems that need to be properly addressed. Advanced materials will continue to be at the forefront of innovation as industries shift toward increased automation, digitization, and sustainability. This will enable new features, increased productivity, and transformative applications across a variety of industries. However, overcoming hurdles in science, technology, economics, and ethics is essential to realizing this promise. There are numerous prospects for future materials improvement. The identification and optimization of new compounds will continue to accelerate using high-throughput simulations, ML, and AI in materials science. Wearable technology, biomedical implants, robotics, and aerospace systems will increasingly rely on smart materials with adaptable, self-healing, or stimuli-responsive properties. Advanced materials will be fundamental to the development of next-generation solar cells, high-capacity batteries, and efficient hydrogen storage systems, all of which will directly contribute to the global transition to carbon neutrality and renewable energy [42–46]. With the emergence of specialized digital materials, additive manufacturing will grow significantly, enabling distributed production, mass customization, and reduced dependence on the supply chain. Furthermore, sustainable materials such as recyclable polymers and biocomposites are expected to accelerate, balancing industrial expansion with environmental responsibility. Through structural engineering and optimization, the electrochemical performance of MXenes can be improved. Both the chemical composition and atomic layer structure of MXenes are of great importance in this case. For example, compared to higher-order MXenes such as M3X2 or M4X3, MXene structures with fewer atomic layers, such as M2X (Figure 1.15), are assumed to have better bulk capacities, as evidenced by higher specific surface areas.

Even with these encouraging trends, many challenges remain. One major obstacle is the challenge of transferring laboratory discoveries to industrial production while maintaining cost-effectiveness, quality, and reproducibility. Due to their complexity, nanomaterials and hybrids often require more complex manufacturing processes, which are also expensive and energy-intensive. As the success of digital transformation depends on standardized platforms, readily available databases, and secure information exchange across sectors and research institutions, data management and interoperability pose another obstacle. Furthermore, the safety, recyclability, and long-term reliability of new materials, particularly nanomaterials, raise health and environmental issues that require comprehensive regulatory frameworks. Socioeconomic issues also remain. While advanced materials can increase competitiveness and open new markets, small businesses and emerging economies may not be able to afford them, potentially widening the technology gap between countries. Furthermore, the adaptability of the workforce is critical, as the transition to digital, materials-based sectors requires continuous education and training programs, as they require new competencies in advanced manufacturing, data analytics, and computational modeling [12–14,47,48].

The future environment is increasingly complex due to issues of sustainability and ethics. Issues such as the environmental impact of production, the potential risks of nanoscale waste, and the responsible procurement of scarce raw materials highlight the need for open and sustainable processes. To avoid societal disparities in the benefits of innovation, it is essential to ensure equitable access to the latest technologies. Innovative Industry 4.0 materials have revolutionary potential to improve mobility, healthcare, energy, and sustainable manufacturing. However, the

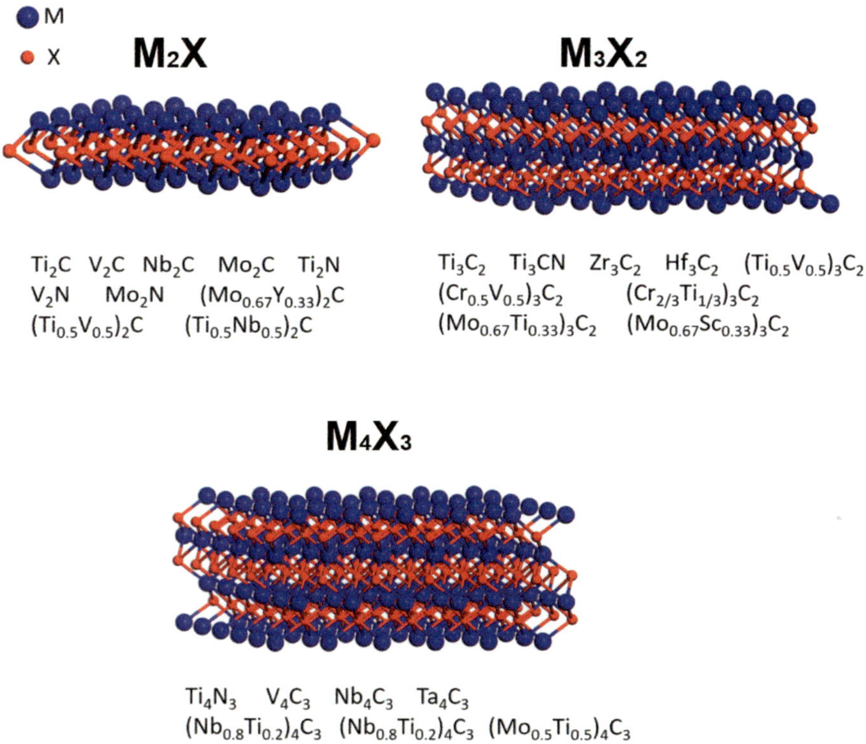

FIGURE 1.15 Bare MXenes structures with a summary of all corresponding experimentally synthesized samples [10] [Open access].

TABLE 1.2
General Information on Publication Selection

Total Number of Records	SCOPUS: 2,496; WOS: 1,954
Document types included	Article, proceedings paper, conference paper, review
Records after filtering by type (before removing duplicates)	SCOPUS: 2,124; WOS: 1,903
Records after filtering by type (after removing duplicates)	SCOPUS: 1,485; WOS: 1,366
Total records after filtering by type	2,851

Source: From Ref. 1 [Open access].

effectiveness of multinational companies, governments, and the research community in addressing issues of scalability, pricing, data integration, environmental responsibility, and workforce readiness will determine the future of the industry, along with technological advancements. Unlocking the full potential of new materials to build a smarter, cleaner, and more inclusive future requires a balanced strategy that combines scientific advancements, ethical responsibility, and sustainable practices [16–20,49]. Table 1.2 shows the total number of publications, article types, and the number of articles after considering the article type filter before and after deleting duplicates, and the total number of publications in SCOPUS and WOS.

PROBLEMS

1. What are the key elements that distinguish the properties of nanomaterials from those of bulk materials?
2. How do reinforcement and matrix influence the performance of composite materials?
3. How can advanced materials help achieve the goals of Industry 4.0?
4. How can digital twins accelerate materials development?
5. What are the major barriers to advanced materials manufacturing?

REFERENCES

1. Dehghan, S., Sattarpanah Karganroudi, S., Echchakoui, S., & Barka, N. (2025). The integration of additive manufacturing into industry 4.0 and industry 5.0: A bibliometric analysis (trends, opportunities, and challenges). *Machines*, 13(1), 62. [Confirmed]
2. Bahrin, M. A. K., Othman, M. F., Azli, N. H. N., & Talib, M. F. (2016). Industry 4.0: A review on industrial automation and robotic. *J. Teknol.*, 78, 137–143.
3. Adel, A. (2022). Future of industry 5.0 in society: Human-centric solutions, challenges and prospective research areas. *J. Cloud Comput.*, 11, 40.
4. Mourtzis, D., Angelopoulos, J., & Panopoulos, N. (2022). A literature review of the challenges and opportunities of the transition from industry 4.0 to society 5.0. *Energies*, 15, 6276.
5. Chatziparaskeva, G., Papamichael, I., Voukkali, I., Loizia, P., Sourkouni, G., Argirusis, C., & Zorpas, A. A. (2022). End-of-life of composite materials in the framework of the circular economy. *Microplastics*, 1(3), 377–392.
6. Bilal, M., & Iqbal, H. M. (2020). New insights on unique features and role of nanostructured materials in cosmetics. *Cosmetics*, 7(2), 24.
7. Çınar, Z. M., Zeeshan, Q., & Korhan, O. (2021). A framework for industry 4.0 readiness and maturity of smart manufacturing enterprises: A case study. *Sustainability*, 13(12), 6659.
8. Moghrabi, I. A., Bhat, S. A., Szczuko, P., AlKhaled, R. A., & Dar, M. A. (2023). Digital transformation and its influence on sustainable manufacturing and business practices. *Sustainability*, 15(4), 3010.
9. Terra dos Santos, L. C., Frimaio, A., Giannetti, B. F., Agostinho, F., Liu, G., & Almeida, C. M. (2023). Integrating environmental, social, and economic dimensions to monitor sustainability in the G20 countries. *Sustainability,* 15(8), 6502.
10. Meseguer-Sánchez, V., Abad-Segura, E., Belmonte-Ureña, L. J., & Molina-Moreno, V. (2020). Examining the research evolution on the socio-economic and environmental dimensions on university social responsibility. *International Journal of Environmental Research and Public Health*, 17(13), 4729.
11. Yang, J., Liu, T., Liu, Y., & Morgan, P. (2022). Review of human–machine interaction towards industry 5.0: Human-centric smart manufacturing. In Proceedings of the International Design Engineering Technical Conferences and Computers

and Information in Engineering Conference, St. Louis, MO, USA, 14–17 August 2022; American Society of Mechanical Engineers: New York, NY; Volume 86212, p. V002T02A060.

12. Alojaiman, B. (2023). Technological modernizations in the industry 5.0 ERA: A descriptive analysis and future research directions. *Processes*, 11 , 1318.
13. Grosse, E. H., Sgarbossa, F., Berlin, C., & Neumann, W. P. Human-centric production and logistics system design and management: Transitioning from Industry 4.0 to Industry 5.0. *Int. J. Prod. Res.*, 2023, 61, 7749–7759.
14. Renda, A., Alcidi, C., Baiocco, S., Corti, F., Schaus, M., Lannoo, K., Sipiczki, A. (2021). The new industrial strategy for Europe. *Intereconomics Rev. Eur. Econ. Policy*, 56, 132.
15. Gródek-Szostak, Z., Siguencia, L. O., Niemczyk, A., & Szeląg-Sikora, A. (2023). From industry 4.0 paradigm towards industry 5.0. In Environment Technologies. Resources. Proceedings of the International Scientific and Practical Conference, Poland, 15–16 June 2023; Rezekne Academy of Technologies: Rezekne; Volume 2, pp. 46–49.
16. Breque, M., De Nul, L., & Petridis, A. (2021). Industry 5.0: towards a sustainable, human-centric and resilient European industry (No. KI-BD-20-021-EN-N). Directorate General for Research and Innovation (DG RTD) of the European Commission. Luxembourg, 2021; p. 46..
17. Pizoń, J., Witczak, M., Gola, A., & Świć, A. (2023). Challenges of human-centered manufacturing in the aspect of Industry 5.0 assumptions. *IFAC-Pap*, 56, 156–161.
18. Rani, S., Jining, D., Shoukat, K., Shoukat, M. U., & Nawaz, S. A. (2024). A human–machine interaction mechanism: Additive manufacturing for Industry 5.0—Design and management. *Sustainability,* 16, 4158.
19. Khan, M., Haleem, A., & Javaid, M. (2023). Changes and improvements in Industry 5.0: A strategic approach to overcome the challenges of Industry 4.0. *Green Technol. Sustain.*, 1, 100020.
20. Maddikunta, P. K. R., Pham, Q. V., Prabadevi, B., Deepa, N., Dev, K., Gadekallu, T. R., Ruby, R., & Liyanage, M. (2022). Industry 5.0: A survey on enabling technologies and potential applications. *J. Ind. Inf. Integr.*, 26, 100257.
21. Zizic, M. C., Mladineo, M., Gjeldum, N., & Celent, L. (2022). From industry 4.0 towards industry 5.0: A review and analysis of paradigm shift for the people, organization and technology. *Energies*, 15, 5221.
22. Eriksson, K. M., Olsson, A. K., & Carlsson, L. (2024). Beyond lean production practices and Industry 4.0 technologies toward the human-centric Industry 5.0. *Technol. Sustain.*, 3, 286–308.
23. Granata, I., Faccio, M., & Boschetti, G. (2024). Industry 5.0: Prioritizing human comfort and productivity through collaborative robots and dynamic task allocation. *Procedia Comput. Sci.*, 232, 2137–2146.
24. Zafar, M. H., Langås, E. F., & Sanfilippo, F. (2024). Exploring the synergies between collaborative robotics, digital twins, augmentation, and industry 5.0 for smart manufacturing: A state-of-the-art review. *Robot. Comput. Integr. Manuf.*, 89, 102769.
25. Longo, F., Padovano, A., & Umbrello, S. (2020). Value-oriented and ethical technology engineering in industry 5.0: A human-centric perspective for the design of the factory of the future. *Appl. Sci.*, 10, 4182.
26. Wolniak, R. (2023). Industry 5.0–characteristic, main principles, advantages and disadvantages. *Zeszyty Naukowe. Organizacja i Zarządzanie/Politechnika Śląska*, 170, 663–678.
27. Ghobakhloo, M., Iranmanesh, M., Mubarak, M. F., Mubarik, M., Rejeb, A., & Nilashi, M. (2022). Identifying industry 5.0 contributions to sustainable development: A

strategy roadmap for delivering sustainability values. *Sustain. Prod. Consum.*, 33, 716–737.

28. Keleko, A. T., Kamsu-Foguem, B., Ngouna, R. H., & Tongne, A. (2022). Artificial intelligence and real-time predictive maintenance in industry 4.0: A bibliometric analysis. *AI Ethics*, 2, 553–577.
29. Gorecky, D., Schmitt, M., Loskyll, M., & Zühlke, D. (2014). Human–machine-interaction in the industry 4.0 era. In Proceedings of the 2014 12th IEEE International Conference on Industrial Informatics (INDIN), Porto Alegre, 27–30 July 2014; pp. 289–294.
30. Haleem, A., & Javaid, M. (2019). Additive manufacturing applications in industry 4.0: A review. *J. Ind. Integr. Manag.*, 4, 1930001.
31. Butt, J. (2020). Exploring the interrelationship between additive manufacturing and Industry 4.0. *Designs*, 4, 13.
32. Bhardwaj, A. R., & Ulhe, P. (2024). Rapid prototyping technology: A driving force in product design innovation. *Int. J. Res. Appl. Sci. Eng. Technol.*, 12, 1001–1002.
33. Narkhede, G., Pasi, B., Rajhans, N., & Kulkarni, A. (2023). Industry 5.0 and the future of sustainable manufacturing: A systematic literature review. *Bus. Strategy Dev.*, 6, 704–723.
34. Nahavandi, S. (2019). Industry 5.0: A human-centric solution. *Sustainability*, 11, 4371.
35. Rahmani, R., Karimi, J., Resende, P. R., Abrantes, J. C. C., & Lopes, S. I. (2023). Overview of selective laser melting for Industry 5.0: Toward customizable, sustainable, and human-centric technologies. *Machines*, 11, 522.
36. Kumar, S., Bongale, A., Patil, S., Bongale, A. M., Kamat, P., & Kotecha, K. (2020). Demystifying artificial intelligence based digital twins in manufacturing: A bibliometric analysis of trends and techniques. *Libr. Philos. Pract.*, 2020, 1–21.
37. Bibby, L., & Dehe, B. (2018). Defining and assessing industry 4.0 maturity levels–case of the defence sector. *Prod. Plan. Control,* 29, 1030–1043.
38. Gardan, J. (2017). Additive manufacturing technologies: State of the art and trends. In *Additive Manufacturing Handbook* (pp. 149–168). CRC Press: Boca Raton, FL. www.taylorfrancis.com/chapters/edit/10.1201/9781315119106-10/additive-manufacturing-technologies-julien-gardan
39. Hull, C. W. (1986). *Apparatus for production of three-dimensional objects by stereolithography*. U.S. Patent 4,575,330, 11 March 1986.
40. Moghaddam, A. O., Shaburova, N. A., Samodurova, M. N., Abdollahzadeh, A., & Trofimov, E. A. (2021). Additive manufacturing of high entropy alloys: A practical review. *J. Mater. Sci. Technol.*, 77, 131–162.
41. Kagermann, H., & Wahlster, W. (2022). Ten years of Industrie 4.0. *Science*, 4, 26.
42. Raja Santhi, A., & Muthuswamy, P. (2023). Industry 5.0 or industry 4.0 S? Introduction to industry 4.0 and a peek into the prospective industry 5.0 technologies. *Int. J. Interact. Des. Manuf.*, 17, 947–979.
43. Mukherjee, D., Gupta, K., Chang, L. H., & Najjaran, H. (2022). A survey of robot learning strategies for human–robot collaboration in industrial settings. *Robot. Comput. Integr. Manuf.*, 73, 102231.
44. Rodríguez-Guerra, D., Sorrosal, G., Cabanes, I., & Calleja, C. (2021). Human–robot interaction review: Challenges and solutions for modern industrial environments. *IEEE Access*, 9, 108557–108578.
45. Bragança, S., Costa, E., Castellucci, I., & Arezes, P. M. (2019). A brief overview of the use of collaborative robots in industry 4.0: Human role and safety. In *Occupational and Environmental Safety and Health* (pp. 641–650). Springer: Cham. https://link.springer.com/chapter/10.1007/978-3-030-14730-3_68

46. Gualtieri, L., Rauch, E., & Vidoni, R. (2021). Emerging research fields in safety and ergonomics in industrial collaborative robotics: A systematic literature review. *Robot. Comput. Integr. Manuf.*, 67, 101998.
47. Alves, J., Lima, T. M., & Gaspar, P. D. (2023). Is industry 5.0 a human-centred approach? A systematic review. *Processes*, 11, 193.
48. Yang, J., Liu, T., Liu, Y., & Morgan, P. (2022). Review of human–machine interaction towards industry 5.0: Human-centric smart manufacturing. In Proceedings of the International Design Engineering Technical Conferences and Computers and Information in Engineering Conference, St. Louis, MO, 14–17 August 2022; American Society of Mechanical Engineers: New York, NY; Volume 86212, p. V002T02A060.
49. Gródek-Szostak, Z., Siguencia, L. O., Niemczyk, A., & Szeląg-Sikora, A. (2023). From Industry 4.0 paradigm towards Industry 5.0. In Environment. Technologies. Resources. Proceedings of the International Scientific and Practical Conference, Krakow, 15–16 June 2023; Rezekne Academy of Technologies: Rezekne; Volume 2, pp. 46–49.

2 Mechanical and Physical Properties of Advanced Materials

2.1 INTRODUCTION

The technological importance of advanced materials and their industrial uses lies in their mechanical and physical properties. Unlike conventional materials, advanced materials at the micro- or nanoscale are designed to perform better under harsh service conditions. These materials are known for their high strength-to-weight ratios, exceptional stiffness, enhanced toughness, and high fatigue resistance. These properties make them highly valuable in critical fields where reliability and durability are critical, such as biomedical engineering, the automotive industry, and aerospace. For example, fiber-reinforced composites and nanoalloys combine exceptional resistance to mechanical stress with lightweight properties, allowing for increased component efficiency and extended service life. Furthermore, ceramic or nanoscale coatings offer improved resistance to wear and tear, reducing maintenance costs and enhancing sustainability [1–4].

Physically, modern materials exhibit special thermal, electrical, magnetic, and optical properties that significantly expand their applications beyond structural roles. While electrically conductive polymers, graphene, and nanowires enable the development of flexible electronics and the miniaturization of electronic devices, materials with specially designed thermal conductivity are essential for energy storage devices, heat exchangers, and electronics cooling systems. Highly stable magnetic materials are essential for energy conversion and data storage technologies, while photonic nanoparticles and photonic structures pave the way for more advanced sensors, coatings, and communication systems. Furthermore, smart materials that bridge the gap between passive structures and active functional systems include self-healing polymers, shape-memory alloys, and piezoelectric ceramics. These materials react dynamically to environmental stimuli.

Together, these physical and mechanical properties demonstrate how advanced materials are transforming modern industry. Their ability to combine strength, toughness, and adaptability with multifunctional physical behaviors makes them important drivers of Industry 4.0 and sustainable technological advancement. However, continuous research and innovation are required to solve environmental issues, improve their properties for mass production, and ensure their long-term reliability to fully realize their promise [1–3]. In this regard, advanced characterization

DOI: 10.1201/9781003758556-2

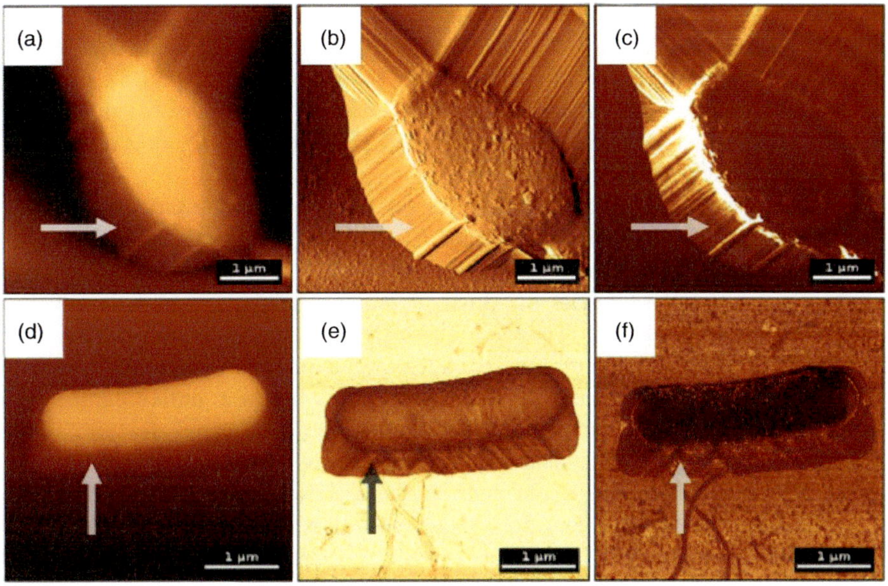

FIGURE 2.1 Cantilever slippage and sample movement cause artifacts in AFM topographic images of (a–c) *Candida albicans* measured in contact mode (where the probe actually touches the surface and scans the sample horizontally) and (d–f) *Bacillus subtilis* measured in quantitative (QI) mode (where force–distance curves are recorded at each pixel of the image, meaning the probe moves vertically towards the surface). Topographic image (a); error signal (b); lateral deflection (c) corresponding to frictional forces (d); stiffness map (e) derived from the slope of the force–distance curve as the object approaches and impacts the surface; adhesion map (f) derived from the force–distance curve as the object deflects from the surface. Arrows show artifacts captured using AFM [From Ref. 1] [Open access].

methods are essential for understanding the structural, mechanical, and functional behavior of these materials at the micro- and nanoscale. Because it provides high-resolution surface imaging and characterization of properties at the nanoscale, atomic force microscopy (AFM) stands out among these methods. However, to effectively collect and analyze data, it is essential to be aware of the limitations and challenges of this technique. While capturing AFM images, users may encounter some defects, often caused by incorrect cantilever tip selection or scanner instability (Figure 2.1). These defects can lead to misleading analyses and incorrect conclusions.

2.2 FUNDAMENTAL MECHANICAL PROPERTIES

As the fundamental mechanical properties of advanced materials determine how they react to external stresses, forces, and long-term service conditions, they form the basis of their use in contemporary engineering systems. One of their most important properties is toughness, which describes a material's ability to withstand deformation or failure under applied loads. Advanced alloys, composites, and nanomaterials are well suited for structural, automotive, and aerospace applications due to their often higher tensile and compressive strengths compared to their conventional counterparts. Toughness and stiffness are closely related. Toughness is a material's resistance to elastic deformation and is usually referred to as its modulus of elasticity. High-strength materials maintain their shape under extreme stresses, which is essential for load-bearing structures and precision parts.

Toughness, which describes a material's resistance to surface scratches or dents, is another essential characteristic. In applications requiring durability and long service life, such as cutting tools, protective surfaces, and corrosion-resistant coatings, toughness is critical. Materials designed to increase toughness include advanced ceramics and nanocoatings. Toughness, which measures a material's ability to absorb energy and resist fracture under shock or sudden loads, is equally important. Toughness is essential for safety-critical components because, unlike hardness and strength alone, it ensures that a material does not suffer catastrophic fracture.

Another important mechanical property is fatigue resistance, particularly for materials such as turbine blades, bridges, and aircraft wings, which are subjected to cyclic stresses. Fatigue-resistant composites and alloys are essential for extending the life of critical systems, where fatigue failure often occurs below the material's maximum strength. Similarly, in high-temperature environments, such as power plants, jet engines, and nuclear reactors, creep resistance—the ability of a material to withstand time-related deformation under sustained load and high temperatures—is essential. Advanced ceramic–matrix composites and superalloys have been specifically designed to meet this need.

A comprehensive understanding of material performance can be achieved by combining these fundamental mechanical properties: strength, stiffness, hardness, toughness, fatigue resistance, and creep resistance. Because of the tremendous advances in these properties, enabled by advanced materials through nanoscale design, hybrid structures, and customized microstructures, it is now possible to create engineering solutions that are lighter, more durable, and more reliable. The improvement of these properties continues to drive innovation across various sectors, highlighting the importance of mechanical performance in creating next-generation products [3]. The movement of pre-sorted cells between layers is used to measure cell migration. There will be cells in other layers that were pre-sorted in the previous layer if migration occurs. In our case, there was essentially no cell movement between layers regardless of whether the scaffolds were equipped with hubs or not, or whether they were cultured for 3 or 6 days (Figure 2.2). Scaffold cell growth is further demonstrated by scanning electron microscopy (SEM). Rare cells cultured in the second scaffold layer migrated to other layers after 3 or 6 days of incubation. According to these results, very little migration occurs in the scaffolds, and this is mostly determined by the pore size of the scaffold.

The survival of scaffold cells with and without spindles was examined, specifically whether the expanded Polytetrafluoroethylene (ePTFE) spindles were removed during scaffold incubation in culture medium. The LIVE/DEAD kit was used to stain cells on uncoated membranes in each layer after 3 and 6 days of static culture. Confocal microscope images show that almost all cells are live, calcein-positive cells

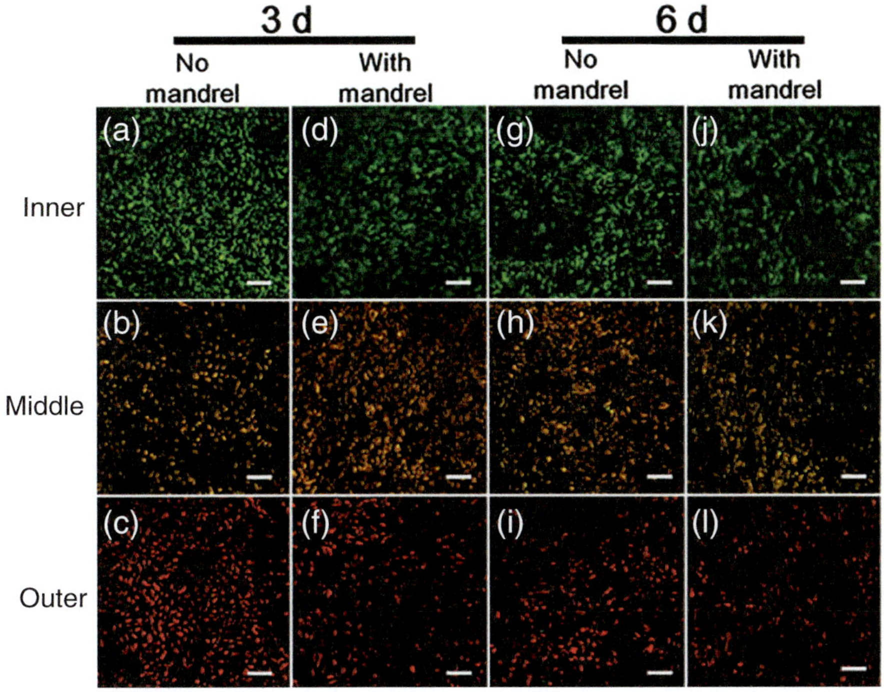

FIGURE 2.2 At days 3 (a–f) and 6 (g–l), cell migration was observed with (d–f, j–l) and without (a–c, g–i) a filament. Inner layer (a, d, g, j); central layer (b, e, h, k); outer layer (c, f, i, l). A typical cell used is C2C12. Each image is randomly selected to represent a layer. Cells in each layer were pre-labeled with different dyes. Each image is randomly selected to be the representative image in its respective layer after unwinding. Cells previously labeled will appear on a different layer if there is inter-level migration. Scale bars are 50 μm [From Ref. 3] [Open access].

(green), with very few cells exhibiting Propidium Iodide (PI)-positive cells (red, indicating dead cells) (Figure 2.3) [3].

2.3 PHYSICAL PROPERTIES OF ADVANCED MATERIALS

The physical properties of advanced materials provide a critical foundation for understanding their applicability and functionality in a range of industrial and engineering applications. Unlike conventional materials, advanced materials are often manufactured at the micro- and nanoscale to produce enhanced or entirely new physical properties. Their interactions with their surroundings and their use in high-performance technologies depend largely on their properties, which include density, optical performance, magnetic response, electrical behavior, and thermal conductivity.

Thermal conductivity is one of the most important physical properties, directly affecting heat dissipation in turbines, electronic devices, and aerospace structures.

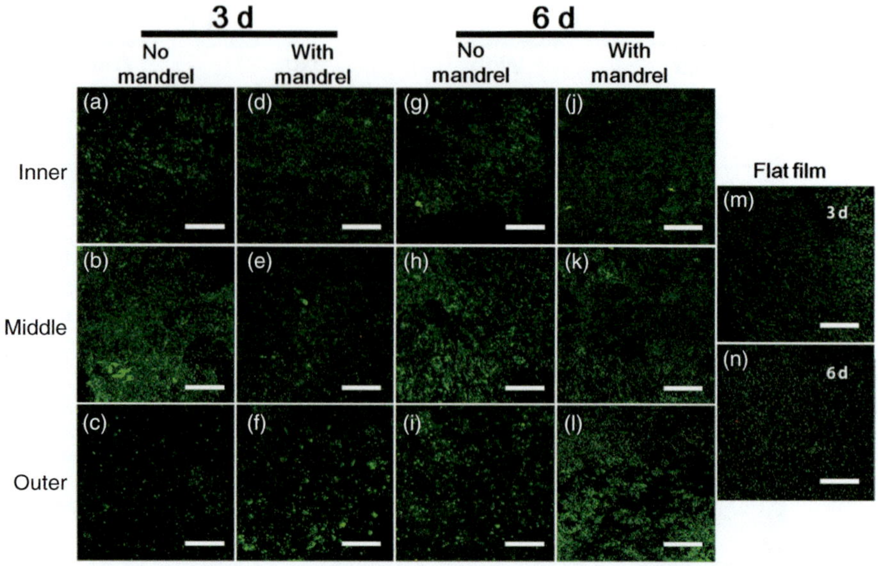

FIGURE 2.3 Cell viability at 3 days (a–f) and 6 days (g–l) with (d–f, j–l) and without (a–c, g–i) filament. Inner layers (a, d, c, j); middle layer (b, e, h, k); outer layer (c, f, j, l). Cell viability in 3D culture on a flat film is shown in (m) and (n). Green represents live cells. Red represents dead cells. The model cell used is a C2C12 cell. Each image in the corresponding layers is randomly selected to be the representative image. Scale bars are 200 μm [From Ref. 3] [Open access].

Depending on their intended use, advanced ceramics, carbon nanomaterials, and metallic composites often have either extremely low conductivity for thermal insulation or extremely high conductivity for rapid heat transfer. Similarly, the use of advanced materials in semiconductors, superconductors, and energy storage devices depends largely on their electrical properties, including conductivity, electrical resistance, and insulating behavior. Because of quantum confinement effects, nanomaterials in particular exhibit tunable electrical properties, making them essential in contemporary microelectronics.

The application possibilities of advanced materials expand because of their optical and magnetic properties. While photonic materials and nanoglasses enable precise control of light propagation, reflection, and absorption, soft and hard magnetic composites are commonly used in data storage, sensors, and electromagnetic shielding. These properties support technologies such as lasers, fiber optics, and energy-efficient lighting systems. Furthermore, the mechanical strength-to-weight ratio and energy storage capacity of modern materials are influenced by their density and specific heat capacity, both of which are critical for lightweight structural elements in aircraft and vehicles.

The versatility of physical properties is also crucial. Advanced materials are often designed to combine multiple physical functions in a single system, unlike standard materials that exhibit a limited set of properties. For example, shape-memory alloys feature special thermomechanical conductivity that allows them to change shape upon heating and return to their original shape upon cooling, while piezoelectric ceramics exhibit both mechanical and electrical interactions. This integration of properties greatly expands the design potential for innovative applications.

Modern advanced materials are remarkably versatile and tunable, as well as possessing improved physical properties compared to conventional materials. Given the ability of researchers and engineers to precisely alter material performance through improved processing and formulation, these properties directly contribute to the development of next-generation technologies, enabling high-value, efficient, and sustainable applications in a wide range of industries [5].

The surface and structural properties of photonic crystal fibers (PCFs) provide a concrete example of these material advancements. Untreated PCFs have smooth surfaces, thin walls, are elongated, and are devoid of twists, resembling the microstructure of wood and cotton (Figure 2.4a and b). However, PCFs initially exhibit an irregular texture due to the waxy components present on the surface. After treatment, removing this wax reveals clusters of microfibrils and fiber bundles, creating a rougher surface (Figure 2.4c). Furthermore, cross-sections of treated PCFs have a circular appearance (Figure 2.4d).

As evidenced by its moisture absorption capacity, which is nearly twice that of conventional textile fibers, and its recovery rate of approximately 18% under normal conditions, this enhanced microfiber organization significantly improves its

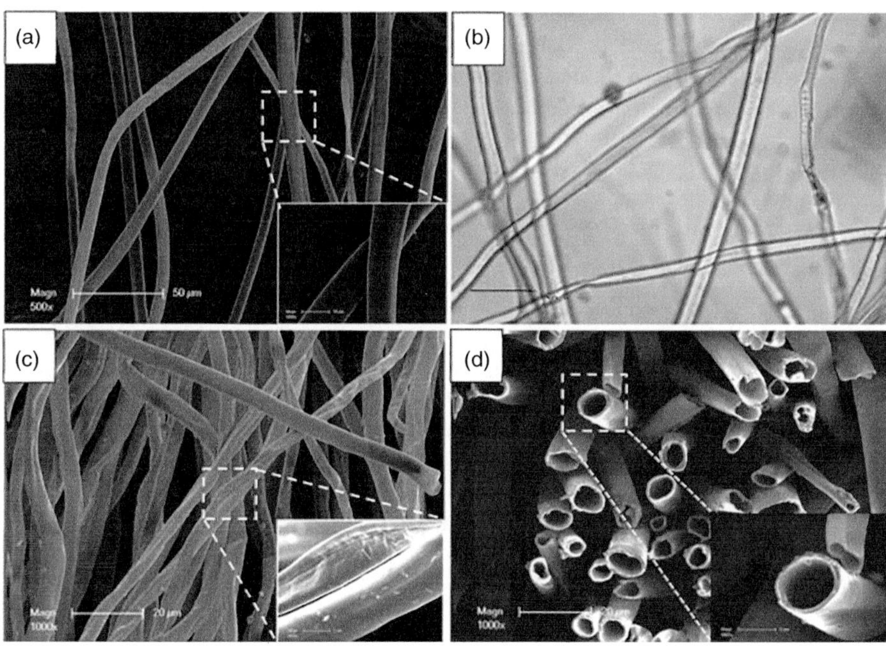

FIGURE 2.4 Scanning electron microscope images of poplar fibers (a, b) show the longitudinal surface before waxing, (c) shows the longitudinal surface after waxing, and (d) shows the cross-section after waxing; the inset images in (c) and (d) are high magnification images [From Ref. 4] [Open access].

functionality. The potential of PCFs as advanced functional materials is demonstrated by the microscopic features shown in Figure 2.4c and d.

Nanomaterials exhibit distinctive optical, electrical, and magnetic properties due to size effects, macroscopic quantum tunneling, and limited surface effects at the nanoscale. Numerous applications are available in cutting-edge energy technologies, such as catalysts, fuel cells, gas sensors, photovoltaic devices, energy storage systems, supercapacitors, and lithium-ion batteries [1]. The production of high-quality nanopowders with precise size, shape, and structure has been a major focus in this field. This has been achieved using a variety of techniques, including carbonylation, solvothermal synthesis [2], anode arc plasma methods [3], co-precipitation [4], microwave irradiation [6], microemulsion sonochemistry [7], ultrasound [8], and sol–gel processes [9]. The most economical and adaptable of these methods is co-precipitation, which allows precise control of physical and chemical parameters without the need for sophisticated equipment.

Nickel-based nanomaterials have received particular attention due to their excellent catalytic activity, stability, low toxicity, and environmental friendliness. In addition to catalysis [2], nickel nanoparticles have been used in electrochemical capacitors [3], solar cells [4], batteries [6], fuel cell electrodes [7], and electrochemical devices [8]. This study designed silicate-supported nickel nanocatalysts with a crystalline width of approximately 30 nm using a modified Stöber co-precipitation method [9]. The effectiveness and stability of these catalysts in hydrogen production were then evaluated using them in methane decomposition processes.

The depletion of fossil fuel supplies and the urgent need to reduce greenhouse gas emissions have led to increased interest in hydrogen, one of the most promising energy sources, on a global scale. Hydrogen is an environmentally friendly and sustainable solution to global energy problems, as it produces no harmful emissions when used [10]. In this regard, the expansion of hydrogen technologies and the transition to clean energy systems are directly supported by the development of highly efficient nanocatalysts.

Advanced characterization tools were used to confirm these results. Zinc oxide (ZnO) and silver-doped zinc oxide (Ag–ZnO) nanorods grown on silicon (Si) substrates exhibited homogeneous morphologies, smooth surfaces, and vertical alignment along the C-axis, according to field-emission scanning electron microscopy (FESEM) analysis), which corroborated the X-ray diffraction (XRD) results [4]. Similarly, an AFM study of untreated transparent carbon fibers (PCFs) revealed a waxy surface layer that affects fiber morphology and performance (Figure 2.5). These results highlight the potential benefits of surface modifications of both natural fibers and manufactured nanomaterials to maximize their functionality in sustainable and energy-related applications.

Furthermore, hydrogen can be produced using renewable raw materials such as water, biomass, and biogas [11]. Environmentally friendly methane separation technology can be considered a sustainable means of hydrogen production, as methane constitutes a large proportion of biomass and is abundant in industrialized countries and in the deep oceans [12]. In short, governments, academic institutions, and the automotive industry have been paying increasing attention to the potential of hydrogen as an alternative fuel, both in terms of production and widespread adoption.

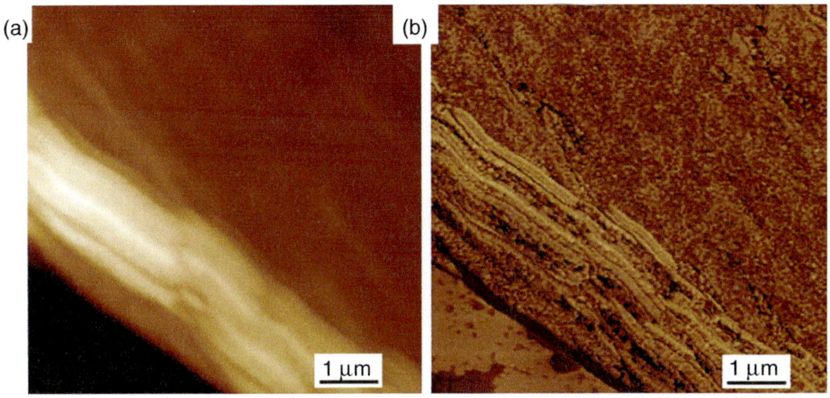

FIGURE 2.5 The longitudinal surface of poplar fibers is imaged using an AFM. (a) height map; (b) phase diagram [From Ref. 4] [Open access].

2.4 SMART AND FUNCTIONAL MATERIALS

A growing class of advanced materials, known as "smart and functional materials," is capable of sensing, reacting, and adapting to changes in their surroundings. Unlike conventional materials, which offer only passive mechanical or physical performance, smart materials are designed to actively respond to external stimuli, such as light, humidity, temperature, pressure, electric and magnetic fields, and chemical conditions. For example, shape memory alloys can recover their previous shape after deformation in response to a stimulus such as heat, while piezoelectric materials have the ability to convert mechanical energy into electrical impulses, and vice versa. These amazing properties demonstrate how smart materials can combine sensing, actuation, and control in a single physical system, simplifying the system and increasing its overall efficiency. Biological systems demonstrate a similar functional response principle. For example, using AFM research, a model of the poorly silicified cell wall of the marine diatom *Cylindrotheca closterium* was created. The researchers gained insights into the material's distinctive mechanical and surface properties by using AFM imaging to observe and characterize the cell wall's nanostructure and structural organization (Figure 2.6). This biological example inspires the development of the next generation of smart, bio-inspired materials and further illustrates how structural function at the micro- and nanoscales contributes to adaptive behavior [6].

Because of their responsiveness, these materials can undergo controlled and reversible changes in their properties, making them essential for the development of next-generation technologies in diverse fields. Their key feature is their multifunctionality. Materials that can perform sensing or actuation tasks while simultaneously providing structural support include shape memory alloys, piezoelectric ceramics, magnetic composites, ferroelectric polymers, and self-healing composites. In addition, energy-related applications, such as energy conversion, storage, and harvesting, rely heavily on functional materials. For example, thermoelectric materials can convert heat directly into electrical energy, helping to recover waste heat and provide

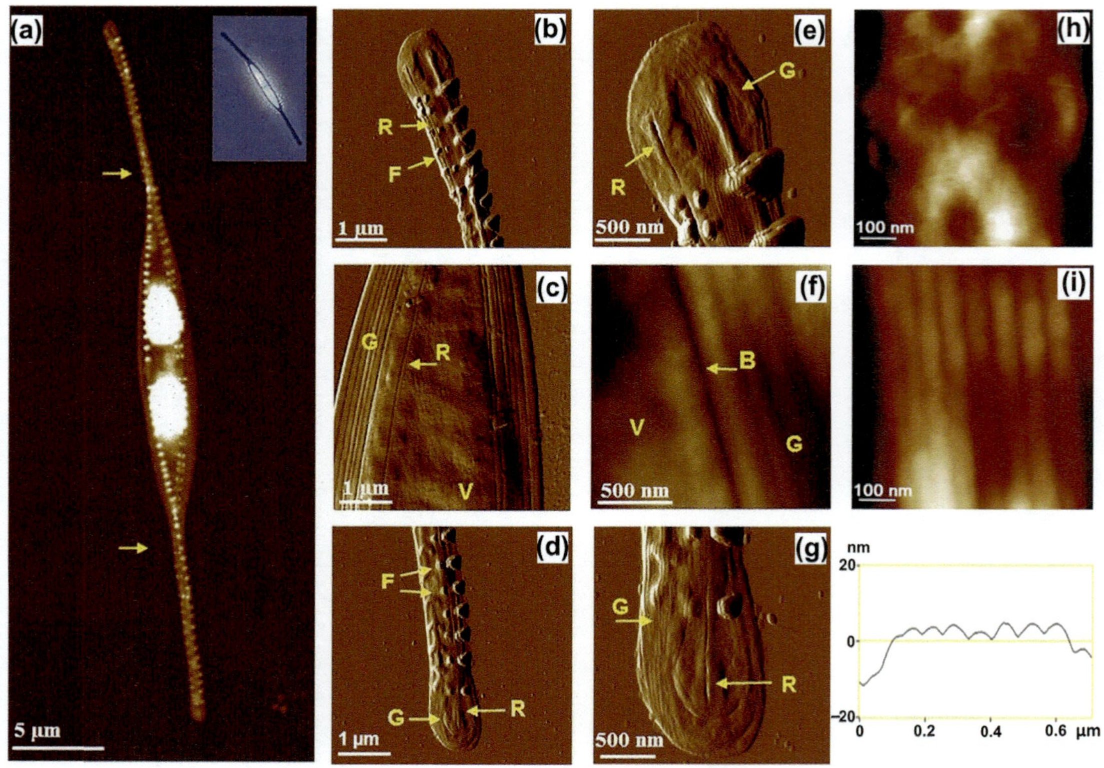

FIGURE 2.6 AFM image of a whole *Cylindrotheca closterium* cell (a) with an inset optical micrograph (b) and morphological details (i). The valve (h) and girdle (i) regions are compared at high resolution with a height profile of the girdle. Using vertical scales of 700 nm (a), 250 nm (f), 25 nm (h), and 250 nm (i), the images were taken in air using contact mode. Images are presented as height data (a, f, h, i) and diffraction data (b–e, g). The helical twisting about the longitudinal axis of the cell in (a) is indicated by arrows. Features are denoted by V, G, R, F, and B, representing the valve, girdle, raphe opening, fragments, and the boundary between them, respectively [From Ref. 6] [Open access].

sustainable energy solutions. Similarly, functional nanomaterials with tunable electrical, optical, and catalytic properties—such as graphene, carbon nanotubes, and metal-organic frameworks (MOFs)—are highly useful in hydrogen storage systems, supercapacitors, and fuel cells. Their applications in green technologies and renewable energy are expanding due to their ability to customize their performance through surface modification, nanostructure, or hybridization with other materials.

The application of smart materials in biomedical engineering is another emerging field of study. By reacting to changes in pH, temperature, or biological signals, stimuli-responsive hydrogels, bioactive ceramics, and nanoscale drug delivery systems can stimulate tissue regeneration or provide controlled drug release. Additionally, functional biomaterials have been integrated into biosensors and wearable technologies, enabling personalized medical care and real-time health monitoring. The impact of smart and functional materials is further enhanced by their integration into Industry 4.0 and digital manufacturing systems. Advanced processing techniques, including nanofabrication, thin-film deposition, and additive manufacturing (AM), enable the design of these materials with unprecedented precision and scalability. This opens new opportunities for smart devices, self-monitoring systems, and adaptable structures that bridge materials science and artificial intelligence.

In addition to serving as structural elements, smart and functional materials are powerful enablers of technological innovation. Advanced industrial applications, smart healthcare solutions, and sustainable energy systems are all made possible by their ability to sense, adapt, and interact with changing environments. To realize their full potential and accelerate the transition to a more technologically advanced and sustainable society, further research in this area is essential.

2.5 CHARACTERIZATION AND TESTING TECHNIQUES

Understanding the structural, physical, chemical, and functional properties of modern materials requires the use of characterization and testing techniques in materials science and engineering. Accurate characterization establishes a link between processing techniques, microstructure, and performance, as well as provides a deeper understanding of the internal structure and surface morphology of materials. Using these methods, engineers and researchers can ensure quality control, optimize materials design, and predict their long-term behavior under specific service conditions. Among the most common techniques is microscopic analysis, which includes AFM, transmission electron microscopy (TEM), light microscopy, and SEM. From the micron level to the atomic level, these methods enable direct imaging of material surfaces and microstructures. For example, the TEM enables observation of crystal structures and defects at the nanoscale, while the SEM provides accurate surface topography and composition through the interactions of electrons with the material. On the other hand, AFM is ideal for investigating thin films, coatings, and biological materials, allowing for three-dimensional mapping of surface properties at sub-nanometer resolution.

Spectroscopic methods analyze the chemical composition and bonding of materials to provide additional information. X-ray photoelectron spectroscopy (XPS), Raman spectroscopy, and Fourier-transform infrared spectroscopy (FTIR) are among the most widely used techniques for investigating surface chemistry and molecular structures. Meanwhile, XRD remains the primary technique for determining phase identification, lattice parameters, and crystal structures. These methods are essential for understanding how processing and composition affect a material's structure, which in turn impacts its performance. Mechanical testing techniques are essential for determining a material's strength, elasticity, hardness, and fracture toughness, in addition to structural and chemical investigation. Standard methods for assessing a material's mechanical behavior include tensile, compression, and fatigue tests. At the microscopic level, nanocavity techniques provide information on hardness and elastic modulus. These tests are essential to ensure that advanced materials meet the mechanical specifications for critical applications, including structural engineering, biomedical devices, and aircraft [7].

To determine their usability, functional materials must also be characterized electrically, magnetically, and thermally. Electrical conductivity, magnetic behavior, and thermal stability are typically evaluated using methods such as differential scanning calorimetry (DSC), vibrating sample magnetometry (VSM), Hall effect analysis, and four-point probe conductivity studies. The analysis of energy storage and conversion devices, such as batteries, fuel cells, and supercapacitors, requires electrochemical techniques such as cyclic voltammetry and impedance spectroscopy.

Finally, nondestructive testing techniques are essential to ensure reliability and safety without causing physical damage. Internal defects, cracks, or porosity in components intended for high-performance applications can be detected through ultrasonic testing, radiography, eddy current analysis, and acoustic emissions monitoring. A material's structure and properties cannot be fully understood through a single characterization or testing method. Whether structural, chemical, mechanical, or functional, each approach provides unique insights, but when combined, their full potential is revealed. Researchers can provide a comprehensive picture of materials behavior at various scales by combining microscopic, spectroscopic, mechanical, thermal, electrical, and nondestructive techniques. Rational design, optimization, and application of new materials in high-performance and sustainable applications depend on this comprehensive understanding [13]. Figure 2.7 shows a schematic diagram of this method. The modified AFM device used in this method is shown in Figure 2.7a, and the tip position when scratching a bulk PC sample under a light normal load is shown in Figure 2.7b. By comparing the measured depth with the machining depth, the elasticity of the sample was determined.

The idea behind the hydrogen economy is its potential to be a clean and efficient energy source. Because it burns to produce only water vapor, without harmful greenhouse gas emissions, it is considered one of the most promising fuels. Furthermore, it has a higher energy density than conventional fossil fuels, making it ideal for energy-intensive applications, such as the use of internal combustion engines.

Hydrogen is essential for converting industrial carbon dioxide emissions into useful chemicals, in addition to its direct applications as a fuel. For example, the German Carbon2Chem® program seeks to convert exhaust gases from steel mills into basic products such as ammonia, methanol, and other basic chemicals [8]. Combining carbon capture and utilization technologies with hydrogen is crucial for the development of hydrogen technology.

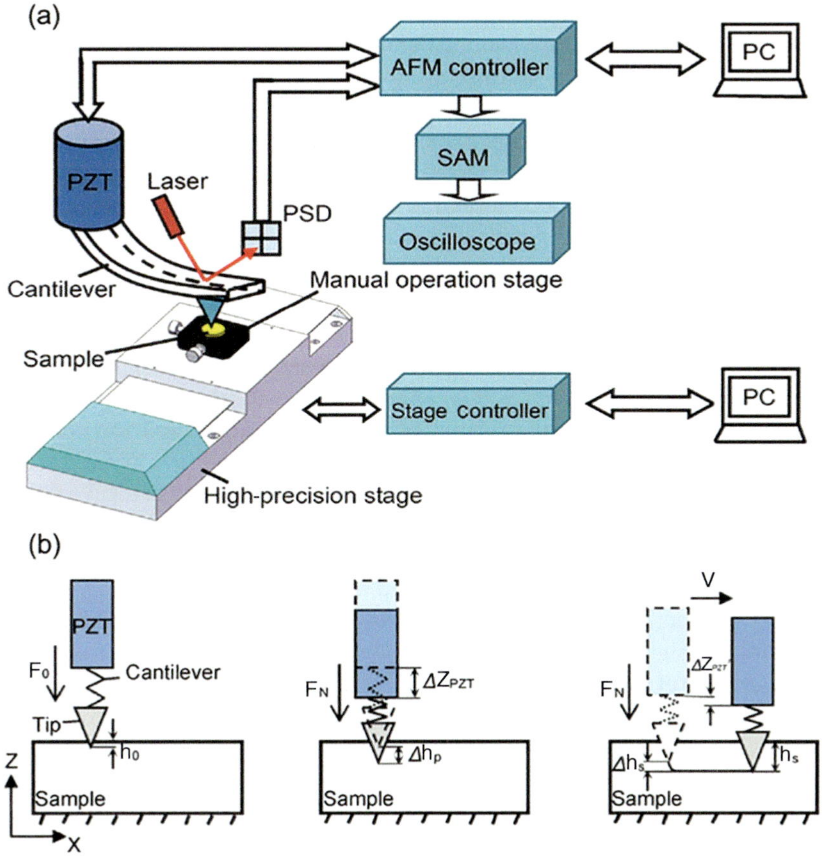

FIGURE 2.7 The modified AFM apparatus utilized in this method is shown in (a) and the tip position when scratching the PC bulk sample with a light normal load is shown in (b). The machining depth and the measured depth were compared to determine the sample's elastic recovery. In this work, the machined depth was obtained by utilizing an oscilloscope to observe the vertical signal change of the piezoceramic tube (PZT) in a modified AFM system. Using AFM tapping mode to image the groove, the measured depth was obtained. [From Ref. 7] [Open access].

The production of nanomaterials that act as carriers, storage media, and catalysts is critical for the development of hydrogen technology. Due to their ability to enhance hydrogen adsorption/dehydrogenation mechanisms, accelerate electrochemical reactions, and enhance electron transfer, zinc oxide (ZnO), nickel oxide (NiO), palladium nanoparticles, and other nanocomposites have been the subject of extensive research. While engineered nanomaterials, such as metal hydrides, carbon nanotubes, and porous frameworks (e.g., MOFs), offer higher storage capacities and faster release rates in hydrogen storage systems, nanocatalysts enhance electrode mobility and energy conversion efficiency in fuel cells. These developments are essential for overcoming the current limitations of hydrogen infrastructure.

The use of characterization and testing techniques is essential to ensure and improve the reliability of these nanomaterials. The shape, grain size, and alignment of nanostructures have been studied using TEM and SEM, while XRD provides important details about phase purity and crystal orientation. As demonstrated in zinc oxide tetrapods (T-ZnO), zinc oxide nanorods (ZnO-NRs) exhibit well-defined hexagonal structures and interconnected porous networks, which were verified by XRD and SEM investigations. Furthermore, the homogeneous distribution of low-agglomeration palladium nanoparticles was revealed by SEM combined with energy-dispersive spectroscopy (EDS), indicating the effectiveness of nanostructure design in catalytic activity [9]. Hydrogen technology is a key pillar of the global energy transition. In addition to efficient fuel cell devices and storage technologies, its success depends on the development of advanced nanomaterials that have been thoroughly characterized. By combining nanotechnology with advanced testing methods, researchers can accelerate the creation of reliable, scalable, and environmentally friendly hydrogen-based energy systems, opening the door to a truly low-carbon economy [11]. It is generally accepted that hydrogen technology is a key component of the global energy transition. Its success depends not only on reliable hydrogen storage technologies and efficient fuel cell systems, but also on the development of advanced nanomaterials specifically designed to improve stability, energy conversion efficiency, and catalytic activity. The distribution of active sites, surface properties, and structural design of these nanomaterials significantly influence their performance and require careful characterization. By combining nanotechnology with advanced testing techniques, researchers can accelerate the development of reliable, scalable, and environmentally sustainable hydrogen-based energy systems, opening the door to a truly low-carbon economy [12].

SEM images of T-ZnO and ZnO-NR films at different magnifications, as shown in Figure 2.8, provide an illustrative example of this characterization. The two images (Figure 2.8A) show the unstructured surfaces of the original films, and (Figure 2.8B) the surfaces after functionalization with palladium nanoparticles. The uniform distribution of palladium (green) was confirmed by complementary elemental x-ray (EDX) mapping, shown in the right column. This indicates successful surface modification and nanoparticle incorporation [8]. These results demonstrate how both palladium functionality and its nanostructure directly influence catalytic activity, increasing the efficiency of processes involving hydrogen. Ultimately, these developments support the development of high-capacity hydrogen storage systems and improved fuel cell electrodes, highlighting the critical role that nanomaterials characterization plays in enabling the next generation of clean energy technologies.

Although hydrogen does not emit any pollutants, its usefulness depends on its energy efficiency [14]. Water electrolysis is the most environmentally responsible method for producing hydrogen using electricity from renewable sources because it has no carbon footprint. Global consumption of renewable energy is expected to be low for the near future, although it is expected to rise in the coming years [15]. Regional limitations, the intermittent nature of renewable energy sources, and the need to resolve energy storage concerns are some of the barriers [14]. For example, up to 2,000 people can be supplied with electricity annually from onshore wind and solar energy resources in Germany, which is far below the amount needed to meet

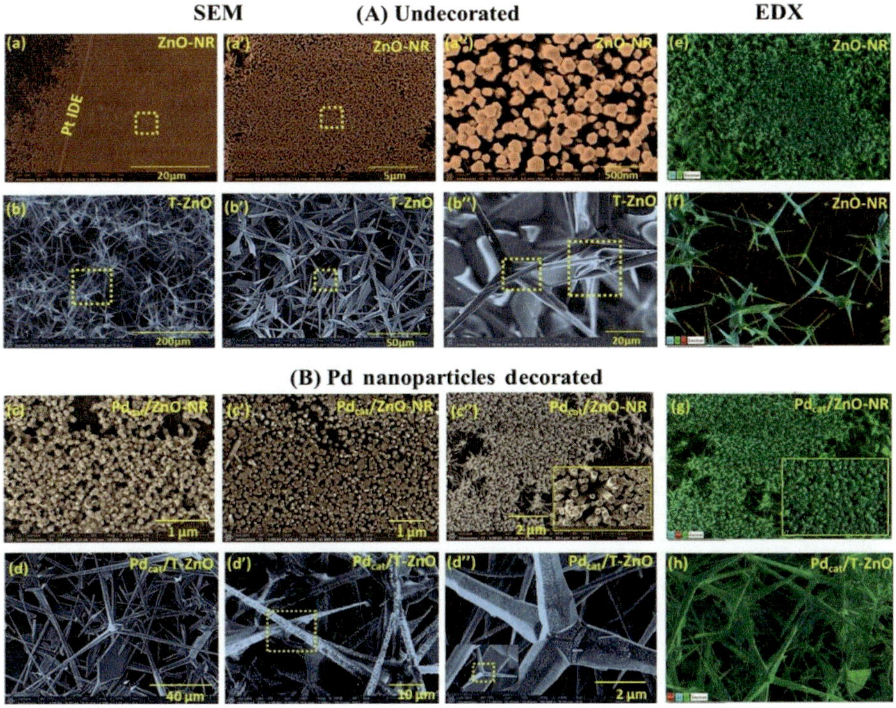

FIGURE 2.8 SEM images of T-ZnO and ZnO-NR films at different magnifications show (A) their unstructured surfaces and (B) their palladium-functionalized surfaces. EDX images showing the elemental mapping and distribution of palladium nanoparticles are shown in the right column (green) [From Ref. 8] [Open access].

the country's annual electricity consumption. [16]. More electricity may be produced from offshore wind power, but its transmission from the coast to inland areas remains challenging. Energy fluctuations will alternate periods of energy scarcity with seasons of abundant production. By storing excess energy, the intermittent characteristics of renewable energy can be mitigated. However, large-scale energy storage remains challenging [15].

The catalysts studied have proven more robust and capable of producing hydrogen after being reinforced with porous supports [8,11]. The effect of additional catalytic supports, such as MgO, Al_2O_3, SiO_2, TiO_2, ZrO_2, $MgOSiO_2$, $Al_2O_3SiO_2$, and H+-ZSM-5, etc., has been studied [15]. SiO_2 is the most efficient thermal conductivity detector (TCD) catalyst support for nanostructured hydrogen generation. Therefore, a highly stable nickel nanocatalyst was produced on porous silicon dioxide [17]. The catalyst was produced by co-precipitation, a modified technique. The lowest deactivation rate (after 300 min) occurred at 500 °C, and the highest hydrogen production (40.4%) occurred at 600 °C. This conclusion was reinforced by the fact that the nickel/silicon dioxide catalyst produced more carbon than the iron/silicon dioxide catalyst [8]. The results of comprehensive studies have shown the following hierarchy of catalytic properties in terms of stability and activity [7,8]. When methane is decomposed according to a temperature schedule, hydrogen is produced on different catalysts (n-Ni/SiO_2 > n-Co > n-Fe; n-Fe/SiO_2 > n-Ni > n-Co > n-Fe).

2.6 APPLICATIONS IN INDUSTRY

Advanced materials, which offer innovative solutions that enhance performance, efficiency, and sustainability across various industries, have emerged as a crucial component of contemporary industrial development. In the context of the Fourth Industrial Revolution and the global shift toward cleaner technologies, their distinctive mechanical, physical, and chemical properties—such as high strength-to-weight ratios, corrosion resistance, exceptional thermal stability, and versatility—make them essential for meeting the challenges of the 21st century. Advanced composites, ceramics, and lightweight alloys are used in the aerospace, automotive, and aviation industries to reduce the weight of structures while maintaining or enhancing mechanical strength and integrity. This results in better overall performance, lower emissions, and higher fuel efficiency. For example, improved steel and aluminum alloys help build high-performance cars, while titanium alloys and carbon fiber-reinforced polymers (CFRPs) are widely used in jet engines and airframes [10,16,18–20]. Table 2.1 shows that the period covered by this work is from 2005 to 2024. In this study, 1290 and 2851 sources (journals, etc.) and other research papers were obtained, respectively. The number of publications increased at an annual rate of 31.58% during this period.

TABLE 2.1
Main Information on AM and Industry 4.0 and 5.0 from SCOPUS and WOS Databases

Category	Parameter	Results
General information	Timespan	2005–2024
General information	Sources (journals, books, etc.)	1,290
General information	Documents	2,851
General information	Annual growth rate (%)	31.58
General information	Document average age (years)	3.21
General information	Average citations per document	15.57
General information	References	117,470
Document contents	Keywords plus (ID)	9,531
Document contents	Author's keywords (DE)	6,860
Authors	Total authors	8,246
Authors	Authors of single-authored documents	178
Authors collaboration	Single-authored documents	197
Authors collaboration	Co-authors per document	4.06
Authors collaboration	International co-authorship (%)	11.36

Source: From Ref. 21 [Open access].

This indicates that researchers' interest in this field increased annually between 2005 and April 2024. Another crucial factor is also shown in Table 2.1.

Figure 2.9 shows, as an example from our study, AFM measurements of poly(acrylonitrile) (PAN) nanofiber mats obtained under identical conditions, after electrospinning (Figure 2.9a and c) and after hot pressing at 180 °C (Figure 2.9b and d). The topography images show clear morphological changes: the nanofibers appear significantly thicker after hot pressing, which is consistent with the expected structural densification. However, the fibers appear to form abnormally high walls in the electrospinning condition (Figure 2.9a), a scanning defect resulting from subtle fiber displacements during imaging. The accompanying phase image (Figure 2.9c), which excludes true structural defects and shows well-separated fiber surfaces with clear

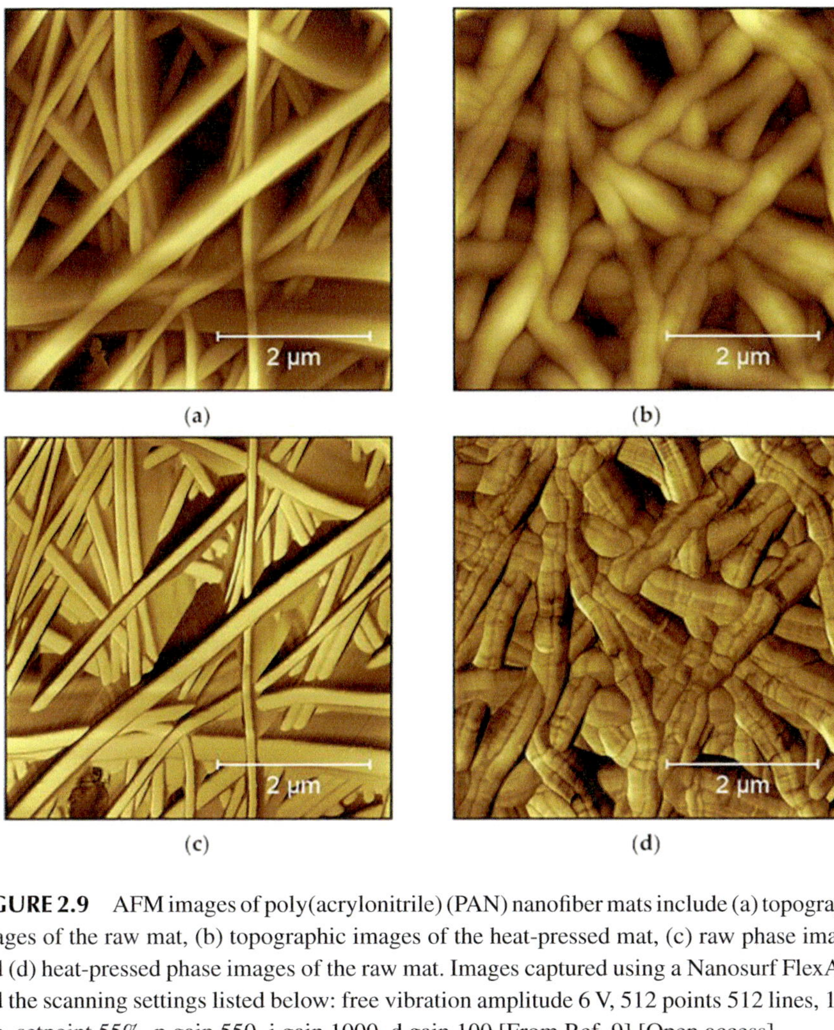

FIGURE 2.9 AFM images of poly(acrylonitrile) (PAN) nanofiber mats include (a) topographic images of the raw mat, (b) topographic images of the heat-pressed mat, (c) raw phase images, and (d) heat-pressed phase images of the raw mat. Images captured using a Nanosurf FlexAFM and the scanning settings listed below: free vibration amplitude 6 V, 512 points 512 lines, 1.4 s/line, setpoint 55%, p gain 550, i gain 1000, d gain 100 [From Ref. 9] [Open access].

phase contrasts, supports this conclusion. On the other hand, the phase image of the heat-pressed sample (Figure 2.9d) shows distinct dark lines, which are also slightly distinguishable in the topography (Figure 2.9b). These patterns reflect heating-induced topographic constrictions, not compositional differences. These constrictions, which validate the structural changes induced by heat compression, are typical of the heat-stabilization process of PAN nanofiber mats.

Advanced materials are essential for creating efficient energy storage devices and renewable energy systems in the energy sector. The use of nanoscale catalysts, membranes, and electrodes improves the performance of fuel cells, solar panels, supercapacitors, and lithium-ion batteries, opening the door to more reliable and scalable sustainable energy solutions. Furthermore, corrosion-resistant materials and advanced thermal insulation coatings extend the life of reactors and turbines, reducing operating costs and improving sustainability. Advanced materials have also made significant contributions to the biomedical sector. Medical implants, targeted drug delivery systems, and diagnostic devices utilize biocompatible polymers, smart hydrogels, and nanomaterials. Bioresorbable structures and shape-memory alloys enable the development of new treatments that improve patient outcomes and reduce long-term risks.

Advanced semiconductors, nanomaterials, and functional coatings contribute to the miniaturization, speed, and energy efficiency of devices in electronics and information technology. Materials such as graphene, quantum dots, and high-K (High-K) insulators have fueled the development of artificial intelligence, the Internet of Things, and 5G and 6G communications technologies, opening new horizons for next-generation transistors, sensors, and flexible electronic systems. The chemical and manufacturing industries rely on advanced materials to produce specialized catalysts, membranes, and coatings that increase reaction efficiency, reduce energy consumption, and minimize waste. Meanwhile, rapid prototyping and customization, enabled by advanced metal powders and polymers, are revolutionizing industrial production through AM (3D printing) technologies.

Overall, by reducing carbon emissions, conserving resources, and promoting circular economy models, the integration of advanced materials across industries directly supports sustainability goals, while also enhancing product performance and energy efficiency. Digital manufacturing, nanotechnology, and the ongoing development of materials science ensure that industrial applications continue to evolve toward smarter, more resilient, and more sustainable systems [22–26].

2.7 CHALLENGES AND LIMITATIONS

Despite their rapid development and promising applications, numerous barriers and limitations still prevent the widespread use of advanced materials in industry and research. These challenges are complex and include issues related to cost, scalability, durability, and the environment, as well as limitations on synthesis and processing. Understanding these limitations is essential to guiding future research efforts and bridging the gap between laboratory innovation and industrial applications. One of the most significant barriers is the complexity of synthesis and processing. The production of many innovative materials, including shape-memory alloys, functional coatings, and nanocomposites, requires highly regulated conditions. These conditions often require the use of expensive raw materials, strict environmental regulations, or specialized equipment. Therefore, achieving consistency and reproducibility in large-scale production remains a significant hurdle.

Scalability and cost are equally important issues. Despite the exceptional performance of new materials in the laboratory, scaling up manufacturing without sacrificing quality can be difficult. Large-scale industrial use is severely hampered by high production costs and the scarcity of raw materials, such as rare earth elements needed in high-performance catalysts or magnets. Long-term stability and durability are another challenge. When many nanomaterials and functional materials are exposed to realistic operating environments, such as varying temperatures, humidity, or corrosive atmospheres, they gradually deteriorate despite exhibiting exceptional performance under tightly controlled experimental conditions. This discrepancy between short-term laboratory performance and long-term industrial reliability raises questions about their practical utility in demanding applications, such as energy storage, biomedical implants, and aerospace.

Figure 2.10a shows the AFM morphology of silver nanoparticles (Ag NPs) deposited on Poly[cyclopenta[2,1-b;3,4-b′]-dithiophene-alt-benzothiadiazole] (PCPDTBT) membranes. The average particle size of these particles is approximately 120 nm (Figure 2.10d). Additional information about the electronic behavior of the Ag-PCPDTBT composite can be obtained from related SKPM measurements.

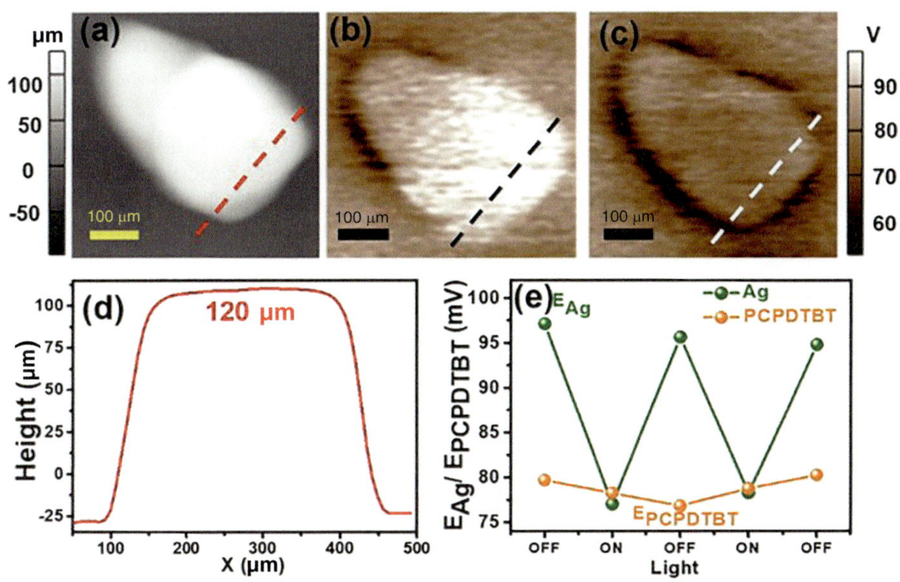

FIGURE 2.10 (a) AFM topographic images of silver nanoparticles on PCPDTBT films; (b) AFM images of the surface potential of silver nanoparticles on PCPDTBT films with and without light excitation at a wavelength of 488 nm; (c) table 120 nm; (d) surface potential of photo-modified silver nanoparticles on PCPDTBT films [From Ref. 11] [Open access].

Figure 2.10b shows the surface potential distribution without light excitation, while Figure 2.10c shows the surface potential over the same area when illuminated at a wavelength of 488 nm. These results demonstrate that the silver nanoparticles completely cover the conductive polymer matrix. Even when the illumination raises the membrane potential, this enhancement alone cannot alleviate the charge separation at the Ag-PCPDTBT interface. Light-free SKPM images (Figure 2.10b) reveal that the silver nanoparticles exhibit a higher potential than the surrounding polymer, as evidenced by the brighter regions in the image. The average surface potential of the silver nanoparticles, measured quantitatively, is 96 ± 3 mV, while that of the PCPDTBT polymer is slightly lower, 79 ± 2 mV. Although slight variations are attributed to local heterogeneity, the surface potential patterns plotted along the dotted lines in Figure 2.10b and c (see Figure 2.10a and b) further support these patterns by showing significant changes in the potential distribution of the silver nanoparticles and the polymer matrix upon illumination. Considered together, these results demonstrate the vital role of the metal–polymer interfaces in enabling light-induced charge dynamics, highlighting their potential.

Health and environmental issues also pose a significant obstacle. In particular, we still do not fully understand the toxicity and environmental impact of nanomaterials. Uncontrolled release of nanoparticles during production, use, or disposal could pose a risk to ecosystems and human health. Furthermore, if not properly controlled, the energy-intensive processes required to manufacture some new materials could outweigh their environmental benefits. The lack of standardized testing and classification methods is another obstacle. Comparing the results of different studies becomes more challenging due to the diversity of modern materials, which often requires specialized measurement methods. Industries find it difficult to evaluate the performance, safety, and reliability of these materials in the absence of globally recognized standards, thereby delaying their adoption for commercial applications.

There are several advantages and disadvantages of the interdisciplinary nature of advanced materials research. Differences in terminologies, methodologies, and technical priorities can hinder innovation, even though collaboration between materials scientists, engineers, chemists, and physicists is crucial. Increased funding for knowledge transfer systems, training, and collaboration platforms is essential to bridge these gaps [27–29].

Finally, while new materials offer opportunities for industrial transformation, issues of synthesis, cost, scalability, durability, environmental safety, and standardization limit their full potential. In addition to technological innovation, international collaboration, sustainable production methods, and regulatory frameworks will be essential to overcome these limitations. Only by removing these barriers can advanced materials fully contribute to the development of high-performance, affordable, and environmentally friendly technologies [5,30–33].

2.8 FUTURE PROSPECTS

The field of advanced materials is expected to have a significant impact in shaping future technologies and industries. The production of next-generation materials will become increasingly important as the world faces more pressing issues, including resource scarcity, climate change, and the growing need for high-performance electronics. In addition to improving the intrinsic properties of materials, the prospects of this field enable multifunctionality, sustainability, and comprehensive integration into industrial applications.

Figure 2.8 shows an example of a MATLAB-based segmentation method. Identifying the extracellular matrix (ECM) and extracting the relevant values from the selected location are the two main operations in this procedure. The stiffness map image generated by the JPK software is initially used to detect the extracellular matrix. The stiffness map and its corresponding scale are shown in this figure (Figure 2.11a). The scale is cropped to create the appropriate image region because the study only requires stiffness data (Figure 2.11b). The green channel of the cropped image is then separated and passed through a thresholding process, producing a binary mask that highlights the extracellular matrix regions (Figure 2.11c).

Using a TSV file containing a list of (x, y) coordinates and their corresponding Young's modulus, the values of the selected regions are retrieved in the second stage. A contour map similar to the original JPK image is produced by mapping the Young's modulus values to a mesh grid constructed using these location values, corresponding to the image format (Figure 2.11d). To isolate regions of interest, a contour plot is finally applied to the previously created binary mask. The resulting plot displays only the ECM or selected regions, allowing for interactive searching. The user can either generate a comprehensive list of values for quantitative study of the selected region or hover the cursor over specific parts to obtain local Young's modulus values.

One promising trend is the development of smart and adaptable materials capable of dynamically responding to external stimuli, such as light, pressure, temperature, and magnetic fields. These materials, which include self-healing polymers, piezoelectric ceramics, and shape-memory alloys, are expected to spur innovation in diverse fields, from wearable electronics and biomedical engineering to robotics and aircraft.

Equally important are advances in nanotechnology and nanomaterials, which will radically transform energy conversion, storage, and stimulation. To achieve higher efficiency, greater stability, and lower costs, future energy systems, such as perovskite solar cells, hydrogen fuel cells, and high-capacity lithium-sulfur batteries, will rely primarily on synthetic nanomaterials. Furthermore, opportunities now exist to create materials with unprecedented electrical, optical, and magnetic capabilities, because of the convergence of nanomaterials and quantum technologies. The development of innovative materials will also be influenced by the circular economy and environmental concepts. Initiatives are already underway to replace key raw materials with readily available alternatives on Earth, produce recyclable composite materials, and adopt greener synthetic processes that consume less energy and produce fewer chemicals. Advanced computer-aided design, nanotechnology, and sustainable production methods will play a greater role in the development of materials in the future.

Because of high-throughput computational modeling, machine learning, and artificial intelligence, digital materials design holds great promise. By predicting structure-property connections and screening large material libraries prior to experimental validation, these approaches will accelerate materials discovery and optimization. The design-to-implementation process will be streamlined by combining advanced

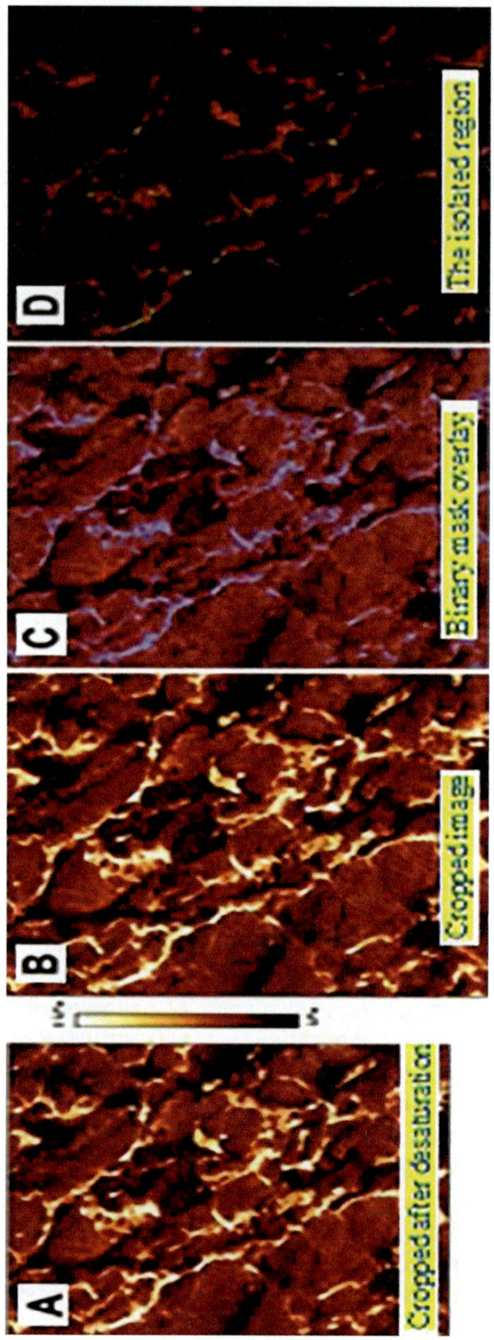

FIGURE 2.11 Representative image of the Matlab segmentation approach. Panels from left to right show: (a) original JPK image, (b) cropped image to removed scale, (c) cropped image with binary mask overlayed (blue), and (d) contour plot of isolated region. [From Ref. 12] [Open access].

experimental methods, such as optical microscopy and real-time spectroscopy, with computational tools.

Scalability and commercialization remain important goals at the industrial level. Advances in automated processing, standardized testing procedures, and AM are essential to bridge the gap between laboratory-scale innovation and large-scale manufacturing. To establish robust frameworks that ensure the reliability, affordability, and safety of new materials in practical applications, governments, businesses, and academics will need to collaborate. In short, the ability of innovative materials to combine cost-effectiveness, sustainability, adaptability, and high performance will determine their future. The next generation of materials will transform industries and help address some of the most important global issues of our time by combining advances in digital design, smart materials, nanotechnology, and green production [34–38].

2.9 MOTIVATIONAL PERFORMANCE MONITORING

Owing to the discovery that nanostructured materials have unique features compared to their bulk counterparts, nanoscience and nanotechnology have received much attention in recent decades [37]. Due to size effects, the macroscopic quantum tunneling effect, and finite surface effect at nanoscale, nanostructured materials have unique optical, electrical, and magnetic properties. Therefore, the creation of catalysts, fuel cells, gas sensors, photoelectronic devices, energy storage devices, supercapacitors, and lithium-ion batteries becomes possible due to the wide range of applications that nanostructured materials open [38]. Currently, the subject of close attention is the production of fine nanopowders of excellent quality in size, shape, and structure. Various methods have been proposed to create materials at the nanoscale, including solvothermal synthesis [21], carbonylation and the following: anode-arc plasma method, coprecipitation, microwave irradiation, microemulsion sonochemistry, ultrasound and sol-gel procedures. Flexibility in changing synthesis parameters and controlling the physicochemical properties of nanopowders without the need for expensive chemicals or complex equipment is a significant advantage of the co-precipitation approach. Due to their exceptional activity, stability, low toxicity, and environmental friendliness, nickel-based nanostructured materials have been widely studied in the field of catalysis [31–33]. In addition, various materials other than catalysts have been developed using nickel-based nanomaterials, including electrochromic materials [25], p–n heterojunctions, gas sensors and magnetic materials [32], optical materials [30] and fuel cell electrodes [31], solar cells [25,26], batteries [33] ,and electrochemical capacitors. In the present study, a modified Stöber co-precipitation method is used to create silicate-supported nano nickel catalysts with crystal diameters of about 30 nm [19]. Methane decomposition is carried out using synthesized nickel catalysts supported on silicate, and then the activity and stability of methane conversion are assessed. One of the newest products in the fields of materials science and sustainable energy is hydrogen. In modern conditions of significant depletion of fossil fuel reserves and degradation of other energy sources, hydrogen can be considered a viable environmental energy carrier as a result of increasing greenhouse gas emissions [10].

Because it produces no greenhouse gas emissions when burned, hydrogen plays a critical role in solving current energy problems.

The morphological shapes of TiO_2 nanorod arrays (NRAs), TiO_2 NP/NRA composites, and branched TiO_2 nanorod arrays are shown in Figure 2.12(a–f). While the side-view SEM image in Figure 2.12 (inset) shows the nanorods arranged almost perpendicularly to the substrate with an average length of approximately 2.5 µm, the surface-view SEM image in Figure 2.12(a) reveals regular, rod-like structures. The original square-shaped nanorods transformed into spherical shapes after the deposition of TiO_2 nanoparticles onto the TiO_2 nanorod arrays and subsequent heat treatment to create TiO_2 NP/NRA composites (Figure 2.12c,

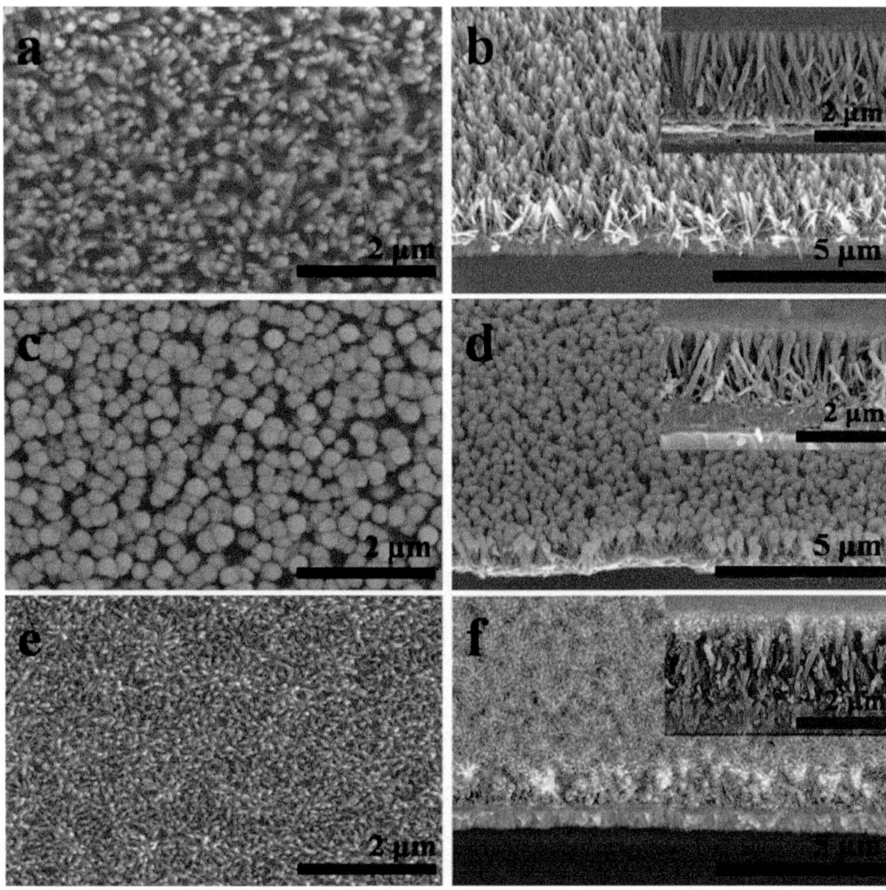

FIGURE 2.12 Surface and side-view SEM images showing the morphological evolution of TiO_2 nanorod arrays: (a,b) pristine TiO_2 NRAs, where (a) represents the surface view and (b) the side view; (c,d) TiO_2 nanoparticle-decorated nanorod arrays (NPs/NRAs), with (c) surface view and (d) side view; and (e,f) branched TiO_2 NRAs, where (e) shows the surface morphology and (f) the side-view structure. The insets in (b), (d), and (f) display high-magnification SEM images of the corresponding structures [39] [Open access].

d). Tree-like titanium dioxide (TiO_2) nanotubes were efficiently produced after a second hydrothermal growth process (Figure 2.12e, f). The nanotubule tips were densely and uniformly covered with branches. These branches grew primarily from the nanotube tips and branched out in all directions, resulting in a spherical shape, as shown in Figure 2.12f. This significantly increased the surface area of the TiO_2 nanotubes.

Moreover, renewable feedstocks, including water, biomass, and biogas, can be used to create hydrogen [16,18]. As methane constitutes a significant portion of biomass and is available in huge quantities both in industrialized countries and on the deep ocean floor, an all-green methane-splitting technology can be considered a sustainable method for hydrogen production [21]. Therefore, to simplify matters, the automotive industry, academic institutions, and governments have focused more attention on the potential use of hydrogen as an alternative fuel, both for widespread adoption and production.

PROBLEMS

1. What is the effect of microstructure on the mechanical properties of advanced materials?
2. How do conventional and nanomaterials differ in their physical properties?
3. What is the effect of surface morphology on the performance of fibers and composites?
4. How do hardness and wear resistance change when nanoparticles are added?
5. Why are characterization methods such as AFM and SEM so important for understanding material properties?

REFERENCES

1. Cieśluk, M., Deptuła, P., Piktel, E., Fiedoruk, K., Suprewicz, Ł., Paprocka, P., ... & Bucki, R. (2020). Physics comes to the aid of medicine—Clinically-relevant microorganisms through the eyes of atomic force microscope. *Pathogens*, 9(11), 969.
2. Ntasiou, P., Kerou, A. K., Karamanidou, T., Vlachou, A., Tziros, G. T., Tsouknidas, A., & Karaoglanidis, G. S. (2021). Synthesis and characterization of novel copper nanoparticles for the control of leaf spot and anthracnose diseases of olive. *Nanomaterials*, 11, 1667.
3. Wang, N., Zheng, W., Cheng, S., Zhang, W., Liu, S., & Jiang, X. (2017). In vitro evaluation of essential mechanical properties and cell behaviors of a novel polylactic-co-glycolic acid (PLGA)-based tubular scaffold for small-diameter vascular tissue engineering. *Polymers*, 9(8), 318.
4. Wu, Y., Wu, X., Shi, T., Chen, H., Wang, H., Sun, M., & Zhang, J. (2019). The microstructure and mechanical properties of poplar catkin fibers evaluated by atomic force microscope (AFM) and nanoindentation. *Forests*, 10(11), 938.
5. Hernán, D. M. V., Aida, M. E., & Gilberto, B. G. (2020). Detailed study of the electrochemical behavior of low-reflectivity TiAlN coatings. *Surf. Innov.*, 9, 293–307.
6. Mišić Radić, T., Vukosav, P., Čačković, A., & Dulebo, A. (2023). Insights into the morphology and surface properties of microalgae at the nanoscale by atomic force microscopy (AFM): A review. *Water*, 15(11), 1983.

7. Yan, Y., Chang, S., Wang, T., & Geng, Y. (2019). Scratch on polymer materials using AFM tip-based approach: A review. *Polymers*, 11(10), 1590.
8. Knoepfel, A., Poudel, B., & Gupta, S. (2023). Surface-catalyzed zinc oxide nanorods and interconnected tetrapods as efficient methane gas sensing platforms. *Chemosensors*, 11(9), 506.
9. Joshi, J., Homburg, S. V., & Ehrmann, A. (2022). Atomic force microscopy (AFM) on biopolymers and hydrogels for biotechnological applications—Possibilities and limits. *Polymers*, 14(6), 1267.
10. Łątka, P., Olszański, B., Żurowska, M., Dębosz, M., Rokicińska, A., & Kuśtrowski, P. (2024). Spherical lignin-derived activated carbons for the adsorption of phenol from aqueous media. *Molecules,* 29(5), 960.
11. Wang, J., Wang, X., Yan, J., Tan, Q., Liang, G., Qu, S., & Zhong, Z. (2018). Enhanced photoelectrochemical properties of Ti3+ self-doped branched TiO2 nanorod arrays with visible light absorption. Materials, 11(10), 1791..
12. Peña, B., Adbel-Hafiz, M., Cavasin, M., Mestroni, L., & Sbaizero, O. (2022). Atomic force microscopy (AFM) applications in arrhythmogenic cardiomyopathy. *Int. J. Mol. Sci.*, 23(7), 3700.
13. Yasakau, K. (2020). Application of AFM-based techniques in studies of corrosion and corrosion inhibition of metallic alloys. *Corros. Mater. Degrad.* 1(3), 345–372.
14. Vobornik, D., Chen, M., Zou, S., & Lopinski, G. P. (2023). Measuring the diameter of single-wall carbon nanotubes using AFM. *Nanomaterials*, 13(3), 477.
15. Lin, Y. H., Sivakumar, C., Balraj, B., Murugesan, G., Nagarajan, S. K., & Ho, M. S. (2023). Ag-decorated vertically aligned ZnO nanorods for non-enzymatic glucose sensor applications. *Nanomaterials,* 13(4), 754.
16. Zaki, S. A., Ouf, S. A., Albarakaty, F. M., Habeb, M. M., Aly, A. A., & Abd-Elsalam, K. A. (2021). Trichoderma harzianum-mediated ZnO NPs: A green tool for controlling soil-borne pathogens in cotton. *J. Fungi,* 7, 952.
17. Hassan, Q., Ahmadi, S., & Kerman, K. (2020). Recent advances in monitoring cell behavior using cell-based impedance spectroscopy. *Micromachines,* 11(6), 590.
18. Iliger, K. S., Sofi, T. A., Bhat, N. A., Ahanger, F. A., Sekhar, J. C., Elhendi, A. Z., Al-Huqail, A. A., & Khan, F. (2021). Copper nanoparticles: Green synthesis and managing fruit rot disease of chilli caused by *Colletotrichum capsici*. *Saudi J. Biol. Sci.,* 28, 1477–1486.
19. Abdelaziz, A. M., Salem, S. S., Khalil, A. M., El-Wakil, D. A., Fouda, H. M., & Hashem, A. H. (2022). Potential of biosynthesized zinc oxide NPs to control Fusarium wilt disease in eggplant (*Solanum melongena*) and promote plant growth. *Biometals,* 35, 601–616.
20. Abdelkhalek, A., Qari, S. H., Abu-Saied, M. A. A.-R., Khalil, A. M., Younes, H. A., Nehela, Y., & Behiry, S. I. (2021). Chitosan nanoparticles inactivate alfalfa mosaic virus replication and boost innate immunity in *Nicotiana glutinosa* plants. *Plants*, 10, 2701.
21. Dehghan, S., Sattarpanah Karganroudi, S., Echchakoui, S., & Barka, N. (2025). The integration of additive manufacturing into industry 4.0 and industry 5.0: A bibliometric analysis (trends, opportunities, and challenges). *Machines*, 13(1), 62.
22. Abdelrhim, A. S, Mazrou, Y. S. A., Nehela, Y., Atallah, O. O., El-Ashmony, R. M., & Dawood, M. F. A. (2021). Silicon dioxide nanoparticles induce innate immune responses and activate antioxidant machinery in wheat against *Rhizoctonia solani*. *Plants*, 10, 2758.
23. Hashem, A. H., Abdelaziz, A. M., Askar, A. A., Fouda, H. M., Khalil, A. M. A., Abd-Elsalam, K. A., & Khaleil, M. M. (2021). *Bacillus megaterium*-mediated synthesis of

selenium nanoparticles and their antifungal activity against *Rhizoctonia solani* in faba bean plants. *J. Fungi,* 7, 195.
24. Kondo, Y., Yoshikawa, H., Awaga, K., Murayama, M., Mori, T., Sunada, K., ... & Iijima, S. (2008). Preparation, photocatalytic activities, and dye-sensitized solar-cell performance of submicron-scale TiO2 hollow spheres. *Langmuir*, 24(2), 547–550.
25. Pei, L., Lin, D., Yuan, S., Lu, R., Bai, Z., Sun, Y., Zhu, Y., Jiang, Y., Zhu, J., & Wang, H. (2023). A multifunctional and long-term waterborne anti-corrosion coating with excellent 'hexagonal warrior' properties. *Chem. Eng. J.,* 457, 141158.
26. Zhang, Y., Wu, H., & Guo, S. (2022). Sandwich-structured surface coating of a silver-decorated electrospun thermoplastic polyurethane fibrous film for excellent electromagnetic interference shielding with low reflectivity and favorable durability. *ACS Appl. Mater. Interfaces,* 14, 40351–40360.
27. Kabeerdass, N., Al Otaibi, A., Rajendran, M., Manikandan, A., Kashmery, H. A., Rahman, M. M., Madhu, P., Khan, A., Asiri, A. M., & Mathanmohun, M. (2021). Bacillus-mediated silver nanoparticle synthesis and its antagonistic activity against bacterial and fungal pathogens. *Antibiotics,* 10, 1334.
28. Mostafa, Y. S., Alamri, S. A., Alrumman, S. A., Hashem, M., & Baka, Z. A. (2021). Green synthesis of silver nanoparticles using pomegranate and orange peel extracts and their antifungal activity against *Alternaria solani*, the causal agent of early blight disease of tomato. *Plants,* 10, 2363.
29. Zhao, Y., Min, X., Ding, Z., Chen, S., Ai, C., Liu, Z., & Fang, M. (2020). Metal-based nanocatalysts via a universal design on cellular structure. *Adv. Sci.,* 7(3), 1902051.
30. Picas, J., Menargues, S., Martin, E., & Baile, M. (2023). Cobalt free metallic binders for HVOF thermal sprayed wear resistant coatings. *Surf. Coat. Technol.,* 456, 129243.
31. Zudyte, B., & Luksiene, Z. (2021). Visible light-activated ZnO nanoparticles for microbial control of wheat crop. *J. Photochem. Photobiol. B. Biol.*, 219, 112206.
32. Dawoud, T. M., Yassin, M. A., El-Samawaty, A. R. M., & Elgorban, A. M. (2020). Silver nanoparticles synthesized by *Nigrospora oryzae* showed antifungal activity. *Saudi J. Biol. Sci.,* 28, 1847–1852.
33. Sofy, A. R., Sofy, M. R., Hmed, A. A., Dawoud, R. A., Alnaggar, A. E.-A. M., Soliman, A. M., & El-Dougdoug, N. K. (2021). Ameliorating the adverse effects of tomato mosaic tobamovirus infecting tomato plants in Egypt by boosting immunity in tomato plants using zinc oxide nanoparticles. *Molecules,* 26, 1337.
34. Helal, H., Ahrouch, M., Rabehi, A., Zappa, D., & Comini, E. (2024). Nanostructured materials for enhanced performance of solid oxide fuel cells: A comprehensive review. *Crystals,* 14(4), 306.
35. Rauschenbach, B., Lotnyk, A., Neumann, L., Poppitz, D., & Gerlach, J. W. (2017). Ion beam assisted deposition of thin epitaxial GaN films. *Materials,* 10(7), 690.
36. Gibson, C. T. (2020). The attachment of carbon nanotubes to atomic force microscopy tips using the pick-up method. *Appl. Sci.*, 10(16), 5575.
37. Wang, D., Zhao, Y., Zhang, S., Bao, L., Li, H., Xu, J., & Hou, X. (2022). Reporter molecules embedded Au@ Ag core-shell nanospheres as SERS nanotags for cardiac troponin I detection. *Biosensors,* 12(12), 1108.
38. Chen, S., Yang, Y., Zhang, M., Ma, X., He, X., Wang, T., & Mao, X. (2022). Bioinspired Pd-Cu alloy nanoparticles as accept agent for dye degradation performances. *Int. J. Mol. Sci.,* 23(22), 14072.
39. Wang, J., Wang, X., Yan, J., Tan, Q., Liang, G., Qu, S., & Zhong, Z. (2018). Enhanced photoelectrochemical properties of Ti3+ self-doped branched TiO2 nanorod arrays with visible light absorption. *Materials*, 11(10), 1791.

3 Processing and Fabrication Techniques

3.1 INTRODUCTION

The structural, mechanical, and functional performance of advanced materials is largely determined by the processing and manufacturing processes, which form the basis of contemporary materials research. The need for high-performance materials with superior properties has driven the development of both traditional and innovative production technologies over the past few decades. Industry continues to use traditional techniques, including casting, forging, extrusion, rolling, and welding, due to their cost-effectiveness, scalability, and reliability. These technologies are essential for industries such as heavy machinery, construction, and automotive, as they are particularly suitable for producing large-scale components with predictable mechanical properties [1–3]. The development of contemporary production technologies has greatly expanded the possibilities of materials engineering. Three-dimensional (3D) printing, also known as additive manufacturing (AM), enables the creation of lightweight structures and complex geometries that are difficult or impossible to create using conventional methods. AM also enables the design of functionally graded structures and multi-material systems with customized features, along with the use of advanced materials, including metal–matrix composites, polymer-based nanocomposites, and biocompatible ceramics. Material behavior can also be further controlled through surface engineering, including coatings and thin-film deposition. This improves resistance to fatigue, oxidation, corrosion, and damage. Heat treatment procedures remain essential to reduce residual stresses, improve hardness, and optimize microstructure, especially in alloys and high-strength steels. A comprehensive and focused analysis model was constructed using traditional probabilistic statistics and a data mining algorithm for the collected wheel production process data. Correlation analysis was then used to identify the critical factors influencing wheel hub quality, further improving production and quality. The data were then reasonably supported by process optimization. Figure 3.1 illustrates the structure of the product quality analysis process based on manufacturing data.

Manufacturing processes now operate with unprecedented precision because of recent advances in sensor-integrated manufacturing, computational modeling, and nanotechnology. Technologies including chemical vapor deposition (CVD), electrospinning, and powder metallurgy enable the production of lightweight,

DOI: 10.1201/9781003758556-3

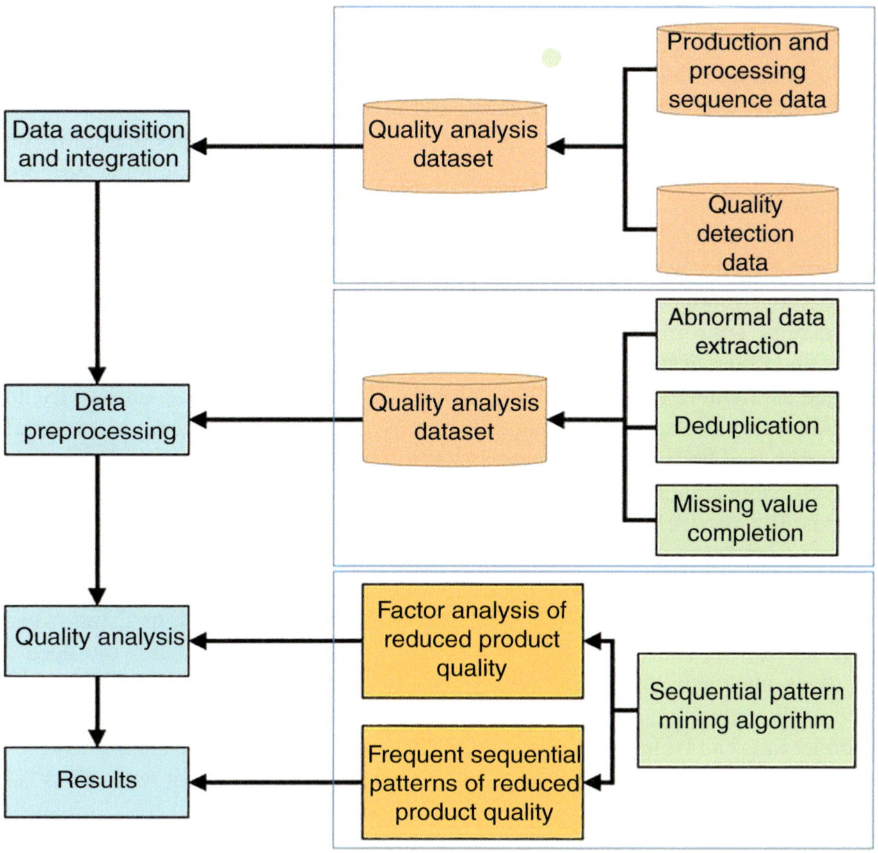

FIGURE 3.1 The process of analyzing product quality using manufacturing data [1] [Open access].

high-strength materials with controlled porosity and uniform nanophase distribution. Industry 4.0 concepts include predictive control, reduced material waste, improved repeatability, and increased overall process efficiency because of integration with artificial intelligence, machine learning, and real-time process monitoring. The importance of understanding manufacturing and processing methods goes beyond just material performance. These technologies are essential for future industrial applications due to their direct impact on cost-effectiveness, energy consumption, and sustainability. While simultaneously considering economic and environmental factors, advanced manufacturing methods provide the flexibility needed to manufacture highly specialized components for electronics, biomedical devices, aerospace systems, and renewable energy systems.

In short, the study of manufacturing and processing technologies combines the principles of materials science with practical engineering applications. By mastering these techniques, engineers and researchers can control material structures, optimize their properties, and create innovative solutions to meet the changing needs of contemporary technology. As this topic evolves, combining traditional and modern technologies will remain essential to produce next-generation materials characterized by sustainability, exceptional performance, and useful functionality [3,4]. Due to their unique properties that distinguish them from their bulk counterparts, including high surface area, increased reactivity, and the ability to interact with biological systems at the molecular level, nanomaterials have become a particularly interesting category in this context. These properties enable them to be used in a variety of industries, such as drug delivery, energy storage, and environmental remediation. However, conventional nanomaterial production processes often involve the use of hazardous chemicals, high energy consumption, and nonrenewable resources, which have detrimental effects on human health and environmental sustainability. Green nanotechnology has emerged as a viable solution to these problems, providing environmentally friendly methods for synthesizing nanomaterials that use less energy, produce fewer harmful byproducts, and help create safer and more sustainable applications.

Green nanotechnology aims to provide economical and environmentally friendly methods for producing nanomaterials. These techniques, which use biological processes involving plant extracts, microorganisms, and enzymes, offer a more environmentally friendly alternative to traditional chemical synthesis. For example, the process of producing nanoparticles involves reducing metal ions using organic materials from plants or microorganisms. This technology eliminates the need for hazardous chemicals and significantly reduces energy consumption. Furthermore, green nanotechnology encourages the recycling and reuse of materials, ensuring that nanomaterials can be integrated into the circular economy [4,5].

The production of nanomaterials using green nanotechnology not only offers safer and more environmentally friendly technologies but also improves the performance of materials in specific applications. Waste management, water purification, solar energy conversion, and reducing the environmental impact of nanomaterial production are just a few examples of the energy-efficient and environmentally friendly applications of nanomaterials enabled by green nanotechnology. With the growing need for greener solutions, the shift to green nanotechnology will be essential to ensure that the benefits of nanomaterials are realized without jeopardizing human health or the environment [6].

3.2 TRADITIONAL PROCESSING METHODS

With their proven reliability, scalability, and affordability, traditional processes such as casting, forging, extrusion, rolling, and welding remain vital to modern industry. For applications requiring long-term performance and structural integrity, these methods have been developed over decades to produce large components with reliable and repeatable mechanical properties. For example, casting uses liquid metal poured into molds and subjected to controlled solidification to produce large components and complex shapes. Components with complex geometries, which are typically difficult to achieve through mechanical deformation, benefit greatly from this method.

By contrast, forging involves plastic deformation of metals under compressive stresses, improving the material's mechanical strength and internal grain structure. This process is widely used to produce high-strength components where durability and fatigue resistance are essential, such as gears, shafts, and automotive components. Similarly, metals can be formed into thin sheets or homogeneous cross-sections using extrusion and rolling, providing excellent surface quality and dimensional accuracy. These techniques are essential for the production of sheet metal, pipes, and structural components for use in industry and construction.

Welding is another essential method for joining metal parts and producing connected structures without the use of fasteners. Incorporating heat treatment procedures into these traditional techniques is critical to improving material quality. Heat treatment improves hardness, toughness, and corrosion resistance while improving microstructure and mitigating residual stresses by regulating temperature, cooling rates, and phase transformations. For example, procedures such as annealing, quenching, and tempering are used to balance strength and ductility according to the specific requirements of each application. The cytoskeleton of cells determines their shape and elastic properties. Figure 3.2 shows images of the cytoskeletal network of red blood cells. The height of the holes above the substrate, measured experimentally, indicated the thickness of the cytoskeletal network, which ranged from 5 to 8 nm.

However, traditional processing techniques remain essential for producing large-scale components with consistent performance and form the basis of contemporary and hybrid manufacturing methods. Their continued application in modern industry underscores the importance of materials engineers and researchers understanding the fundamentals and limitations of these techniques [7].

3.3 MODERN FABRICATION TECHNIQUES

Materials engineering has been radically transformed by modern manufacturing technologies, enabling the production of complex structures, multipurpose parts, and high-performance systems that are difficult or impossible to achieve using conventional techniques. AM, also known as 3D printing, enables the production of layers of materials and components with precise geometries with minimal material waste. This is one of the most significant innovations. The technology's adaptability is evident in industries including aerospace, automotive, and healthcare, supporting the production of lightweight structures, functionally graded materials, and even patient-specific biomedical implants. Figure 3.3a–e illustrates a typical procedure for combining TBN technology with conventional photolithography to create nano-electromechanical resonators (NEMS). This TBN technique is easy to use, highly reliable, and capable of operating in ambient environments. Importantly, this TBN approach is ideal for prototyping nanodevices due to its compatibility with conventional microfabrication techniques.

Another advanced method is powder metallurgy, which enables the production of parts with enhanced mechanical properties, controlled porosity, and customized microstructures. When combined with sintering and hot isostatic pressing, this process produces materials with exceptional strength and reliability, widely used in the manufacture of high-performance gears, engine parts, and turbine blades. Similarly, advanced coating technologies, such as CVD and physical vapor deposition (PVD), enable the creation of surface layers and thin films with superior electrical performance, corrosion resistance, and high hardness. Due to their superior mechanical and

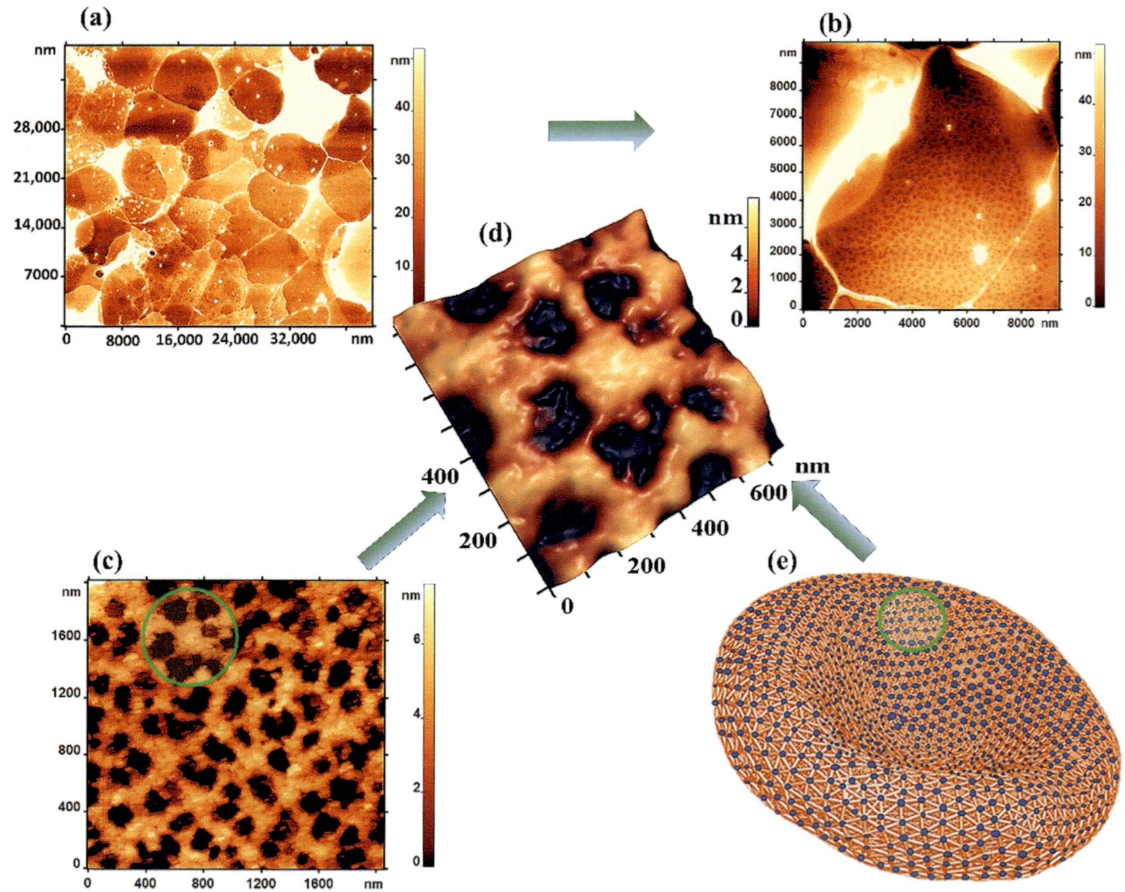

FIGURE 3.2 Atomic electron microscopy images of the cytoskeleton: (a) single red blood cell phantom ($10 \times 10\ \mu m^2$); (b) monolayer of control cell phantom ($35 \times 35\ \mu m^2$); (c) part of the cytoskeleton ($1.8 \times 1.6\ \mu m^2$); (d) 3D atomic electron microscopy image of part of the cytoskeleton of an RBC ($0.8 \times 0.7\ \mu m^2$); (e) model of the RBC cytoskeleton [2] [Open access].

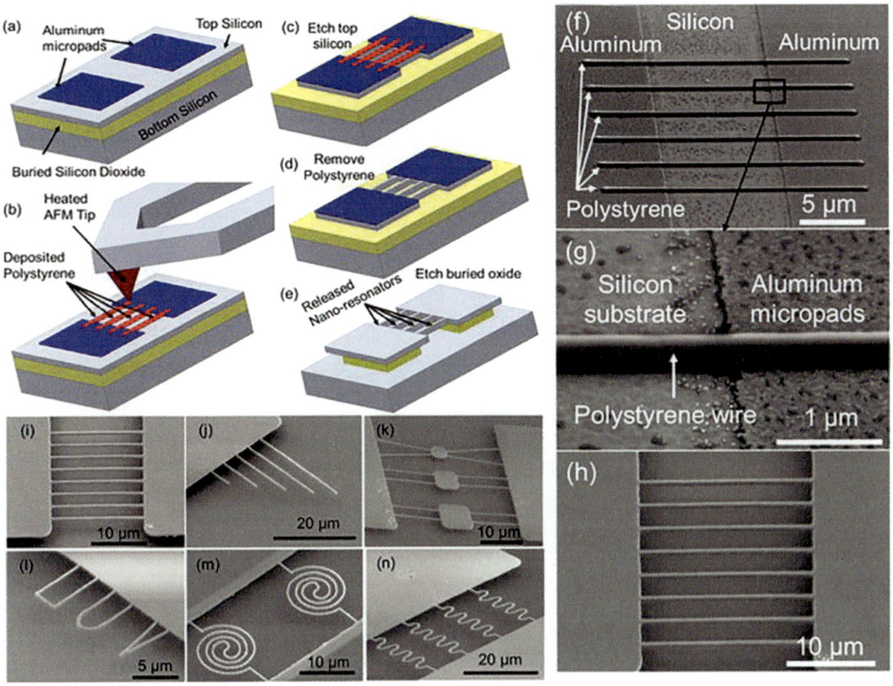

FIGURE 3.3 (a–e) Summarize the process of producing nanomechanical resonators using tip-based nanofabrication (TBN). Aluminum micropads are first printed onto a silicon-on-insulator (SOI) wafer (a), and polystyrene nanowires are then deposited across the pads using a heated atomic force microscope (AFM) tip (b). These wires serve as an etch mask to create silicon nanobundles (c). The nanowires are then removed (d) and the buried oxide is etched to produce hanging bundles (e). Scanning electron microscope (SEM) images confirm the deposited nanowires (f, g) and the evacuated nanobundles (h). Figures (i–n) illustrate the six types of silicon nanomechanical resonators available with this technique [3] [Open access].

functional properties, these coatings are widely used in cutting tools, solar cells, microelectronics, and protective coatings for aerospace applications. In this context, incorporating graphene with silicon substrates is an effective strategy for enhancing device performance. Consequently, using micro-mechanical exfoliation techniques, graphene samples were fabricated on a silicon substrate coated with a 285 nm thick layer of silicon dioxide (SiO_2), as illustrated in Figure 3.4. An enhanced scanning atom microscopy (AFM)-based photolithography technique was used to create high-quality graphene constrictions (Figure 3.4a). AFM was used to determine the crystalline orientation of the graphene lattice before the forming process. Contact-mode imaging utilizes atomic potential-induced frictional force modulation to reveal lattice symmetry orientations (Figure 3.4b, c). This allows for precise selection of the cutting direction, which is essential for producing superior edge quality. With the feedback loop switched off, the graphene was cut by pressing the AFM tip against the surface until a predetermined deflection (corresponding to an apparent tip drop of

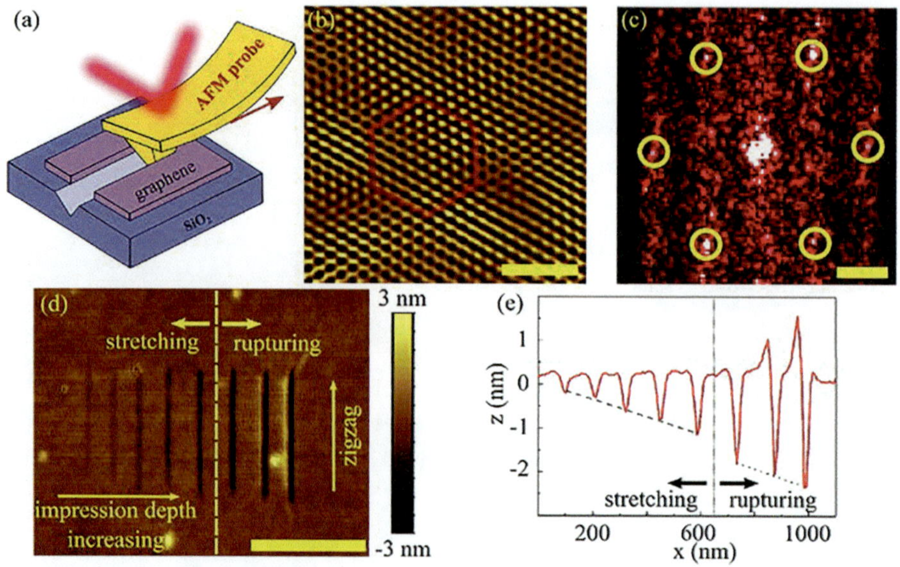

FIGURE 3.4 Figure 3.4 (a) Schematic of the AFM cleavage lithography process. (b) Lattice resolution on graphene acquired in contact mode, (FFT filtered for clarity), revealing the main crystallographic directions. Scale bar: 2 nm. (c) Fourier transform of the original lattice resolution image. Scale bar: 2 nm^{-1}. (d) Linear AFM indentations in graphene along zigzag direction with gradually increasing depth. Scale bar: 500 nm. (e) Depth profiles of the patterned lines from panel (d), revealing the border between stretching and cutting of the graphene sheet [3] [Open access].

100–200 nm), and then moving the tip along the desired cutting path (Figure 3.4d). (Figure 3.4e) depth profiles of the patterned lines from panel d. While graphene is stretched and adheres to the plastically deformed silica substrate at shallow indentation depths, the probability of tearing increases dramatically at indentation depths of approximately 1.5 nanometers..

The scope of design in contemporary engineering is expanding because of other advanced production technologies, such as hybrid systems that combine subtractive and additive processes, laser AM, and nanofiber electrospinning. For example, electrospun nanofibers are used in drug delivery systems and filtration membranes, while laser-based 3D printing is increasingly being used to create strong, lightweight parts for electric vehicles and aircraft.

In line with Industry 4.0 principles, the precision, reproducibility, and efficiency of these technologies are enhanced by combining computational modeling, artificial intelligence, and real-time monitoring. Overall, researchers and engineers have powerful tools to innovate, optimize, and utilize next-generation materials because of modern manufacturing technologies. Its importance in influencing the path of advanced manufacturing is evident through its ability to combine complexity, sustainability, and high performance [1,2]. Nanoparticles are widely used in many different products and applications, including medical devices, cosmetics, industrial catalysis, energy conversion and storage, and imaging technologies, due to their unique physical properties [3–5].

Bacteria are the primary source of natural nanopolymers (bacterial cellulose (BCs)) [11–14]. Due to the wide variety of nanoparticles and their special properties, some occur naturally, such as BC, which is produced by bacteria and has unique properties that distinguish it from ambient and synthetic nanoparticles. Bacteria produce granular cellulose using cellulase–enzyme complexes (Figure 3.5c). BC is between 0.01 and 0.10 μm wide, making it ribbon-like. This is two to three times smaller than the diameter of plant cellulose, which is typically 10 μm. The fibers range in length from a few hundred nanometers to microns and intersect to form a network-like structure (Figure 3.5e and f) [2].

Because cellulose produced by microbial processes is a byproduct of microbial metabolism and not a structural component of the cell wall, it is free of impurities such as hemicellulose and lignin, unlike plant cellulose. Humans are surrounded by large quantities of naturally occurring, random particles, as well as intentionally engineered nanoparticles. These include airborne nanoparticles from industrial combustion and traffic, and nanoparticles from erosion or corrosion in water [6,7]. Human exposure is inevitable given the presence of nanoparticles in many consumer goods and the environment.

However, little is known about the potential health risks that nanoparticles may pose in the short and long term. In this work, we introduce the major classes of nanoparticles and investigate how their physical and chemical properties influence

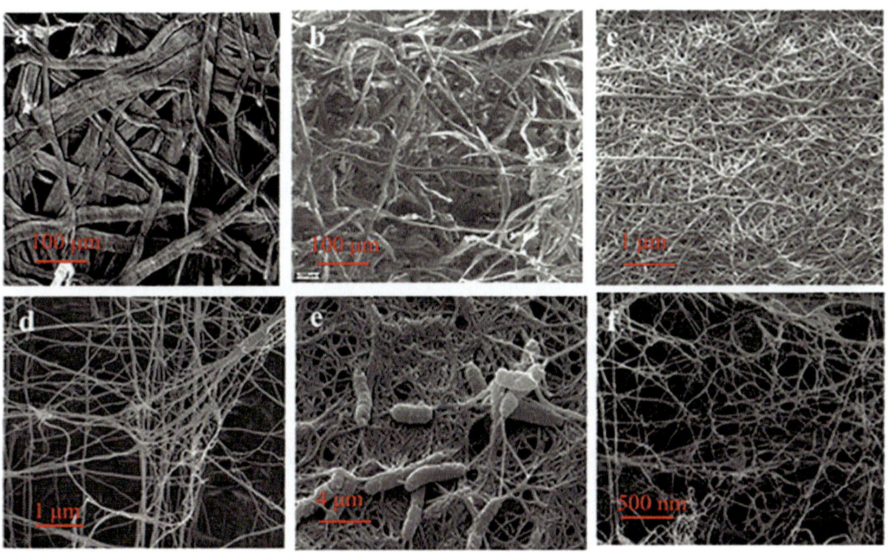

FIGURE 3.5 SEM images of cellulose from different sources: (a) softwood (b) cotton, (c) tunicate cellulose, (d) algae, (e) bacterial cellulose (BC), and (f) regenerated cellulose RSC. (Reprinted with permission from [4] [Open access]).

toxicity. The number of microparticles increased from 55 ± 1% after laccase treatment to 62 ± 1% during cellulose pretreatment after enzyme pretreatment. These results demonstrate that cellulases can promote fiber growth by damaging their internal and external regions, allowing more water molecules to enter the fiber [6].

The fundamental processes of nanotoxicity are discussed, along with an overview of how nanoparticles interact with the body at the organ, tissue, cell, and biomolecular levels. This mechanistic insight is enabled by the numerous experimental and theoretical techniques that have been developed and applied to assess nanotoxicity. To fully leverage nanotechnology for human safety, this review concludes by sharing our views on potential strategies for reducing nanotoxicity. We hope this review will serve as a useful resource to highlight the current state of nanotoxicity research and inspire further studies to deepen our understanding of nanoparticle toxicity. Ultimately, a deeper understanding of nanotoxicity may influence and guide the development of regulatory frameworks to mitigate potential risks to human health. Nanoparticles interact with biological materials in two main ways: internally and externally. Through ingestion, inhalation, or skin absorption, the body can absorb nanoparticles from food, water, air, or other substances. Internally, biologically implanted nanoparticles are injected directly into the body.

In addition to investigating how discrimination affects human health, safer nanoproducts, such as nanocellulose, that can be modified by enzymes to enhance their properties and barriers to impact, should be developed. After enzymatic hydrolysis, the recovered nanocellulose remains almost intact due to the strong selectivity of the enzyme (Figure 3.6h) [2]. The main enzymes used for this are cellulases, which break the β-1,4 glycosidic bonds in the cellulose molecule. More precisely, amorphous cellulose can only be broken down by cellulases, reducing the energy required for mechanical processing [7,8].

Endogenous or exogenous mechanisms can lead to primary effects such as oxidative stress, DNA damage, inflammation, and toxicity. These health effects gradually lead to secondary effects such as tissue and organ deposition, systemic inflammation, dysfunction, and secondary toxicity [4]. As previously noted, several factors, such as genetics, preexisting disease states, exposure level, and nanoparticle characteristics, are associated with the adverse effects of nanomaterials on human health. Studies in animals and in clinical settings have demonstrated the inefficiency of macrophage clearance techniques in removing nanoparticles from the body, which harm the nervous system, lungs, lymphatic system, and some organs. Due to their small size, nanomaterials can pass through the blood–brain barrier, allowing them easy access to most organs.

3.4 SURFACE ENGINEERING AND FUNCTIONALIZATION

Surface engineering and functionalization are critical fields in contemporary materials science and engineering, aiming to improve and modify the surface properties of a material without altering its overall properties. The surface becomes the primary domain for enhancing mechanical, chemical, and physical interactions, as the interaction of engineering components with their external environments often determines their performance and longevity. Materials can be engineered to resist oxidation,

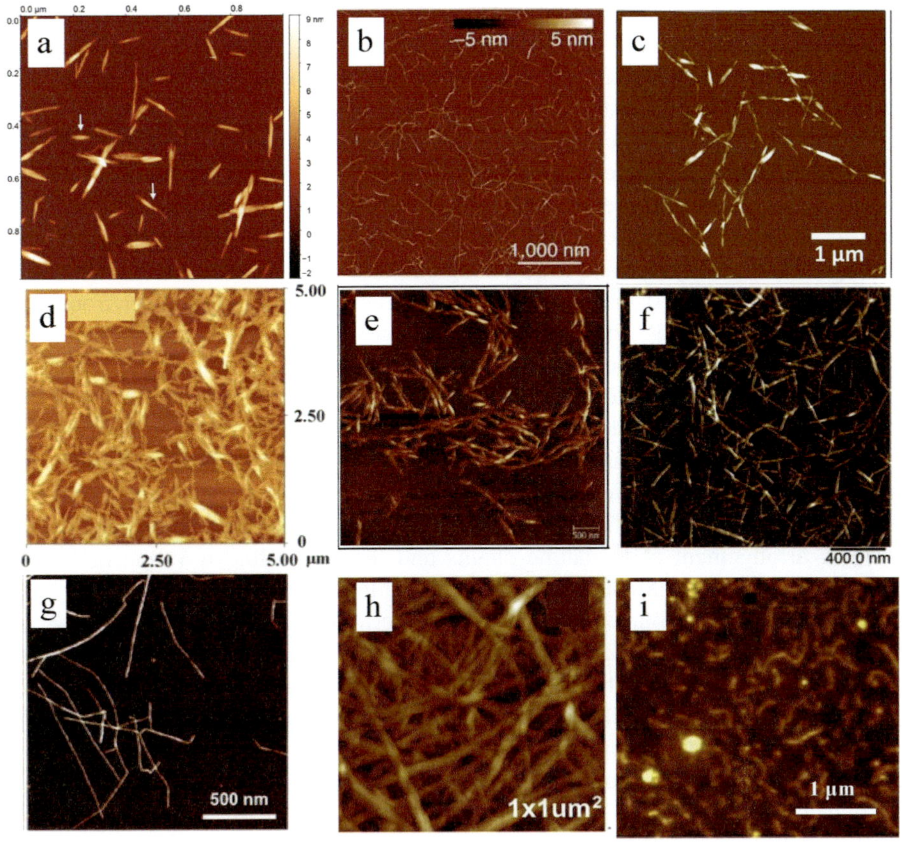

FIGURE 3.6 Nanocellulose prepared by different methods: H_2SO_4 (sulfuric acid, a), TEMPO (2,2,6,6-tetramethylpiperidine-1-oxyl, b), H_3PO_4 (phosphoric acid, c), HCl (hydrochloric acid, d), and maleic acid (e). (f) oxalic acid, (g) TEMPO-oxidized CNF, (h) enzymatic hydrolysis, and (i) carboxymethylation, [4] [Open access].

corrosion, and damage through advanced surface engineering, and functionalization enables the integration of specific properties such as hydrophobicity, biocompatibility, and catalytic activity.

There are many different surface engineering techniques, including laser surface treatment, thermal spraying, ion implantation, and PVD and CVD. These methods enable the production of protective coatings, thin films, and surface modifications that significantly extend the lifespan of parts used in energy, automotive, medical, and aerospace applications. Functionalization, on the other hand, goes beyond protection to include the imparting of new capabilities. For example, adding chemical groups or nanostructures to surfaces can improve adhesion, enable self-cleaning properties, or confer antibacterial activity on biomedical devices. In nanotechnology, where the surface-to-volume ratio is critical, improving surface functionality at the nanoscale is crucial. Functionally enhanced nanoparticles, coatings, and thin films are increasingly used in drug delivery systems, energy storage devices, and smart sensors. Furthermore, the design of multipurpose materials with customized performance becomes possible by combining surface engineering with cutting-edge disciplines such as AM and Industry 4.0 concepts.

Overall, surface engineering and functional optimization facilitate the development of next-generation technologies by linking fundamental materials research with practical industrial requirements. Surface treatments are essential for sustainability, increasing energy efficiency, reducing material use, and enhancing recyclability. Consequently, surface engineering and functionalization are now considered key pillars of advanced manufacturing, biomedical innovation, and environmental conservation. Designing chemical processes that minimize the production of hazardous compounds is a focus of the growing field of green chemistry, sometimes referred to as sustainable chemistry. These approaches, when combined with surface engineering, enable the production of environmentally friendly coatings and high-performance functional materials. For the molecular understanding of these surfaces, advanced characterization methods are essential. For example, AFM has been widely used to visualize nanoscale structures. Molecular features, such as the periodic arrangement of L-alanine chains and the protrusion of valine molecules, can be analyzed in AFM images of hybrid cellulose films and amino acid crystals. Fourier-filtered AFM images reveal enhanced periodic patterns, providing important information about molecular organization, although sub-molecular resolution is difficult to achieve. These results demonstrate how the rational design of sustainable and functional surfaces is directly supported by nanoscale characterization (Figure 3.7).

Figure 3.8a–d shows typical clusters adsorbed on the protein monolayer covering the bottom of an atom probe microscope (AFM) cell. Both clusters under study contained between four and nine molecules. Nucleation theories predict that larger clusters will be closer to a metastable equilibrium with the solution, while smaller clusters will dissipate rapidly [4]. However, after about 20 min of observation, Figure 3.8a–d shows that the smaller clusters retain four molecules after gaining and losing two. On the other hand, over the same time period, the larger clusters gradually dissipate [5].

3.5 SMART AND SUSTAINABLE MANUFACTURING

In the era of Industry 4.0, where digital transformation is coupled with environmentally friendly practices to achieve high efficiency and long-term sustainability, smart and sustainable manufacturing has emerged as a key paradigm. To facilitate real-time monitoring, smart decision-making, and predictive maintenance, this strategy integrates advanced technologies, including automation, big data analytics, the Internet of Things, and artificial intelligence. By integrating smart capabilities into industrial processes, manufacturers can reduce energy consumption and material waste while increasing productivity, improving product quality, and reducing downtime.

The sustainability component of this concept, which focuses on reducing negative environmental impacts, conserving natural resources, and promoting circular economy concepts, is crucial. By facilitating closed-loop systems, maximizing resource utilization, and encouraging energy-efficient processes, smart technologies

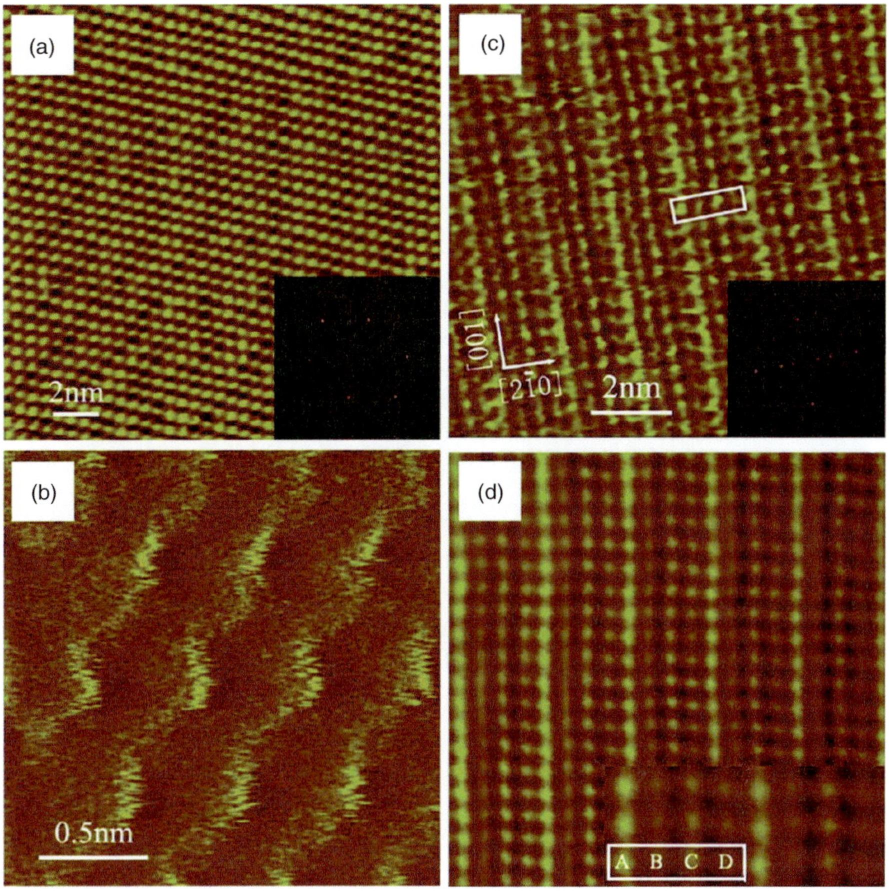

FIGURE 3.7 (a) Atomic force microscope (AFM) image of the DL-valine (001) crystal face with a scan size of 18 nm × 18 nm (input FFT image). (b) AFM image of the DL-valine (001) crystal face at higher resolution, using a scan area of 2 nm × 2 nm. (c) AFM image of the L-alanine (120) crystal face showing a unit cell of four molecules within the white rectangular frame (from input FFT image). Scan area is 11 nm × 11 nm. (d) Rotation and cropping of the original image (a) followed by the Fourier-filtered image. The four molecules of the unit cell, labeled A, B, C, and D in the white box, are enlarged in the inset [5] [Open access].

are essential to achieving these goals. For example, digital twins and AM reduce material waste and support lightweight product designs that extend product lifecycles, while sensors and data-driven models enable continuous tracking of materials and energy across production lines, preventing unnecessary losses and reducing costs. The Fourth Industrial Revolution, or Industry 4.0, is expected to transform not only industrial production but also society as a whole. Figure 3.9 illustrates the four major industrial changes that society has experienced throughout history.

By providing transparency, traceability, and flexibility, this combination of sustainable and smart approaches has revolutionized supply chain management.

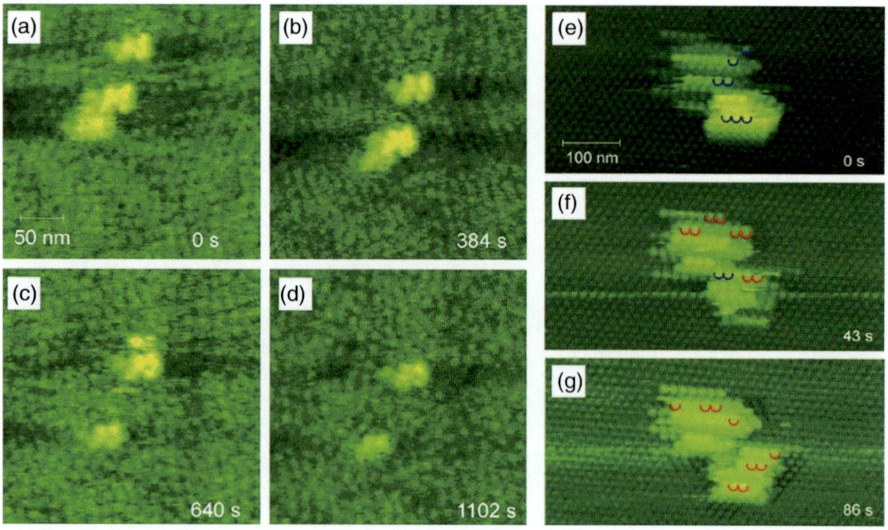

FIGURE 3.8 The progression of two adsorbed protein clusters on the substrate is depicted in AFM images (a–d). AFM images (e–g) illustrate the kinetics of protein exchange between a typical protein cluster and the solution (molecules that appear in or are absent from the cluster with respect to the next frame are shown in red and blue, respectively) [5] [Open access].

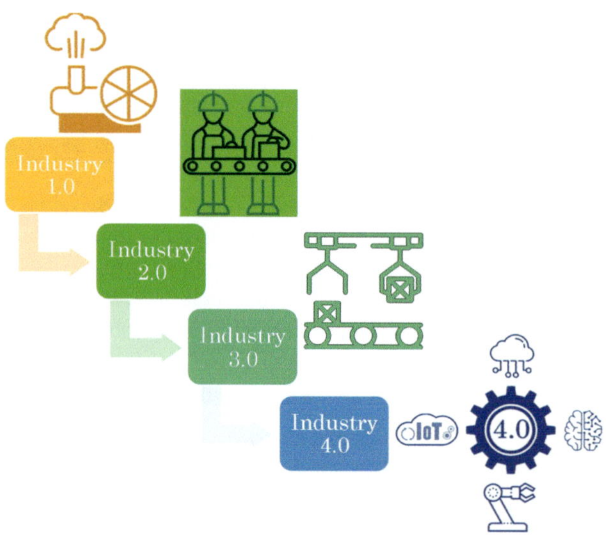

FIGURE 3.9 Revolutions in industry [6] [Open access].

Real-time data systems and blockchain technology ensure ethical sourcing and a low carbon footprint, while dynamic production schedules synchronize industrial production with renewable energy supplies. The application of these approaches has

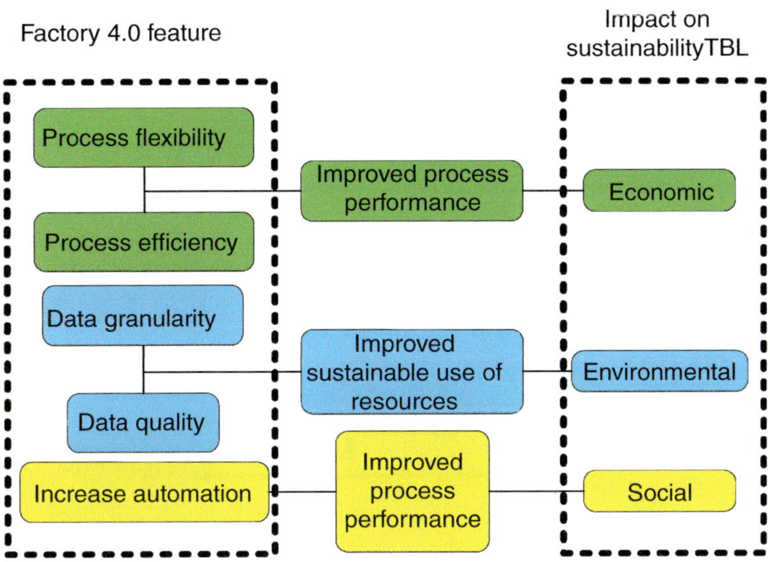

FIGURE 3.10 Factory 4.0 uses the Triple Bottom Line (TBL) concept to promote sustainability [6] [Open access].

led to safer, lighter, and more environmentally friendly goods in fields such as biomedicine, aerospace, and automotive, demonstrating the revolutionary impact of this paradigm.

However, smart and sustainable manufacturing is a cultural and strategic shift that balances economic competitiveness with environmental responsibility, not simply a technological change. This strategy will continue to evolve as a fundamental pillar of future industrial development, encouraging innovation, flexibility, and sustainability across many industries amid escalating global issues such as resource efficiency, energy scarcity, and climate change [4].

As shown in Figure 3.10, a previous study [5] attempted to capture the benefits of the triple bottom line for sustainability using Factory 4.0 concepts. They also noted, as Figure 3.11 illustrates, that the impact of Factory 4.0 on the triple bottom line of sustainability is interactive in nature. This interaction suggests that aspects of the triple bottom line can be improved in conjunction with the implementation of Factory 4.0 technologies. The following section provides a detailed analysis of how Factory 4.0 impacts each of these sustainability factors.

As Industry 4.0 impacts manufacturing and industry in the future, it is essential to understand how it supports sustainable growth. The term "sustainable development" also known as "sustainable development," once referred primarily to environmental issues, but now encompasses the environmental, social, and economic aspects of development.

It is clear that integrating Industry 4.0 principles and sustainability perspectives, when carefully and strategically implemented, can ensure not only continued economic growth but also improved customer relations, enhanced product and facility

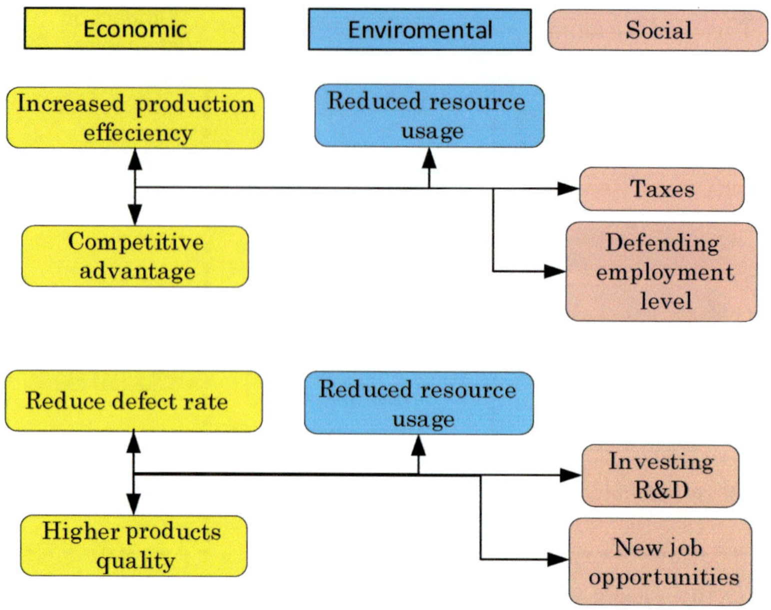

FIGURE 3.11 Benefits of implementing Factory 4.0 in the interactive TBL concept [6] [Open access].

value, enhanced employee safety and well-being, and improved environmental management. This interconnectedness highlights the importance of combining sustainable practices with technological advancements to build resilient and forward-looking industries. Therefore, organizations are under increasing pressure to balance these two perspectives and develop a comprehensive, smart, and sustainable business plan that meets the needs of internal stakeholders such as management and employees, as well as external stakeholders such as consumers, regulators, and society.

Figure 3.12 illustrates a diagram of an integrated, smart, and sustainable business model. This model illustrates a comprehensive framework in which long-term value is generated through the convergence of digital transformation, operational efficiency, and environmental responsibility. When implemented correctly, this model can increase overall profitability, ensure regulatory compliance, and enhance competitiveness in global markets. Additionally, it encourages the innovation of superior products, builds consumer confidence, and enhances worker well-being by ensuring better and safer working conditions.

The effective use of several key elements is essential for the success of this integrated model. These include the application of the latest digital technologies to facilitate real-time data monitoring and predictive decision-making, such as the Internet of Things, artificial intelligence, and cyber–physical systems.

Ultimately, a sustainable and smart business model represents a strategic roadmap that enables sectors to balance accountability and profitability. In addition to enabling companies to easily adapt to changing market needs, its adoption enhances resilience in the face of global concerns, including resource scarcity and climate change.

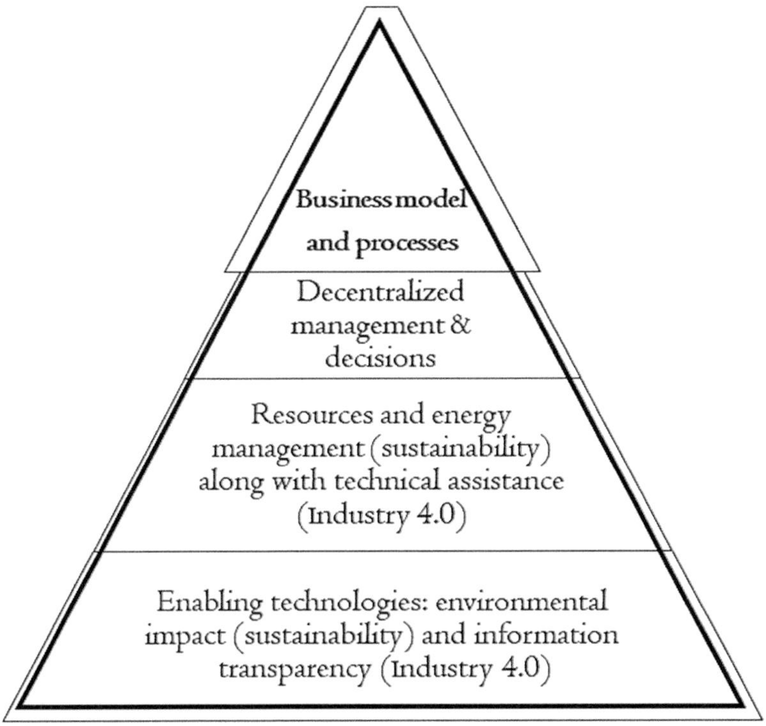

FIGURE 3.12 Components of a sustainable and smart business model [6] [Open access].

By integrating technological innovation with sustainability principles, industries can experience a synergistic transformation that not only enhances economic success but also ensures long-term social and environmental benefits.

Industries can experience a synergistic transformation that not only boosts economic growth but also provides long-term social and environmental benefits by integrating technological innovation with sustainability principles. In this regard, the goals of green chemistry and the vision of smart and sustainable production are closely aligned. Green chemistry promotes safer, more efficient, and more sustainable chemical processes to prevent environmental damage. It focuses on prevention at the design stage of materials and processes, rather than only addressing waste and pollution after they are produced. This proactive strategy complements the broader goals of sustainability-led industrial transformation by emphasizing the importance of mitigating risks before they arise.

Green chemistry helps reduce waste and energy consumption by using renewable raw materials, reducing derivatives, and improving resource utilization. This aligns directly with the efficiency gains facilitated by Industry 4.0 technology. Furthermore, the creation of safer chemicals, solvents, and processes contributes to reducing risks to public health and ecosystems and enhancing the social responsibility component of sustainable business models. Green chemistry concepts become powerful accelerators for cleaner production systems when combined with digital technologies such as life cycle assessment, predictive analytics, and real-time monitoring. Industry's ability to balance innovation, profitability, and environmental stewardship is fundamentally enhanced by integrating green chemistry into smart and sustainable manufacturing frameworks. It provides a scientific understanding of the composition, properties, and behavior of materials at the nanoscale[7,8]. According to these theories, less hazardous chemical synthesis techniques are replacing the nuclear economy, which has historically relied on hazardous materials and procedures. Another crucial element is the concept of real-time analysis, which enables early detection and prevention of contamination during chemical reactions. Reducing environmental impact requires this proactive approach [9].

Moreover, safer chemistry includes preventing accidents and creating safer working conditions, particularly in sectors that manufacture polymers and other complex materials. Sustainability relies heavily on ongoing efforts to replace hazardous solvents with environmentally acceptable ones. Because they provide an environmentally acceptable channel for chemical reactions and eliminate the need for hazardous solvents, solvent-free synthesis techniques have grown in popularity as more sustainable solutions. Using safe reagents, photocatalysts, and natural raw materials, low-energy catalytic technology has been developed to reduce pollution. To reduce waste generation, nontoxic and energy-efficient synthetic methods have been developed [9]. These changes have enabled the next generation of researchers to understand the methods, practices, and principles of environmentally friendly green chemistry, as well as its role in mitigating negative impacts on the environment and public health [10,11]. Twelve fundamental principles of green chemistry guide the design of sustainable and environmentally friendly chemical processes [12] focus on sustainability and environmental responsibility [13]. With the emergence of the concept of green chemistry, which focuses on innovating and promoting industrial processes that reduce health risks while maintaining financial viability, this field has seen significant growth. Sustainable chemistry focuses on the research, development, and application of chemical processes and products that reduce their negative environmental impacts and improve human health and safety. The primary goal of this field is to develop chemical products and processes that reduce or eliminate the use and production of hazardous substances. Green chemistry achieves this by providing a scientific foundation that enhances industrial competitiveness and sustainability goals. Below is an overview of its core ideas, uses, and benefits:

- *Green Chemistry Preventive Principles:* Preventing waste is preferable to treating or cleaning it up after production. Creating techniques that do not produce trash or pollutants is a basic idea.
- *Atom Economy:* By properly designing synthetic processes, all materials employed in a process should be utilized as much as feasible in the finished output. The effective utilization of atoms is highly valued in this concept.
- Developing less dangerous chemical formulations and materials that are less harmful to the environment and the general public's health.
- *Developing Safer Chemicals:* Chemicals should have the least level of toxicity while still being as effective as feasible. Reducing the possible harm from the chemicals utilized is the aim.

- *Safer Solvents and Reaction Conditions:* Avoid any solvents, reactants, or reaction conditions that could be hazardous. When possible, switch to safer solvents or eliminate them.
- *Energy Efficiency:* To save energy, operate processes at as close to ambient temperature and pressure as possible. Energy-efficient processes reduce the overall impact on the environment.
- Instead of exhausting all currently available resources, use renewable commodities or raw materials. This makes chemical processes more sustainable in the long term.
- *Reduce Derivatives:* Avoid unnecessary derivatives and chemical groups whenever possible. Derivatives often require more reactants and produce more waste.
- *Design for Degradation:* Chemicals must degrade into harmless substances after use to minimize their negative impact on the environment.
- *Real-Time Analysis to Prevent Pollution:* To prevent the development of hazardous compounds and reduce pollution, implement real-time monitoring and control techniques in your processes.
- *Use Inherently Safer Materials to Avoid Accidents:* Reduce the likelihood of spills, explosions, and other hazards by improving processes and materials.
- *Reduce Environmental Impact:* Green chemistry methods contribute to reducing waste and pollution, resulting in cleaner air, water, and soil.
- *Improve Human Health:* Green chemistry reduces exposure to and use of hazardous chemicals, protecting human health.
- *Increasing Cost Efficiency:* By reducing waste and using resources efficiently, green chemistry can lead to cost savings.
- *Promoting Green Practices:* Sustainable green practices are facilitated in the long term by using renewable resources and developing products that decompose safely.

The 12 pillars of green chemistry are listed in Table 3.1.

3.5.1 Green Chemistry Scope and Applications

The demand for clean technology in the chemicals sector is well known. A paradigm shift from the traditional perspective of process efficiency, which focuses heavily on chemical outputs, to one that places financial value on source disposal and the avoidance of the use of toxic and/or hazardous materials is essential to keep pace with the trend toward "green chemistry" [14] or "sustainable technologies." The term "green chemistry" was first used by the U.S. Environmental Protection Agency (EPA). Japan responded by establishing the Green and Sustainable Chemistry Network, which has contributed to promoting the global trend toward sustainable and environmentally friendly chemical practices. The first edition of the journal "Green Chemistry" was funded by the Royal Society of Chemistry and published in 1999. Therefore, green chemistry is believed to be here to stay. To study the design principles used in the synthesis of silica structures, atomic-scale microscopy was conducted on the cell walls of sixteen different diatom species. In addition to the

TABLE 3.1
The 12 Main Principles of Green Chemistry

No.	Principle	Remarks
1	Prevention	Better to prevent generating waste than to clean it up.
2	Atom economy	Synthetic methods should maximize the incorporation of all materials used in the process into the final product.
3	Less hazardous chemical syntheses	Alternative synthetic methods should be devised to generate substances with little or no toxicity to health or the environment.
4	Designing safer chemicals	Chemical products should be designed to achieve their functions with minimized toxicity.
5	Safer solvents and auxiliaries	When possible, auxiliary substances should substitute for harsh solvents, separation agents, etc.
6	Design for energy efficiency	Experiments should be designed to be conducted at ambient temperature and pressure to minimize energy use and environmental impact.
7	Renewable feedstocks	Renewable, raw material/feedstocks should be used when technically and economically feasible.
8	Reduce derivatives	Needless derivatizations (physical/chemical processes) should be avoided to prevent the use of additional reagents and generation of waste.
9	Catalysis	Catalytic reagents are superior to stoichiometric reagents and should be used.
10	Design for degradation	Chemical products should be designed to innocuously degrade at the end of their function with minimal environmental impact.
11	Real-time analysis for pollution prevention	Analytical methodologies should be developed for real-time, in-process monitoring and control prior to the formation of hazardous substances.
12	Inherently safer chemistry for accident prevention	Substances used in a chemical process should be chosen to minimize the potential for chemical accidents, including releases, explosions, and fires.

Source: [7] [Open access].

industrial uses of green chemistry, understanding the natural processes that produce materials at the nanoscale, such as the synthesis of silicon structures in diatoms, may inspire new sustainable technologies. Samples are air-cleaned after being treated with acid to remove organic layers, preparing them for atomic-scale microscopy. The silica structures of diatoms vary greatly in the same properties between different species, as well as in distinctive features within a single species and across different aspects of the organism (Figure 3.13) [8].

Since its formal adoption by the Environmental Protection Agency in 1993, the U.S. Green Chemistry Program has spawned new programs, including the Presidential Green Chemistry Challenge Awards and the Green Chemistry and Engineering Conference. Although green chemistry research predates the 1990s, significant progress has been made in the United States, Italy, and the United Kingdom. To produce environmentally friendly processes and products that embody the principle of safety by design, green chemistry has focused since the early 1990s on the efficient use of renewable raw materials, waste disposal, and the avoidance of harmful compounds. This concept is embodied in the Twelve Principles of Green Chemistry [15], which can be summarized as follows:

- Avoiding losses rather than repairing them
- Atomic efficiency

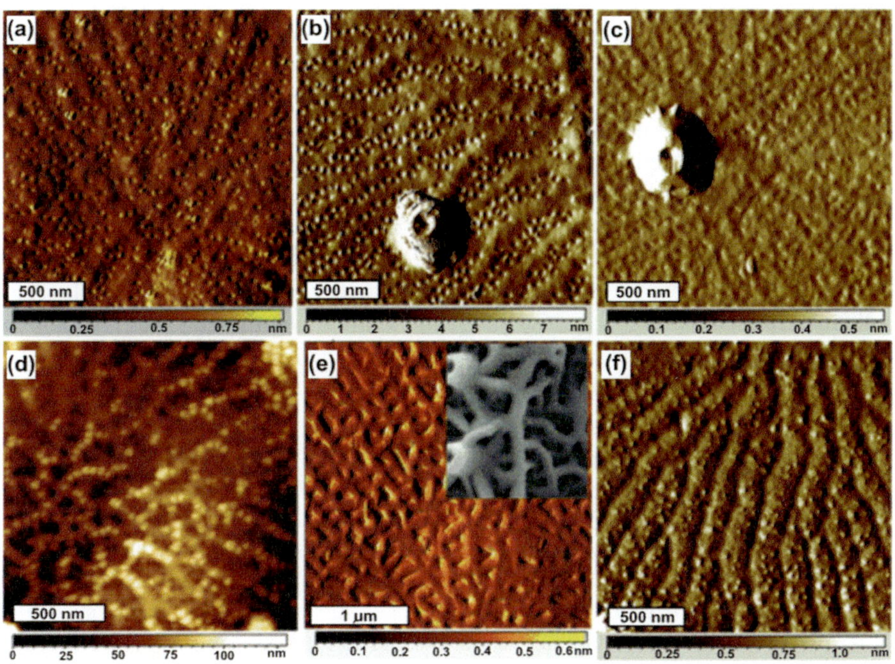

FIGURE 3.13 Three species of *Thalassiosira* have similar proximal valve characteristics and different distal valve characteristics. *T. pseudonana* (a), *T. weissflogii* (b), *T. oceanica* (c) are proximal valve surfaces, *T. pseudonana* (d), *T. weissflogii* with inserted SEM image (e), *T. oceanica* (f) are examples of distal surfaces [8] [Open access].

- Reducing hazardous or toxic materials
- Producing safer products
- Using safe auxiliary materials and solvents.

A key element of green chemistry is the development of innovative solvent systems or solvent-free technologies to produce environmentally friendly chemicals. Furthermore, reducing waste production and energy consumption encourages adherence to the twelve principles of green chemistry. Industries most impacted by green chemistry include pharmaceuticals, reaction solvents, biological transformations, biomaterials, safe alternative energy sources (such as photovoltaics, quantum dots, fuel cells, and batteries), innovative catalyst designs, and low-risk molecular designs, among others.

3.6 GREEN NANOTECHNOLOGY

Green chemistry principles in nanotechnology enable the safe, economical, sustainable, and environmentally friendly production and processing of nanoparticles. As a result of a better understanding of green chemistry, the pharmaceutical industry is using fewer or no nongreen ingredients. In this regard, nanostructure fabrication approaches are essential for achieving green chemistry goals, as both top-down and bottom-up manufacturing strategies enable the sustainable and environmentally friendly synthesis of nanomaterials [16]. Bottom-up technology focuses on producing nanomaterials from atoms and molecules with diameters ranging from 1 to 10 nm, while top-down technology focuses on producing nanomaterials from bulk materials [17]. This comparison between the top-down and bottom-up approaches highlights the importance of selecting the best synthesis techniques to produce nanoparticles in a sustainable and environmentally friendly manner. Magnetite (Fe_3O_4) nanoparticles were biosynthesized using an aqueous extract of papaya leaves at room temperature. Figure 3.14 shows the X-ray diffraction (XRD) pattern and scanning electron microscope image of the produced magnetite nanoparticles. The plate-like particle shape is characterized by large grains and shell structures. Using the Scherrer formula, the magnetite crystal size is determined to be approximately 33 nm [9].

The categories of physical, chemical, and biological treatments are further separated. Techniques used include mechanical milling, laser ablation, spraying, etching, and electro-blasting. The lipid membrane developed massive defects or pores when exposed to 5 micromolar alamethicine, leaving small lipid islands between the larger interconnected lipid bilayer sections (Figure 3.15) [10].

It is impossible to record the early stages of peptide association and membrane dissociation due to the rapid interaction of peptides with membranes. In less than 10 min, indolicidin and alamethicin undergo primary peptide bonding reactions, according to a quartz crystal microbalance with dissipation monitoring (QCM-D) study. Furthermore, a variety of techniques, such as CVD, plant and microbial methods, thermal and hydrothermal methods, sol–gel procedures, and reverse micelle methods, can be used to create nanomaterials [18]. In general, nanotechnology can be divided into two main groups based on its uses and production methods: dry nanotechnology,

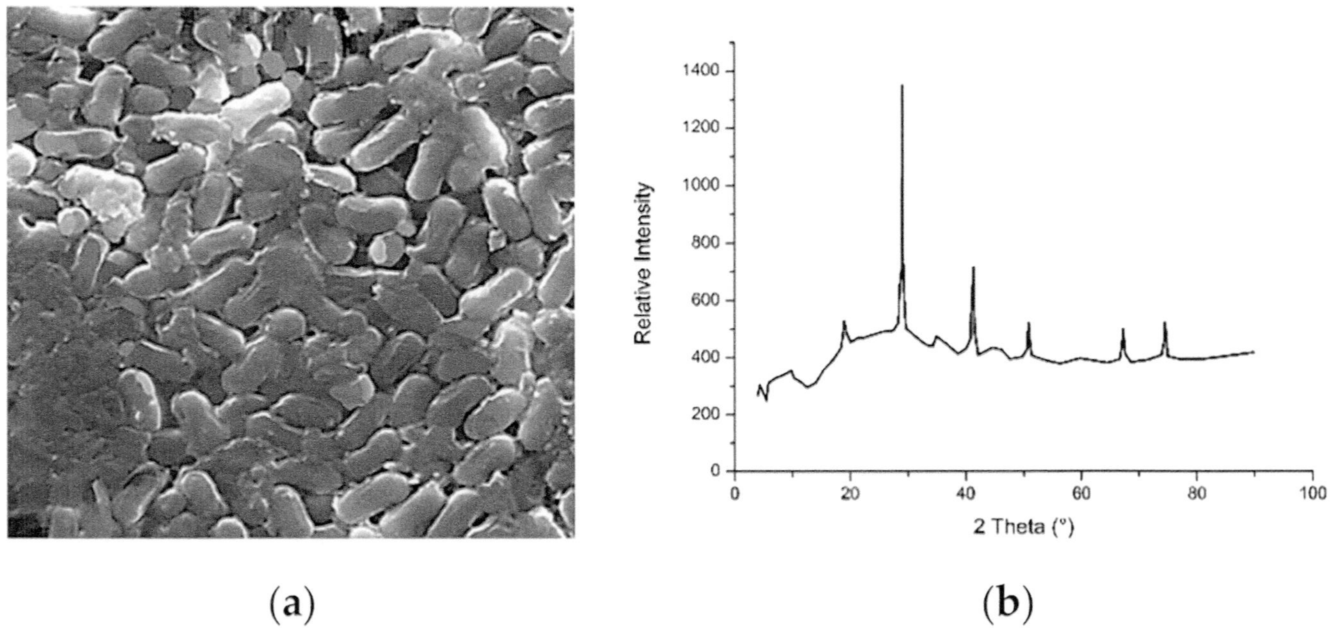

FIGURE 3.14 (a) Fe_3O_4 nanoparticles synthesized with papaya leaf extract; (b) SEM image; and XRD pattern [9] [Open access].

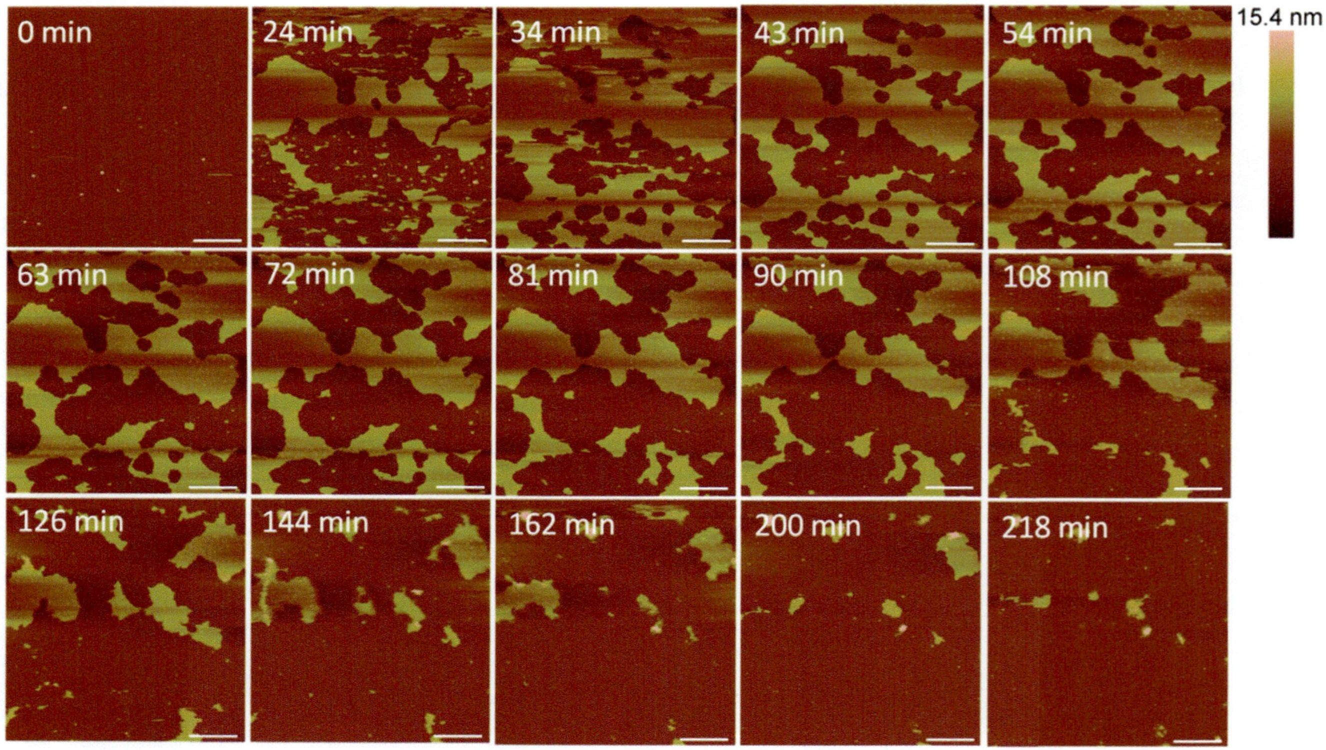

FIGURE 3.15 Defects developed in PC bilayer supported after injection of 5 μM alamethicin for 218 min. On the substrate, each AFM image is scanned at same location. One micrometer is represented by the scale bars [10] [Open access].

which focuses on the production of silicon, metals, and semiconductors, and wet nanotechnology, which works with biological systems such as enzymes and genetic material. Shape, size, topography, agglomeration, solubility, size distribution, composition, surface chemistry, hydrophilicity, and formulation are important factors in the synthesis of nanomaterials.

Graphene carbons, active nanomaterials, semiconductor devices, green nanotechnology, nanocomposites, membranes, sensors, nanoencapsulation, and computational nanotechnology have received increasing attention in recent years [19]. The major advantages of using nanoparticles in applications include the fabrication of finer and more robust composites, larger production scales, improved commercial performance, increased sustainability, and significant implications in the biomedical field [20]. One of the most important benefits of nanoparticles is that they exhibit quantum effects that make them useful in drug delivery, catalysis, and sensors. High-efficiency nanobatteries and storage systems that offer endurance and performance with minimal charging time are being developed by the energy sector. The usage of nanomaterials in a wide range of sectors has shown them to be extremely revolutionary. It is anticipated that additional nanotechnology research will produce creative solutions and address significant global issues.

Nanomaterials can be synthesized from organic or inorganic materials using the abovementioned techniques. Many types of nanomaterials are used for useful purposes. However, their use is limited due to their potential harm and hazardous nature, especially when used in chemical processes. Green nanomaterials are one of the modern synthesis techniques used in recent decades to reduce harmful effects and expand the scope of applications. Scientists are focusing on integrating natural resources to reduce the risks associated with nanoparticles. They can be used in various ways without posing a threat to living organisms [21]. Green technology is environmentally friendly and reduces the risks associated with chemical reactions. The primary goal of green nanotechnology is to synthesize sustainable particles with fewer negative impacts by using green chemicals in the synthesis process. Environmental impact analysis includes studies of the life cycle of green nanoparticles, including their synthesis, use, disposal, and raw material processing. Green nanotechnology significantly reduces waste and utilizes efficient technologies for recycling products. Among other environmental uses, a variety of nanoproducts are used in water treatment, renewable energy, and building materials. Green products ensure innovation, efficiency, sustainability, and the well-being of consumers and the environment. They also offer green solutions to a range of issues facing various sectors [22–24].

3.6.1 Creation of Green Nanomaterials

Nanotechnology-based approaches are gaining popularity in various fields. When nanomaterials are used in environmentally friendly processes and products, from manufacturing to disposal, the term "green nanomaterials" has emerged, describing the nanoscale aspects of green chemistry. The most common technique for producing environmentally friendly green nanomaterials is the bottom-up approach, which uses microorganisms or plant extracts. Green nanomaterials with unique properties, such as antibacterial, catalytic, optical, and magnetic properties, are produced using this method. Because oftheir unique properties, which make them easier to use and less harmful to the environment, nanomaterials have a wide range of uses. In medicine, where nanotechnology is used in multiple applications, green nanomaterials that are safe for human use are being used to accelerate associated technological processes [25]. They can be applied in both environmental and industrial settings. Cost, application resistance, and technological limitations are drawbacks of green nanomaterials. This means that green nanomaterials could find widespread use if more funding for research and development is made available and appropriate laws are enacted to promote their use in processes. This chapter discusses in detail the synthesis, properties, and industrial applications of green nanomaterials. In addition, AFM studies are conducted on silica-supported silver nanoparticles. Nanotechnology-based approaches are gaining popularity in a wide range of industries. The term "green nanomaterials" has emerged to define the properties of nanoparticles in green chemistry due to their applications in environmentally friendly production and waste disposal processes. Touch-mode images are shown in Figure 3.16a. By masking the contact-mode images and determining the maximum height of the selected particles, the heights of 222 silver nanoparticles were measured. Figure 3.16b displays the histogram of the silver nanoparticle height distribution. A mean height of 6.3 nm and a standard deviation of 1.4 nm were obtained by fitting this distribution to a Gaussian. To measure individual silver nanoparticles, quantum information theory and methods are used. Figure 3.16e, d, and c illustrate the values of adhesion, elasticity, and height, respectively, which can be obtained. This single silver nanoparticle is about 5 nm

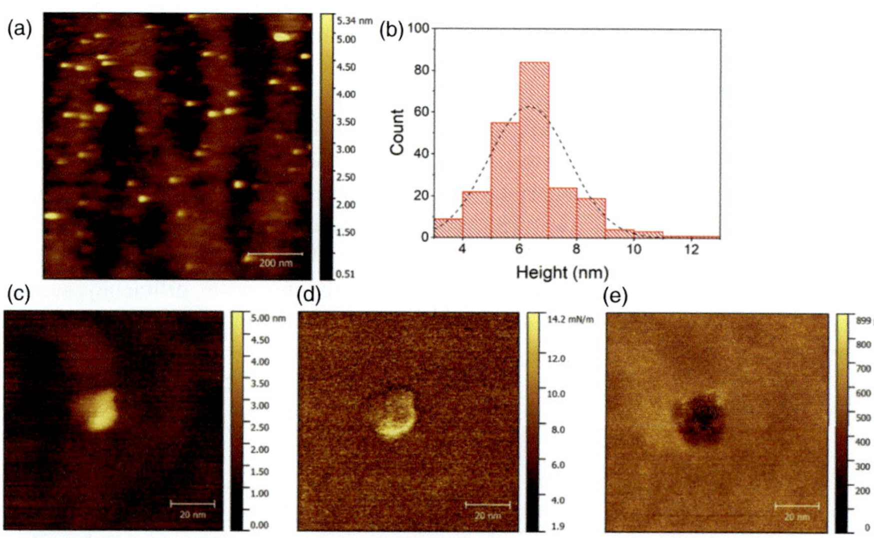

FIGURE 3.16 AFM image of Ag clusters deposit on a SiO2 wafer substrate. (a) Topographical height obtained in contact mode; (b) Histogram of Ag NPs from contact-mode images; (c) Height in QITM mode; (d) Elastic constant in QI mode; and (e) Adhesion in QI mode. [11] [Open access].

long, and its elastic constant is slightly larger than that of the silica background. Furthermore, the adhesion of the silver nanoparticles is lower than that of the silica background. Examining Figure 3.16d and e [7], it is clear that the nanoparticles are composed of a different material than the background.

The following methods are used in the development of green nanomaterials to apply green chemistry concepts:

- Creating safer nanomaterials requires evaluating the material's biological properties (size, structure, and surface function), using safe raw materials, and focusing on essential properties.
- Environmental design includes studying the toxicity of broken subunits, the degradation properties of nanomaterials, and the use of biosynthetic techniques.
- Waste prevention should be considered during design by reducing the production of unwanted derivatives or byproducts, developing alternative purification techniques (such as nanofiltration), and eliminating the need for solvents during purification.
- *Process Safety Design:* replacing hazardous solvents and reagents with safe alternatives and renewable raw materials.
- *Material Performance Design:* using advanced synthesis techniques, optimizing raw materials and advanced green solvents, and tracking nanomaterial production in real time [30].

These design approaches can be used in nanofabrication to create environmentally friendly and sustainable nanomaterials.

3.7 GREEN NANOMATERIALS: APPLICATIONS IN DIAGNOSIS

As green-synthesized nanomaterials have better biocompatibility even at high concentrations and after long incubation times, they are widely used in biomedical disciplines, including diagnostics and therapeutics. This chapter elaborates on the use of green nanomaterials for nanomaterial-assisted diagnostics in the fields of biosensing and bioimaging. Among the green-synthesized nanomaterials reviewed in this chapter, metals and metal oxide nanoparticles such as gold (Au), silver (Ag), and iron oxide (Fe_3O_4) nanoparticles, as well as magnetic nanoparticles, have shown great promise for bioimaging using computed tomography (CT) and magnetic resonance imaging (MRI). In addition to bioimaging, the development of biosensors using environmentally friendly synthesized nanomaterials for diagnostic purposes is also being considered. The conclusion is that some nanomaterials, including gold, silver, and certain metal oxides such as ZnO and CeO_2 nanoparticles, can be used in electrochemical methods for the determination of analytes such as lactose, hydrogen peroxide, and glucose [17], depending on the synthesis methods used [18].

Biosensors made using nanomaterials are used not only for bioimaging but also for diagnostic purposes. To create biosensors capable of detecting glucose levels electrochemically (by electrochemically reducing glucose oxidase), reduced graphene oxide–gold nanocomposites are prepared using rose water [18]. Similarly, another study used a one-pot green synthesis method to create a self-assembled membrane from a combination of reduced graphene oxide and gold nanoparticles with amperometric detection of hydrogen peroxide as a cancer biomarker. Similarly, using gold and silver nanoparticles in quercetin-based synthesis, a third-generation cellobiose dehydrogenase-based lactose biosensor is developed, which shows high stability and sensitivity in detection [14].

3.8 GREEN NANOMATERIALS: APPLICATIONS IN THERAPEUTICS

An emerging area of research in nanomedicine is the use of green chemistry for the synthesis of metal nanoparticles (MNPs) and their biological use. The green route, which uses plants, bacteria, algae, and fungi as biological sources, has several advantages over the current synthetic route. It is a simple, one-step process that is fast, effective, inexpensive, and environmentally friendly. It also does not require any special tools or catalysts. Compared to chemically synthesized nanoparticles, most biologically synthesized metal nanoparticles (b-MNPs) are more therapeutically effective and biocompatible. Due to their unique physicochemical properties, synthesis, and biocompatibility, biosynthesized silver nanoparticles, or b-AgNPs, are one of the most promising options among MNPs in biomedical research. In addition, since ancient times, silver has been widely used as a medicinal substance to treat various diseases. Moreover, the overall therapeutic efficacy of b-AgNPs is enhanced by active compounds derived from biological sources.

The biological use of b-AgNPs for the treatment of various diseases, including cancer, infectious diseases, diabetes, inflammation, wound healing, and neurodegeneration have been investigated by several researchers over the years. This chapter depicts the recent advances of b-AgNPs in biomedical research, including their use in drug transport, bioimaging, anticancer, antiangiogenic, antibacterial, and biosensing applications, as well as their use as therapeutic agents that can replace more traditional methods. approaches to therapy. In addition, we will review the biodistribution, pharmacokinetics, and clearance of b-AgNPs in animal models in light of their encouraging clinical results. Finally, we will discuss the challenges and potential applications of b-AgNPs for future therapeutic applications in healthcare settings. These biocompatible gold particles have demonstrated good radiopaque properties, making them suitable for use in computed tomography [12]

Moreover, photodynamic therapy and in vitro photothermal therapy on HeLa cells are a line of human cervical cancer cells that are widely used in scientific research and effectively tested using gold nanotriangles made from yam bean extract. Using an ascorbic acid-mediated green reduction process, a heterodimeric nanoparticle containing a hybrid of gold, copper, and selenium is created. Plasmonic interaction between copper and gold caused a remarkable response in Hep-2 cells treated with photothermal therapy [13]. Using the mouse fibroblast cell line 3T3, silver nanoparticles with excellent biocompatibility are synthesized using egg white medium. This silver–protein biocomposite exhibited increased radioactivity under X-ray irradiation, making it suitable for radiotherapy in cancer treatment [14]. In addition, iron oxide nanoparticles prepared with chitosan derivative have shown a strong biocompatibility and efficient absorption with natural lactose [15].

These magnetic nanoparticles demonstrate stimulus-responsive cellular localization and enable magnetic hyperthermia for cancer treatment, as demonstrated using A549 human lung cancer cells and L929 mouse embryonic fibroblasts [15]. Ferrite nanoparticles are biogenically generated within magnetotactic bacteria (Geobacter sulfur-reducing) to show good potential for magnetic hyperthermia-induced MRI imaging and cancer therapy [16]. Due to their ability to fight bacteria (both gram-positive and gram-negative), fungi, viruses, and cancer, b-AgNPs have been the subject of numerous studies [14–17]. Another well-established therapeutic approach is drug delivery using green nanomaterials, which uses metal, metal oxide, carbon, liposomes, micelles, quantum dots, dendrimers, and polymer nanoparticles to actively or passively deliver drugs to the affected areas. Green-synthesized gold nanoparticles have found fantastic applications as drug and gene delivery carriers due to their biocompatibility and functionalizability [11,13]. Graphene and its derivatives function as drug and gene carriers due to their huge surface area and high drug-carrying capacity [13]. Another excellent illustration of a green synthetic polymer-based drug delivery system is the Pluronic polymer-based nanomicellar system [14].

To track endothelial cells both in vitro and in vivo, fluorescent and biocompatible gold nanoparticles combined with liposomes have been prepared and used as biomarkers [15]. Similarly, gold nanoparticles modified with polyethylene glycol (PEG) and cell-targeting peptides effectively target specific cell types and are easily internalized [16]. Some more examples of applications of green nanomaterials in biomedicine are shown in Table 3.2. The desire to develop green nanotechnology, that is, to combine nanotechnology with the principles of green chemistry, is driven by concerns about the safety of nanoparticles affecting human health and the environment. Many of the major problems associated with nanotoxicology could be addressed sustainably through green nanotechnology. Significant safety improvements can be achieved through changes in the design of materials synthesis and proper handling [26].

Safer products have been developed while reducing production costs, raw material consumption, harsh chemicals, and energy consumption through the adoption of environmentally friendly chemical technologies. With recycling/reuse provisions, efficient use of resources has been achieved through the application of nuclear economics. Process efficiency is ensured by minimizing unnecessary intermediates and preventing waste generation. When biogenic synthesis methods are used instead of hazardous chemically produced or engineered nanoparticles, product quality is improved to safer levels [27]. More biocompatible products that can be safely administered into the body for diagnostic and therapeutic purposes are being produced using green nanotechnology. Moreover, without jeopardizing the environment, green nanotechnology provides a healthy future for the pharmaceutical, cosmetics, electrical, and other commercial industries [28, 29, 31].

Green nanomaterials, which use environmentally friendly and sustainable technologies to produce nanoparticles, represent a breakthrough in nanotechnology. Green nanotechnology increases sustainability and reduces its environmental impact by using biological processes, such as plant extracts and bacteria, and reducing reliance on hazardous chemicals. In addition to encouraging safer manufacturing

TABLE 3.2
Biomedical Applications of Green Synthesized Nanomaterials

Nanomaterial	Synthesis Approach	Application
Carbon dots	*Ocimum sanctum*-mediated hydrothermal method	Fluorescent sensing of Pb^{2+} ions in breast cancer cells (live imaging) and detecting Pb^{2+} ions in water
Reduced graphene oxide	*Ocimum sanctum* hydroalcoholic-mediated phytochemical synthesis	Biocompatibility in animal models. Potential biocompatible drug/gene carrier
Silver-chondroitin sulfate nanocomposite	Thermal treatment or ultraviolet (UV) irradiation	Antimicrobial scaffolds
Silver/ carboxymethylcellulose—poly(acrylamide) hydrogel nanocomposite	Plant extract based in situ reduction	pH responsive drug delivery
Silver-TiO_2 nanocomposite	Euphorbia heterophylla leaf extract-mediated reduction	Catalytic activity
Iron oxide–gold core–shell nanoparticles	Sonochemical synthesis	Targeted drug delivery and contrast agent for MRI/CT
Fe_2O_3/$CuFe_2O_4$/chitosan nanocomposites	Sol–gel auto-combustion route	MRI imaging
Au@C@Pt core–double shell nanocomposite	Microwave synthesis	Biosensing H_2O_2
ZnSe/Au nanocomposites	Low-temperature–aqueous synthesis and direct coupling approach	Cellular bioimaging
ZnO-carbon dots nanocomposites	Hydrothermal method	Photocatalysis and photoreduction of CO_2
Chitosan gold nanocomposite	Solvothermal synthesis	Antifungal activities
Carbon dots	Egg yolk-mediated solvothermal	Bioimaging
CQD–polymer composite	Microwave-assisted pyrolysis method	Bioimaging and bio-labeling
Fe_3O_4–Ag core shell	*Vitis vinifera* (grape) stem extract based reduction	Reusable antibacterial activity
Iron oxide–pectin hybrid nanocomposite	Co-precipitation method	Targeted drug delivery, molecular separation and ion exchange
Silver–graphene oxide nanocomposite	Glucose-mediated reduction process	Tryptophan biosensor
Fluorescent carbon dots	Garlic extract-mediated hydrothermal synthesis	Cellular fluorescent imaging (multicolor), radical scavenging
Silver-activated charcoal nanocomposite	West Indian cherry extract-mediated synthesis	Microbial filtration efficiency

processes, this move toward greener approaches combines creativity and environmental responsibility, providing long-term solutions for sectors such as technology, energy, and healthcare. Ultimately, green nanomaterials represent the incorporation of environmentally friendly methods into scientific and commercial breakthroughs.

PROBLEMS

1. What is the impact of direct processing techniques on the properties of final materials?
2. How do manufacturing methods relate to product quality?
3. Why is process control critical to the production of advanced materials?
4. How can processing techniques improve the mechanical and physical properties of a material?
5. How can selecting an optimal manufacturing strategy increase product reliability?

REFERENCES

1. Yao, L., Huang, H., & Chen, S. H. (2020). Product quality detection through manufacturing process based on sequential patterns considering deep semantic learning and process rules. *Processes*, 8(7), 751.
2. Sergunova, V., Leesment, S., Kozlov, A., Inozemtsev, V., Platitsina, P., Lyapunova, S., ... & Sherstyukova, E. (2022). Investigation of red blood cells by atomic force microscopy. *Sensors,* 22(5), 2055.
3. Kun, P., Fülöp, B., Dobrik, G., Nemes-Incze, P., Lukács, I. E., Csonka, S., ... & Tapasztó, L. (2020). Robust quantum point contact operation of narrow graphene constrictions patterned by AFM cleavage lithography. *npj 2D Materials and Applications*, 4(1), 43..
4. Li, J., Zhang, F., Zhong, Y., Zhao, Y., Gao, P., Tian, F., & Cullen, P. J. (2022). Emerging food packaging applications of cellulose nanocomposites: A review. *Polymers,* 14(19), 4025.
5. Rotondi, S. M. C., Ailuno, G., Mattioli, S. L., Pesce, A., Cavalleri, O., & Canepa, P. (2023). Morphological investigation of protein crystals by atomic force microscopy. *Crystals*, 13(7), 1149.
6. Abubakr, M., Abbas, A. T., Tomaz, I., Soliman, M. S., Luqman, M., & Hegab, H. (2020). Sustainable and smart manufacturing: An integrated approach. *Sustainability*, 12(6), 2280.
7. Foday, Jr., E. H., Bo, B., & Xu, X. (2021). Removal of toxic heavy metals from contaminated aqueous solutions using seaweeds: A review. *Sustainability*, 13(21), 12311.
8. Mišić Radić, T., Vukosav, P., Čačković, A., & Dulebo, A. (2023). Insights into the morphology and surface properties of microalgae at the nanoscale by atomic force microscopy (AFM): A review. *Water,* 15(11), 1983.
9. Abuzeid, H. M., Julien, C. M., Zhu, L., & Hashem, A. M. (2023). Green synthesis of nanoparticles and their energy storage, environmental, and biomedical applications. *Crystals*, 13(11), 1576.
10. Kareem, M. A., Bello, I. T., Shittu, H. A., Awodele, M. K., Adedokun, O., & Sanusi, Y. K. (2020). Green synthesis of silver nanoparticles (AgNPs) for optical and photocatalytic applications: A review. *IOP Conf. Ser. Mater. Sci. Eng.,* 805, 12020.

11. Mery, M., Orellana, N., Acevedo, C. A., Oyarzún, S., Araneda, F., Herrera, G., & Romero, C. P. (2018). Fabrication of silver nanoparticles using a gas phase nanocluster device and preliminary biological uses. *Materials,* 11(12), 2574.
12. El-Sherbiny, I. M., & Salih, E. (2018). Green synthesis of metallic nanoparticles using biopolymers and plant extracts. In Kanchi, S., & Ahmed, S. (Eds.), *Green Metal Nanoparticles* (pp. 293–319). Scrivener Publishing LLC: Beverly, MA.
13. Adyani, S. H., & Soleimani, E. (2019). Green synthesis of Ag/Fe_3O_4/RGO nanocomposites by Punica Granatum peel extract: Catalytic activity for reduction of organic pollutants. *Int. J. Hydrogen Energy,* 44, 2711–2730.
14. Ahmed, R. H., & Mustafa, D. E. (2020). Green synthesis of silver nanoparticles mediated by traditionally used medicinal plants in Sudan. *Int. Nano Lett.,* 10, 1–14.
15. Ahmed, T., Noman, M., Shahid, M., Niazi, M. B. K., Hussain, S., Manzoor, N., Wang, X., & Li, B. (2020). Green synthesis of silver nanoparticles transformed synthetic textile dye into less toxic intermediate molecules through LC-MS analysis and treated the actual wastewater. *Environ. Res.,* 191, 110142.
16. Banasiuk, R., Krychowiak, M., Swigon, D., Tomaszewicz, W., Michalak, A., Chylewska, A., ... & Krolicka, A. (2020). Carnivorous plants used for green synthesis of silver nanoparticles with broad-spectrum antimicrobial activity. *Arab. J. Chem.*, 13(1), 1415–1428..
17. Braccini, A. M., & Margherita, E. G. (2018). Exploring organizational sustainability of Industry 4.0 under the triple bottom line: The case of a manufacturing company. *Sustainability,* 11, 36.
18. Jabbour, A. B. L. D. S., Jabbour, C. J. C., Foropon, C., & Filho, M. G. (2018). When titans meet—Can industry 4.0 revolutionise the environmentally-sustainable manufacturing wave? The role of critical success factors. *Technol. Soc. Chang*., 132, 18–25.
19. Waibel, M., Steenkamp, L., Moloko, N., & Oosthuizen, G. (2017). Investigating the effects of smart production systems on sustainability elements. *Procedia Manuf*., 8, 731–737.
20. Chin, T., Li, G., Jiao, H., Addo, F., & Jawahar, I. (2019). Career sustainability during manufacturing innovation. *Career Dev. Int*., 24, 509–528.
21. Nahavandi, S. (2019). Industry 5.0—A human-centric solution. *Sustainability,* 11, 4371.
22. Zhou, J., Li, P., Zhou, Y.-H., Wang, B., Zang, J., & Meng, L. (2018). Toward new-generation intelligent manufacturing. *Engineering*, 4, 11–20.
23. Garcia, D. J., Mozaffar, M., Ren, H., Correa, J. E., Ehmann, K., Cao, J., & You, F. (2018). Sustainable manufacturing with cyber-physical discrete manufacturing networks: Overview and modeling framework. *J. Manuf. Sci. Eng*., 141, 021013.
24. Demir, K. A., Döven, G., & Sezen, B. (2019). Industry 5.0 and human-robot co-working. *Procedia Comput. Sci*., 158, 688–695.
25. Zhou, J., Zhou, Y., Wang, B., Zang, J., Ji, Z., Yanhong, Z., Baicun, W., & Jiyuan, Z. (2019). Human–cyber–physical systems (HCPSs) in the context of new-generation intelligent manufacturing. *Engineering*, 5, 624–636.
26. Raj, A., Dwivedi, G., Sharma, A., Jabbour, A. B. L. D. S., & Rajak, S. (2019). Barriers to the adoption of industry 4.0 technologies in the manufacturing sector: An inter-country comparative perspective. *Int. J. Prod. Econ*., 224, 107546.
27. Bathula, C., Subalakshmi, K., Kumar, K. A., Yadav, H., Ramesh, S., Shinde, S., Shrestha, N. K., Mallikarjuna, K., & Kim, H. (2020). Ultrasonically driven green synthesis of palladium nanoparticles by *Coleus amboinicus* for catalytic reduction and Suzuki-Miyaura reaction. *Colloids Surf. B Biointerfaces,* 192, 111026.

28. Kiani, M., Rabiee, N., Bagherzadeh, M., Ghadiri, A. M., Fatahi, Y., Dinarvand, R., & Webster, T. J. (2020). High-gravity-assisted green synthesis of palladium nanoparticles: The flowering of nanomedicine. *Nanomed. Nanotechnol. Biol. Med.,* 30, 102297.
29. Rashmi, B. N., Harlapur, S. F., Avinash, B., Ravikumar, C. R., Nagaswarupa, H. P., Anil Kumar, M. R., Gurushantha, K., & Santosh, M. S. (2020). Facile green synthesis of silver oxide nanoparticles and their electrochemical, photocatalytic and biological studies. *Inorg. Chem. Commun.,* 111, 107580.
30. Foday, Jr., E. H., Bo, B., & Xu, X. (2021). Removal of toxic heavy metals from contaminated aqueous solutions using seaweeds: A review. *Sustainability,* 13(21), 12311.
31. Dehghan, S., Sattarpanah Karganroudi, S., Echchakoui, S., & Barka, N. (2025). The integration of additive manufacturing into industry 4.0 and industry 5.0: A bibliometric analysis (trends, opportunities, and challenges). *Machines*, 13(1), 62.

4 Fundamentals of Nanostructured Materials

4.1 INTRODUCTION

The ability of nanomaterials to combine size-dependent phenomena with special functional properties impossible to achieve in conventional bulk material systems makes them one of the most exciting and revolutionary areas of materials research and engineering today. Materials exhibit unique behaviors at the nanoscale, typically less than 100 nm, due to surface and quantum effects that dramatically alter optical responses, chemical reactivity, mechanical strength, and electrical performance. Due to their unique properties, nanomaterials are essential for fundamental research, as well as for advanced technological applications, such as biomedical advancements and next-generation energy systems. The ability to modify the structure, shape, and surface chemistry of these materials in ways that provide entirely new design paradigms for functional materials garners scientific interest in addition to their intrinsic properties [1–3].

Significant advances in synthesis and processing technologies have made it possible to produce nanostructured materials. While chemical methods, such as sol–gel, hydrothermal, sol-thermal, and reverse micelle, provide flexible pathways for producing nanoparticles (NPs), nanowires, and porous nanocomposites with controlled shapes, physical methods, such as chemical vapor deposition, sputtering, and laser ablation, enable the precise fabrication of thin films and nanocoatings. Furthermore, bioinspired techniques, using microorganisms, plants, or enzymes, provide sustainable approaches for synthesizing nanomaterials, emphasizing the intersection of green chemistry and nanotechnology [3]. The performance and usability of nanomaterials are directly influenced by key factors, such as particle size distribution, crystallinity, agglomeration state, solubility, and surface functionalization, all of which are controlled by the choice of synthesis technique. Nanomaterials can be divided into three categories based on their pore size, according to the International Union of Pure and Applied Chemistry (IUPAC) nomenclature (Figure 4.1).

Classifying nanomaterials based on their dimensions is crucial. Strong quantum confinement effects and enhanced optical properties make zero-dimensional (0D) nanostructures such as quantum dots and NPs of great importance in photonics and imaging. Nanowires, nanotubes (NTs), and nanorods are examples of one-dimensional (1D) systems characterized by their exceptional mechanical flexibility and directional

DOI: 10.1201/9781003758556-4

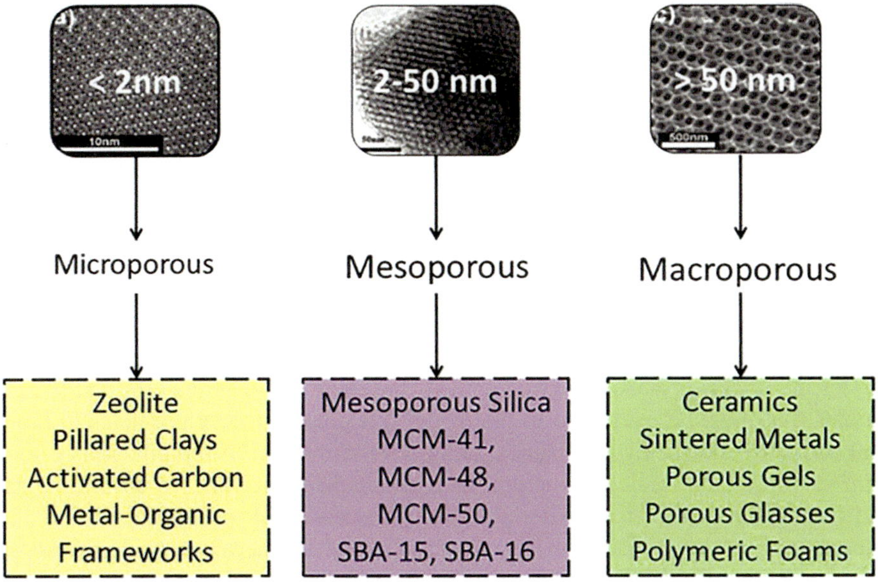

FIGURE 4.1 The International Union of Pure and Applied Chemistry (IUPAC) classifies mesoporous silicate materials into three graded classes based on their porosity. Special forms of ordered mesoporous silicate systems include MCM-41, MCM-48, MCM-50, SBA-15, and SBA-16 [1] [Open access].

conductivity, making them suitable for nanoelectronics and sensors. Graphene, nanosheets, and thin films are examples of two-dimensional (2D) nanostructures that have attracted significant interest due to their excellent electrical conductivity, mechanical robustness, and modifiable surface chemistry. Nanoporous and porous nanomaterials are examples of three-dimensional (3D) designs that combine multiple functions and offer enhanced structural stability, expanding their applications in structural reinforcement, filtration, and catalysis. This classification provides a basis for understanding the diversity and adaptability of nanosystems.

Advanced characterization methods are essential for accurately investigating the properties of nanomaterials. High-resolution spatial imaging of nanoscale morphology and topography is possible using tools such as the atomic force microscopy (AFM), transmission electron microscopy (TEM), and scanning electron microscopy (SEM) are widely used characterization techniques. Spectroscopic techniques that reveal molecular interactions, chemical bonding, and crystallinity include Fourier-transform infrared (FTIR) spectroscopy , Raman spectroscopy, and X-ray diffraction (XRD). Furthermore, surface-sensitive methods, such as dynamic light scattering (DLS) and X-ray photoelectron spectroscopy (XPS), are crucial for examining colloidal stability, particle size distribution, and surface chemistry. Combining these methods ensures a comprehensive evaluation of nanomaterials and establishes relationships between synthesis, structural properties, and functional performance.

The technological uses of nanomaterials continue to grow rapidly, highlighting their critical importance in contemporary innovation. By improving conductivity, charge mobility, and optical sensitivity, nanomaterials in electronics and photonics enable faster, smaller, and lower-power devices. Because nanoscale shapes enhance active surface area, promote electron transport, and enable higher power conversion efficiencies, they are essential for improving the performance of batteries, supercapacitors, fuel cells, and solar cells in the energy sector. NPs are used in medicine as therapeutic agents, drug delivery vehicles, and imaging agents. Their customizable size and surface chemistry enable precise and controlled interactions with biological materials [1–4]. Applications include carbon capture, water purification, and pollution management using nanocatalysts and adsorbents, which offer long-term solutions to pressing global issues.

Despite these promising opportunities, the field of nanomaterials still faces hurdles that must be overcome to realize its full potential. Key obstacles to industrial adoption and public acceptance include issues related to reproducibility, long-term stability, large-scale production, and potential toxicity [4]. To ensure the safe and responsible use of nanomaterials, it remains essential to understand how they interact with biological and environmental systems. Furthermore, bridging the gap between laboratory-scale invention and commercial application requires the development of standardized protocols for characterization, performance evaluation, and risk assessment.

Nanomaterials are a vital field of science and technology, offering exceptional opportunities to create and design materials with unprecedented functionality and precision. Their study includes a comprehensive understanding of synthesis techniques, classification schemes, intrinsic properties, and advanced characterization tools, all of which form the foundation for their diverse applications. Overcoming these obstacles will open the door to the broader integration of nanomaterials into advanced technologies that meet pressing global demands in the areas of sustainability, energy, health, and communications as research advances. Therefore, this chapter aims to provide a comprehensive study of the fundamental ideas behind nanomaterials and build the knowledge necessary to understand their pivotal role in influencing the development of advanced materials and engineering in the future [1]. The energy diagram in Figure 4.2a facilitates understanding the crystallization process in terms of changes in surface and free mass energies during this process. The nucleus does not return to the dissolved gas state and is normally in a stable state [2]. The nuclei multiply and form a stable NP upon reaching a critical threshold. For example, in the formation of metallic NPs, the starting material containing metal cations is reduced or dissociated into metal atoms (zero oxidation state), which subsequently combine to form the NPs. Various crystallization mechanisms have been observed and documented in the scientific literature to date. Figure 4.2b illustrates the crystallization mechanism proposed by La Mere and Denegar.

Numerous drawbacks of enzyme sensors, such as complex synthesis procedures, high operating costs, unstable performance, and critical operating conditions, are the main reasons for their inability to achieve the desired results. Furthermore, numerous environmental factors, such as temperature, humidity, pH, and the instability of

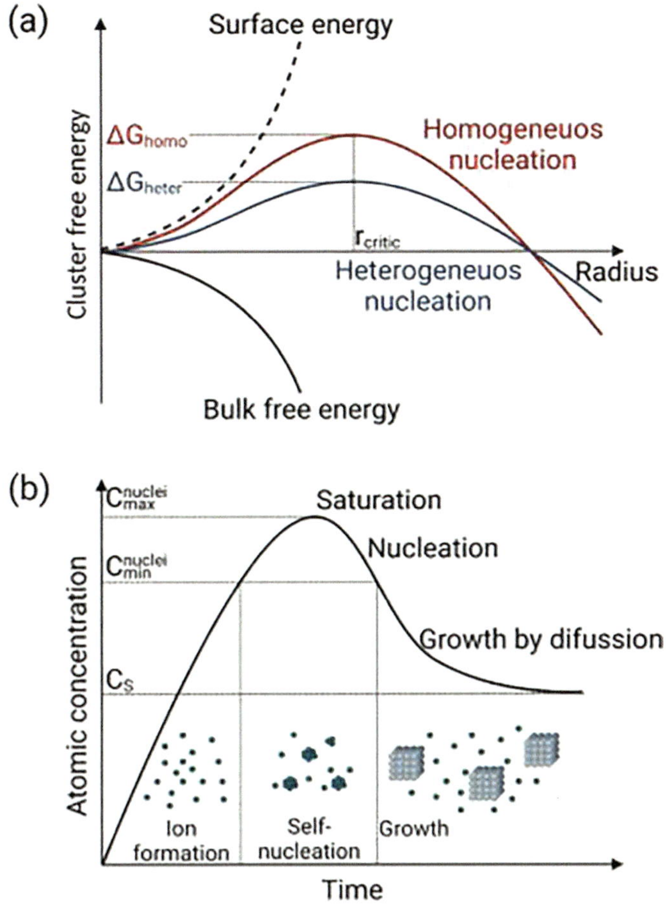

FIGURE 4.2 Schematic diagram of the crystallization steps. (a) The cluster radius (r) affects the free energy of clusters (ΔG) in classical crystallization theory. The formation of the first stable particles occurs at the critical cluster size rc, where the highest value of ΔG is obtained. (b) The crystallization mechanism postulated by Lamier and Denegar shows the production of nanoparticles as a function of time [1] [Open access].

enzyme molecules, affect the repeatability, selectivity, and sensitivity of enzyme electrodes. These drawbacks negatively impact the suitability of enzyme sensors for applications such as food monitoring, glucose sensing, and bioprocess control. Given the challenges facing enzyme glucose sensors, research is currently underway into enzyme-free electrodes. Electrochemical screening of NPs used in enzyme-free sensors results in performance degradation, characterized by surface corrosion or contamination due to the adsorption of intermediates throughout the reaction process. In addition, the high cost of some metal particles increases the cost of sensor assays. Therefore, the development of sensitive, robust, and affordable glucose sensors for enzyme-free electrochemical assays is of paramount importance. Nickel-based NPs are metals with these properties. Therefore, oxidation processes and natural electrocatalytic activities based on the $Ni(OH)_2/NiOOH$ redox reaction in an alkaline environment have been investigated [6,7].

Physical or non-covalent adsorption is the main reason for the electrocatalytic activity of nickel-based catalysts. A fine nickel layer is formed over an electrical mechanism made of gold and glucose NPs. Obtaining materials for research, monitoring, and experiments is challenging [4]. Therefore, it is essential to design new sensor materials. Finding effective ways to create robust connections between the electrode surface and the material under test is also crucial to enhancing electrocatalytic activity. A biosensor is a small analytical device that detects and measures a specific chemical (analyte) via a recognition element connected to a transducer (reader). Numerous studies have focused on developing novel biosensor combinations, immobilization techniques, and material applications that enable these strategies to be translated into useful applications [5]. A reliable biosensor requires an inexpensive, rapid, stable, and inert platform to detect the target molecule accurately and sensitively [6].

The use of noble metals, including gold [7], palladium [8], platinum [9], and their alloys [10], has been extensively studied in the construction of non-enzymatic glucose sensors. However, due to their high cost and rarity, precious metals have very little economic use. Conversely, transition metals, including copper [11], cobalt [12], nickel [13], and related metal hydroxides [14], exhibit remarkable electrocatalytic properties toward glucose in an alkaline environment. Compared with noble metals, transition metals are more active, less expensive, and easier to regulate during synthesis. Given these factors, research on transition metal-based non-enzymatic glucose sensors has attracted much attention [8,9].

4.2 CLASSIFICATION OF NANOMATERIALS

Nanomaterials can be comprehensively classified based on their origin, shape, dimensions, and functional properties. The most common method divides these materials into four groups based on their structural dimensions, which significantly influence their chemical and physical behavior. All dimensions of 0D nanomaterials, including quantum dots, NPs, and nanoclusters, exist at the nanoscale. These materials are critical in imaging, catalysis, and optoelectronics due to their distinctive optical and electrical properties, which are dominated by quantum and surface confinement effects.

Nanowires, NTs, and nanorods are examples of 1D nanomaterials. These materials exhibit anisotropic electrical and thermal conductivity, high aspect ratios, and exceptional mechanical flexibility in one extended dimension, making them suitable for use in energy conversion devices, sensors, and nanoelectronics. Graphene, transition metal dichalcogenides, and nanosheets are examples of 2D nanomaterials characterized by their vast lateral dimensions and atomic or molecular thickness. These materials are pioneers in flexible electronics, photonics, and innovative coatings due to their exceptional mechanical strength, high carrier mobility, and tailorable surface chemistry. Advanced microscopy methods, such as atomic force microscopy (AFM), are essential for accurately assessing nanostructures when characterizing nanomaterials. The use of Peak Force Tapping (PFT) mode has been shown to significantly improve the performance of carbon nanotube (CNT)-based AFM probes, increasing their efficiency in routine high-resolution imaging of nanostructures. For example, a standard tapping probe (Bruker FMV) and a CNT1 probe, a newly developed small-diameter AFM probe from ScanAsyst (Bruker ScanAsyst), were used to image a Nioprobe tip calibration sample. As shown in the final AFM image (Figure 4.3), the comparison demonstrates the high resolution and imaging capability offered by CNT-AFM probes. By providing precise insights into the shape, surface properties, and dimensional features of various classes of nanomaterials, including 0D NPs, 1D NTs, and more complex structures, this development underscores the importance of reliable characterization tools in research.

Networks, porous structures, and nanocomposites that combine multiple nanoscale building blocks into large structures are examples of 3D nanomaterials. These composites are widely used in biomedical structures, filtration, catalysis, and structural reinforcement due to their improved stability, multifunctionality, and synergistic properties. Nanomaterials can also be classified according to their origin, in addition to their dimensions. Nanomaterials specifically designed for specific applications coexist with natural nanomaterials, such as volcanic ash or bio-based NPs.

Furthermore, the functional classification distinguishes between carbon, metallic, ceramic, and polymeric nanomaterials, each with unique performance characteristics. Considering all aspects, this classification provides a framework for understanding the many types of nanomaterials and the connections between their structure and properties, forming the basis for their rational design and application in modern industry and science. In this regard, biosensors are among the most promising uses of nanomaterials, with a variety of applications in clinical diagnosis, prognostication, food analysis, environmental monitoring, bioprocess monitoring, and bioassays [7].

Given their wide range of clinical and experimental applications, glucose detection has received the most attention, prompting many laboratories around the world to focus on developing fast, accurate, and affordable glucose biosensors. The history of blood glucose monitoring strips has been extensively researched [8], and glucose

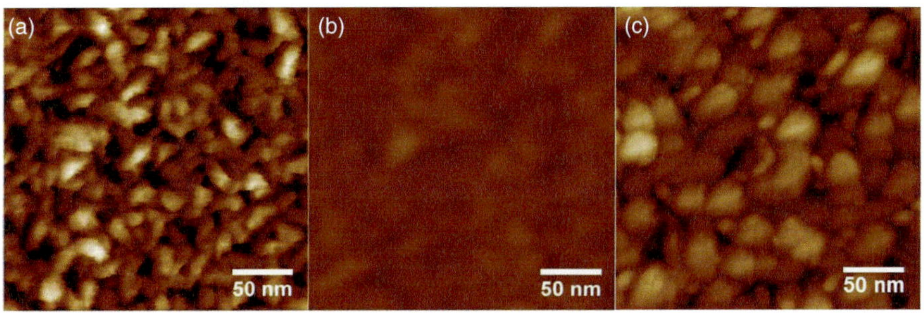

FIGURE 4.3 PeakForce tapping (PFT) mode AFM images of a Nioprobe tip calibration sample using (a) a CNT1 probe, (b) a new silicon FMV probe with a nominal tip diameter of 20 nm, and (c) a new ScanAsyst probe with a nominal diameter of 4 nm. Each image has a height or color scale of 20 nm [2] [Open access].

biosensors are among the few to have achieved commercial success [9]. The performance of these devices has further improved because of recent advances in nanomaterial-based electrode modifications. For example, adding a GO/Co/chitosan composite to gold electrodes has been shown to significantly increase sensing efficiency.

According to experimental results, the transmittance of the proposed structural model gradually decreased with increasing AgNPs at a constant glucose concentration, indicating improved sensor response. A GO/Co/chitosan composite-modified Au film electrode was successfully deposited on the electrode surface (Figure 4.4a) according to FESEM characterization, which also revealed morphologies that were different from those of the bare electrode (Figure 4.4b). Furthermore, consistent with previous research on intermediate and elemental composition analysis (Figure 4.4c–e), the GO/Co/chitosan composite demonstrated beneficial structural properties [3].

There is significant interest in developing accurate, rapid, and affordable sensors for glucose detection, and efforts have focused on improving performance using NPs. The use of silver nanoparticles (AgNPs) has shown great potential in improving glucose concentration detection through advanced techniques such as finite-difference time-domain (FDTD) method. Additionally, the researchers evaluated a 20-micromolar glucose concentration, which included predetermined levels of ascorbic acid (AA) and uric acid (UA), to measure any potential interference. Sensor interference is minimal at 20 and 80 μM AA and UA, respectively. The current–time (I–t) curve responds quickly and sensitively to changes in the glucose concentration in the sensor as it changes gradually.

With a correlation coefficient of $R^2 = 0.9990$ and a detection limit of 0.1 μM at a signal-to-noise ratio (SNR) of 3, the glucose concentration range was found to

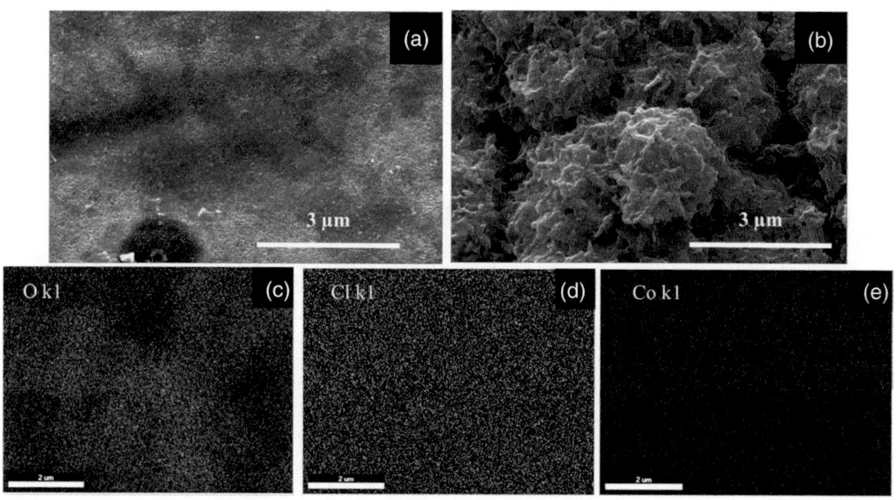

FIGURE 4.4 Surface characterization of GO/Co/chitosan modified electrode in fabricated biosensor using FESEM images and elemental mapping: (a) bare and (b) GO, Co and chitosan treated electrodes. EDS analysis and elemental mapping images of (c) O, (d) Cl, and (e) Co [3] [Open access].

be 0.5–50 μM. The sensor sensitivity was found to be 135.9 μA/mmol. The serum sample had to be diluted appropriately to detect glucose levels above 50 μM. A near-constant value was achieved for the fluctuating glucose concentration and decreasing current. With a correlation coefficient of $R^2 = 0.9990$ and a detection limit of 0.1 μM at an SNR of 3, the estimated linear range for glucose concentration was found to be 0.5–50 μM after optimization. The sensor sensitivity was found to be 135 μA/mmol.

To detect glucose levels above 50 μM, the serum sample must be appropriately diluted. Glucose is used as a source of detection signals for glucose biosensors, either directly from glucose or by catalyzing its conversion to other detectable, electrically active molecules. Both enzymatic and enzyme-independent techniques are currently used to measure glucose levels. However, immobilized glucose oxidase (GOx)-based detection is the most widely used. Due to the impact of enzyme immobilization on biosensor response, new immobilization techniques, electrode surface modifications, and stabilization techniques are required [10].

In this regard, emphasis has been placed on improving sensor performance through the use of advanced techniques such as FDTD and NPs such as AgNPs. Studies have demonstrated that NPs can increase the accuracy of glucose concentration sensing due to their rapid and sensitive response to changes in concentration. Although many conventional techniques, such as enzymatic sensing, attempt to use nanomaterials have seen significant progress has been made in increasing sensing sensitivity and accuracy. Electrode surface modification and immobilization techniques have emerged as key tools for improving the response of biosensors. As these studies progress, we are discovering that combining innovative immobilization techniques, nanomaterials, and new immobilization methods represents a major step toward developing more efficient and stable sensors.

In addition to using AgNPs to enhance sensor sensitivity, research has also been conducted on the nanostructure of materials used in the fabrication of advanced sensors. For example, the morphology of gallium arsenide nanowires, with diameters of up to 200 nm and lengths of approximately 8.5 μm, was studied using an SEM and exhibited some regularity. The crystals' conformity to the zinc-doped gallium arsenide structure and purity were verified using transmission electron microscopy (TEM) and single-area electron diffraction (SAED) images. Furthermore, 2D gallium arsenide nanosheets (2DNLS), produced using atomic-field electron microscopy (AEM), exhibited clean surfaces and a consistent thickness of 18 nm, demonstrating the complete removal of sacrificial aluminum arsenide layers. The p-type semiconductor capabilities of 2D GaAs layers are enhanced by both the GaAs nanowires and their regular crystalline properties, suggesting their potential applications in advanced sensors. Figure 4.5a shows the morphology of GaAs nanowires as measured by an SEM. The nanowires exhibit a clearly uniform size, with a diameter of 200 nm and a length of 8.5 μm. Figure 4.5b shows a TEM and SAED image of the GaAs nanowires. The GaAs nanowires are consistent with zinc-doped GaAs, exhibiting good crystallinity with a lattice pitch of 0.325 nm along the (111) planes. Figure 4.5c shows a GaAs 2DNLS image obtained by AEM. The smooth surface of this sheet, with a thickness of about 18 nm, shows that all sacrificial AlAs layers have been removed. The GaAs 2DNLS layer with zinc-doped crystals, as shown in Figure 4.5d, exhibits a consistent thickness and good

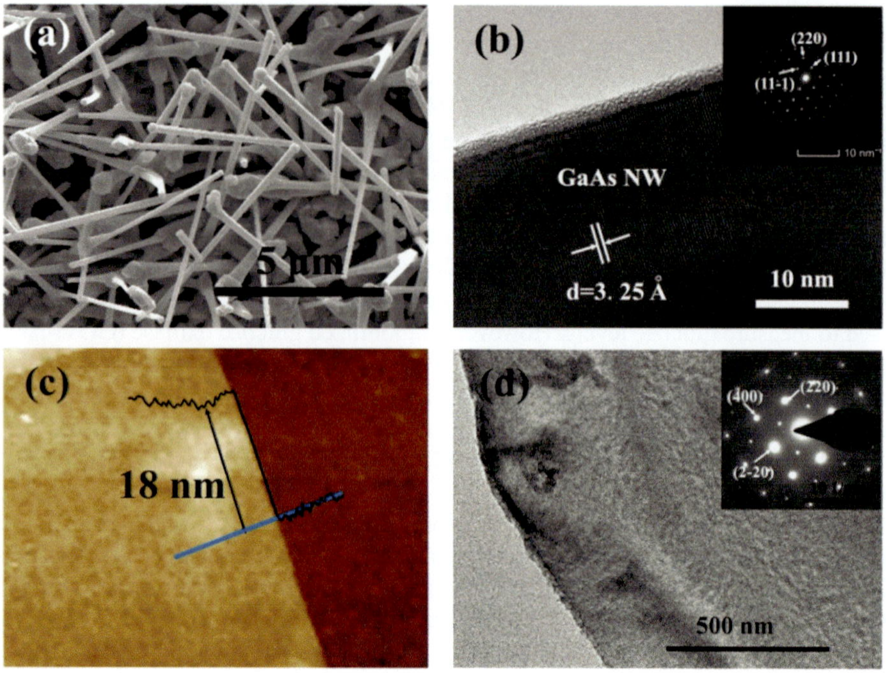

FIGURE 4.5 (a) Top-view SEM image of GaAs NWs, (b) TEM image of GaAs NW; the insert is the SAED (c) AFM image of GaAs 2DNLS, (d) TEM image of GaAs 2DNLS; the insert is the SAED of GaAs 2DNLS [4] [Open access].

crystallinity, as confirmed by SAED and TEM. Both the GaAs 2DNLS layer and the GaAs NW exhibit p-type semiconductor properties [11].

Innovative nanomaterials, such as NTs, quantum dots, nanosilica, nanofibers, NPs, and nanorods, have been made possible by nanotechnology in the past 10 years, opening up promising prospects in the field of biosensing [17]. Due to the many significant advantages of NPs compared to conventional materials, they can be combined.

4.3 SYNTHESIS METHODS OF NANOSTRUCTURED MATERIALS

Because the size, shape, and surface structure of nanomaterials significantly influence their distinct physical, chemical, and electrical properties, their synthesis has emerged as a key area of study in contemporary materials research. To create nanostructures with ordered properties, several techniques have been developed; these can be broadly divided into top-down and bottom-up approaches. Top-down techniques, including laser ablation, ball milling, and lithography, involve reducing bulk materials to nanoscale dimensions. Although well-established and enabling the creation of ordered nanostructures, these techniques often suffer from drawbacks, including surface defects and significant energy consumption. On the other hand, bottom-up strategies, such as hydrothermal synthesis, chemical vapor deposition

(CVD), sol–gel processing, and biological techniques, rely on the assembly of atoms and molecules into nanostructures. For applications in biomedical devices, sensors, and catalysis, these techniques are particularly useful for achieving homogeneity, high purity, and controlled shape. Gold nanoparticles (AuNPs), graphene oxide-based hybrids, and metal-organic frameworks are examples of nanocomposite materials that have recently attracted significant interest due to their tunable properties and suitability for biosensing platforms. For example, AuNPs produced by seed-mediated growth techniques or chemical reduction exhibit superior plasmonic behavior, biocompatibility, and surface functionalization, making them ideal for optical biosensors. Similarly, composite materials combining metal oxides or polymers with carbon nanomaterials can be produced through in situ polymerization, electrodeposition, or solution mixing. These materials exhibit synergistic properties that improve the stability and sensitivity of detection systems.

The case of biosensors for diabetes monitoring demonstrates the clear link between synthesis techniques and their applications. Diabetes is a common medical condition in humans that requires the development of highly sensitive and selective biosensors to accurately assess glucose levels. Nanomaterials are essential for developing next-generation glucose sensors, as their controlled synthesis ensures reliable and reproducible performance. One potential technique that is highly sensitive to changes in local refractive index is localized surface plasmon resonance (LSPR). The use of tapered optical fibers coated with AuNPs is an innovative design, where the NPs are synthesized and applied to the fiber surface via surface functionalization or wet chemical procedures. The LSPR signal is enhanced by the even distribution of NPs on the fiber, allowing for highly accurate monitoring of glucose concentration in real time. This combination of synthesis techniques and device engineering demonstrates the critical role of nanomaterials in improving biological diagnostics. Ultimately, the rational synthesis and assembly of nanostructures not only improves the performance of biosensors but also opens the door to scalable and affordable solutions for major global health issues such as diabetes. AuNPs approximately 10 nm in size were produced for this purpose.

The properties of the AuNPs were characterized using transmission electron microscopy (TEM), scanning electron microscopy (SEM), and UV–Vis spectroscopy. Furthermore, glucose oxidase was used to enhance the sensor's selectivity. Several analytes abundant in blood serum, such as L-cysteine, urea, cholesterol, galactose, and AA, were also used to evaluate the sensor's selectivity. The potential of the AuNP/nanogas hybrid nanocomposite as a practical electrochemical catalyst for glucose sensing was evaluated [12]. The article details an analytical procedure that uses amperometric measurement (amperometric response) of the entire material at a voltage of 0.0 V and a concentration of 4.0 mM. The measured response time is approximately tenfold. At a noise-to-noise ratio of 3 (S/N = 3), the level of detection (LOD) should be 12 μm. AFM data captured in conventional 2 μm × 2 μm fields, where the height change (from red to white) is displayed over a magnification range of 0–125, are compared with the results. The surface volume and coverage of the gold and tin oxide NPs can be calculated using bearing analysis (Supplementary Information). Figure 4.6a–c shows the percentages of 99%, 99%, and 97%, respectively, of AuNPs electrodeposited using processes 1, 2, and 3. AuNP films are produced on the surfaces of tin, gold oxide, and tin oxide, as shown in Figure 4.6a–c [3].

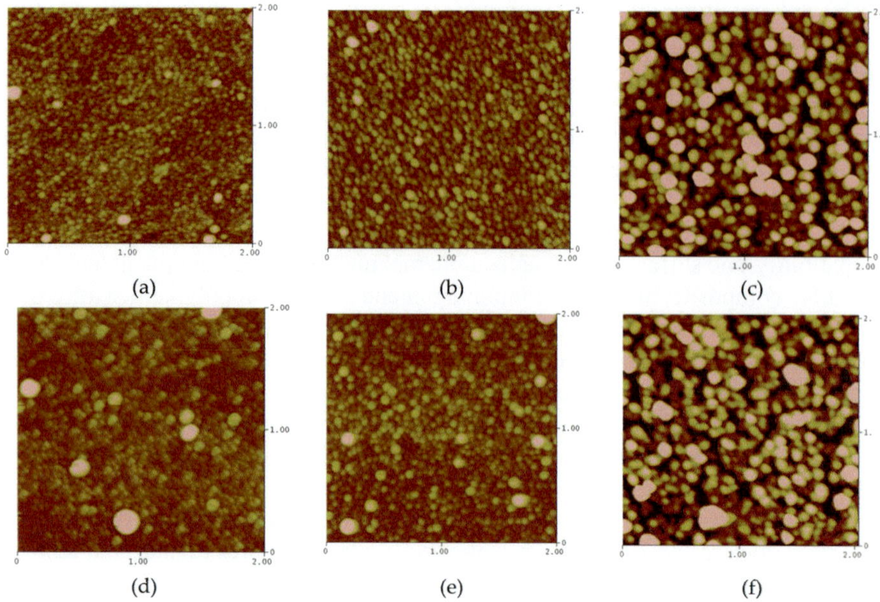

FIGURE 4.6 AFM scans of gold nanoparticles electrospun with (d) BSA, (e) GOx, and (f) casein-coated gold nanoparticles using (a) Method 1, (b) Method 2, and (c) Method 3. AFM scan area is 2 μm by 2 μm with a color scale from 0 to 125 nm [5] [Open access].

The AuNP/NG particle-decorated ethanolamine electrode exhibited remarkable sensitivity and anti-interference behavior towards AA (0.05 mM), UA (0.05 mM), and DA (0.005 mM). The researchers concluded that the fabricated electrode detects glucose non-enzymatically at 0.0 V with very little interference. The long-term stability of the sensor is evaluated by studying the amperometric response. Current losses can reach 4.5% of the original data after 700 s of operation (Figure 4.7c). The patterned AuNP/NG electrodes were drawn into ethanolamine using cyclic scan voltammetry (CV) testing with 4 mM glucose in 0.1 M NaOH. As shown in Figure 4.7c, 90% of the peak current is measured during the saccharification process, indicating good stability of the electrode compared to the initial glucose value. The recovery rate ranged from 93.1% to 94.6% when examining the viability of the indium tin oxide electrode decorated with gold/nickel nanoparticles. With the help of the researchers' innovative glucose sensor, advanced micro- and nanodevices, such as microfluidic sensors and nanoscale biosensors, can accurately detect and quantify glucose levels in real biological and clinical samples.. These features include a large surface-to-volume ratio, small size, variable behavior, strong chemical reactivity, and adjustable properties [17]. Due to their special properties, nanomaterials offer a promising future in biotechnology and bioanalytical chemistry. Figure 4.7a shows the corresponding surface cross-section and atomic energy microscopy (AFM) image of γ-APTES+NPs near the coated tip using the spin-impact method. AFM scan shows that the thickness of γ-APTES+NPs particles coated by the spin-impact method ranges from about 450 to 470 nm [13].

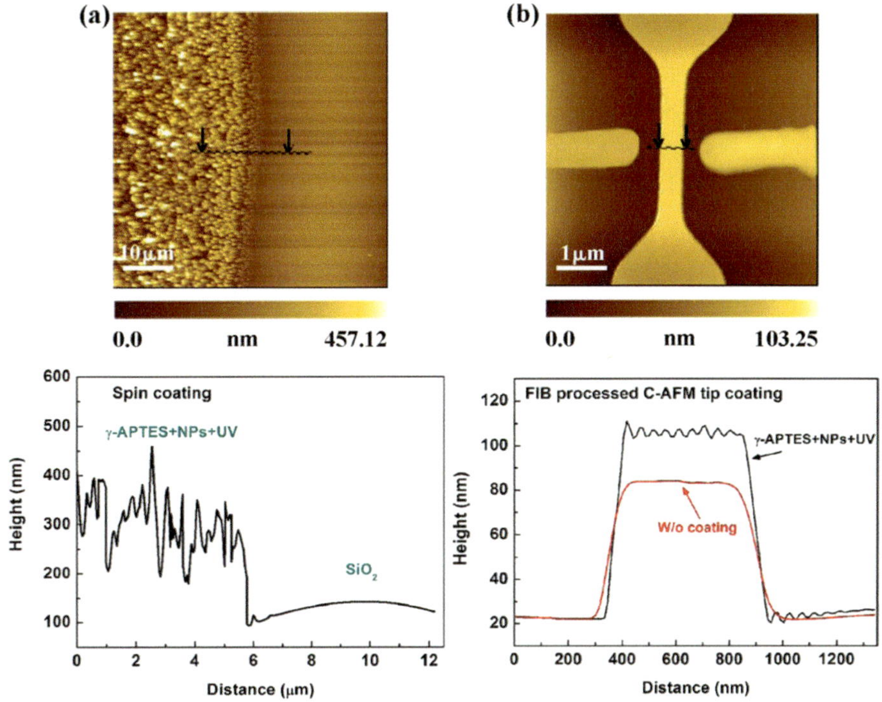

FIGURE 4.7 For PSW coated with γ-APTES+NP film, AFM images, and surface characterization are obtained using (a) micropipette/spin-coating and (b) C-AFM tip/coating scanning using FIB [6] [Open access].

Because it is difficult to generate a smooth film surface profile with a large detection thickness using AFM, the maximum surface profile is used as the film profile. Figure 4.7b shows the cross-sectional surface profile of an AFM image of PSW particles coated with γ-APTES+NPs using a C-AFM probe. For comparison, the PSW surface profile without the γ-APTES+NP coating is shown. It was discovered that the film coated with a C-AFM tip could have a smooth profile and a thickness of 21–23 nm. Metallic poly(NPs) are widely used in many biomedical sectors. The electron flow from the enzyme to the electrode and the expanded surface area of the medical lines led to their subsequent invention in sensors.

Furthermore, the properties of gold nanoparticles (AuNPs) facilitate the assembly of GOx-bound nanoparticles onto the electrode surface [14]. These particles can form conductive wires on the electrode surfaces [15]. As the particles grow, the surface area of the MPTS-nAuE electrode increases in conjunction with the amount of glucose present. It is believed that the novel concept of a neutral atom/water oxide mediator may help explain the catalytic activity of gold nanoparticles (AuNPs). [16]. To ascertain how AuNPs affect the ability of gas chromatography electrodes to adsorb proteins, a potentiometric approach is used to analyze the true discrimination and interference using an AC current measurement. During electrode deposition, a larger

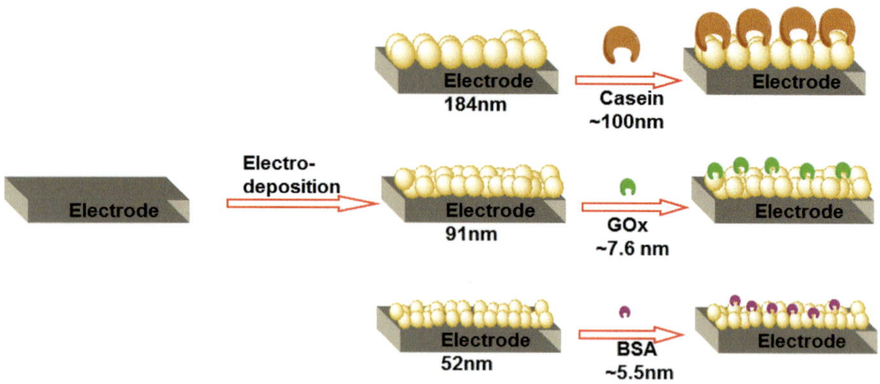

FIGURE 4.8 Schematic representation of protein adsorption on AuNP surfaces of different sizes [7] [Open access].

negative potential is applied to read the binding of the horns to the AuNP surface (as shown in Figure 4.8). As germination and development are essential conditions for deposition independence, the results indicate that the potential drop during deposition is large and severe for AuNPs.

The following section illustrates how printing pressure and its mechanism affect the shape of a polyethylene terephthalate (PET) film. It is well known that the transferred printing pressure causes tensile stresses in the area surrounding the nanopores. Figure 4.9 shows a schematic illustration of the stress process attributed to the microdisc in a thin gold layer on the substrate. The PET film used in this research contains holes approximately 1,000 nm deep.

A 40-nm-thick gold layer is deposited on a quartz glass substrate. A 40-nm-thick layer of deposited gold foil is directly pushed into the 1,000-nm-deep PET holes. After the holes penetrate the gold layer to a certain depth, some gold clusters leak into the PET hole chambers. As a result, during the transfer printing process, the height of the gold layer in the holes gradually increases. It has been confirmed that fine gold disc shapes form on a 50-nm-thick glass substrate.

Several CNT-nanoparticle composites were used in high-throughput electrochemistry in 2007. Following the introduction of CNT devices, the sensitivity of these devices was examined based on the amount of AuNPs on different CNTs. Spin-coated and tip-coated synthetic films differ in their linewidth resolution [19]. Electron microscope images of γ-APTES+NP-coated PSW particles of different diameters after spin-coating and tip-coating after UV exposure for 120 s are shown in Figure 4.10a, b.

Figure 4.10a also shows the large solid particles collected in the rotary press output. In Figure 4.10, the variations in surface-to-volume ratio and ΔJ illustrate the different feature sizes. For the tip-coated film, Figure 4.10b shows the regular distribution of numerous small clusters on the PSW surface. The root mean square deviation of the surface roughness of the UV-modified PSW nanocomposite with varying levels of γ-APTES+NP is shown in the range of 1.67–1.75 nm [4].

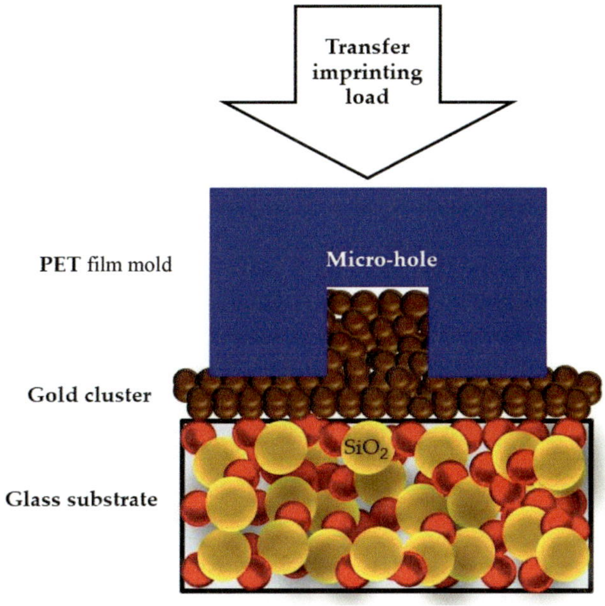

FIGURE 4.9 Schematic illustration of the compression mechanism of transfer printing process [8] [Open access].

The improved sensitivity is attributed to the higher electron transfer rate from GOx to the detection electrode. Figure 4.10 illustrates the production process of AuNP–CNT composites. In the first step, an AuNP–CNT composite is made, and the AuNP surface is modified with cysteamine based on strong thiol–gold interactions. In the second step, the GOx core is attracted at neutral pH to the cysteamine amine group, resulting in a positive surface charge of the composite. In the third step, the modified electrode is spread on a polished graphite disc and allowed to dry at 4 °C [17], the electrode is completely coated in the fourth step. Anodic glucose oxidation peak current data. A solution of 5 mmol/L glucose and 0.1 mol/L sodium hydroxide (NaOH) was used to study the electrochemical reaction of an electrode designed with a scan rate of 0.002–0.100 V/s (Figure 4.10). As the anodic and cathodic glucose peak currents gradually increased with increasing scan speed, the anodic potential changed to a positive value. The scan speed affected the peak current. To achieve linearity, the anode peak was connected [18]. Calculate the current value and the square root of the scan rate using linear regression, the best-fit standard error range (listed in Figure 4.10), and the correlation coefficient $R^2 = 0.996$. In another important study, researchers discovered a method for creating a modified gold electrode that functions as a glucose sensor using covalent linkages between GOx and AuNPs [20]. The fabricated NP-based enzyme sensor demonstrated exceptional electrocatalytic performance and rapid response to changes in glucose concentration, reaching 95% of the steady-state current in just 8 s at an operating voltage of 0.3 V compared to a saturated calomel electrode (SCE) (Figure 4.11). Successive increases in glucose concentrations,

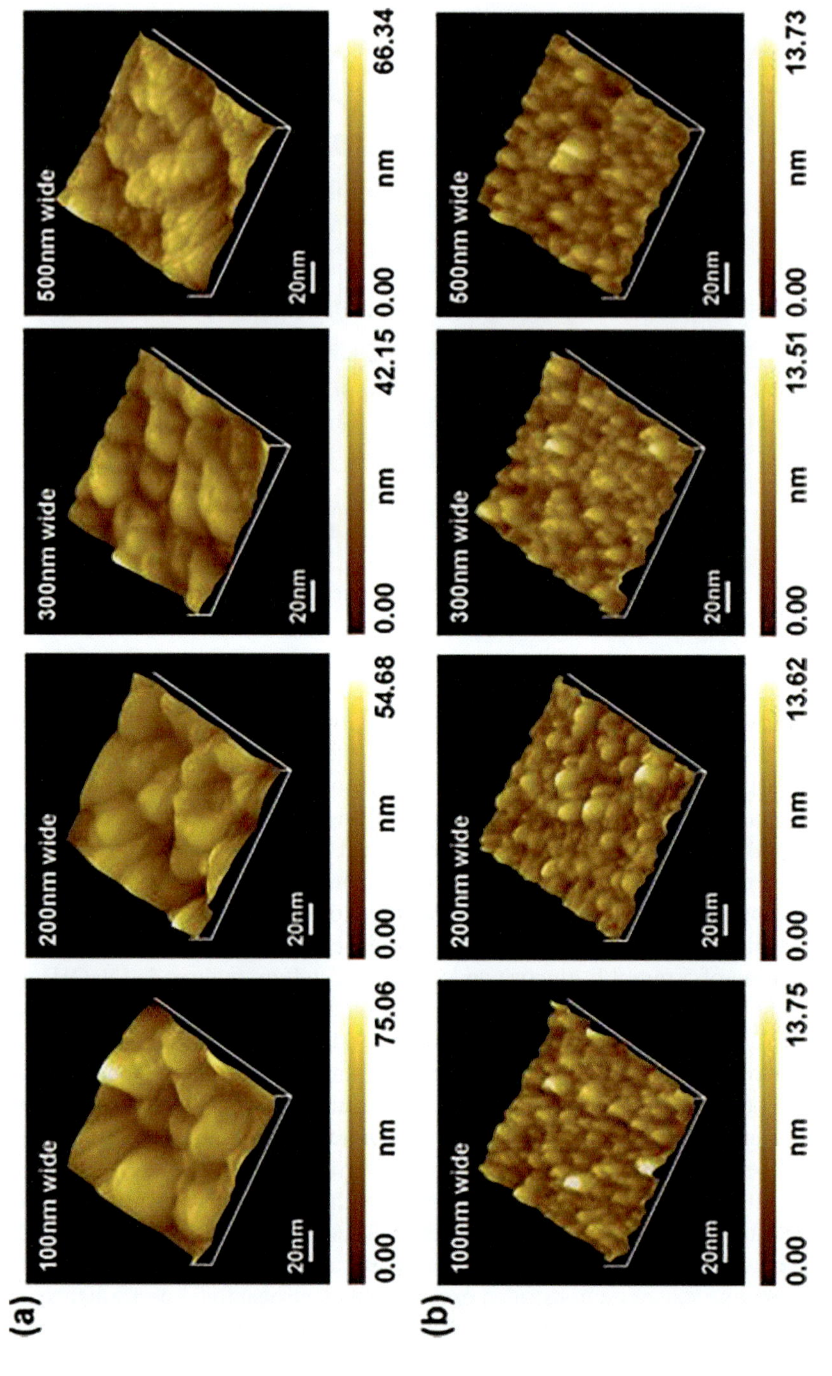

FIGURE 4.10 Using micropipette/spin coating and scanning/probe coating C-AFM, AFM images of surface morphology of γ-APTES+NPs deposited on PSW surface with 100, 200, 300, and 500 nm are obtained under UV illumination for 120 s [9] [Open access].

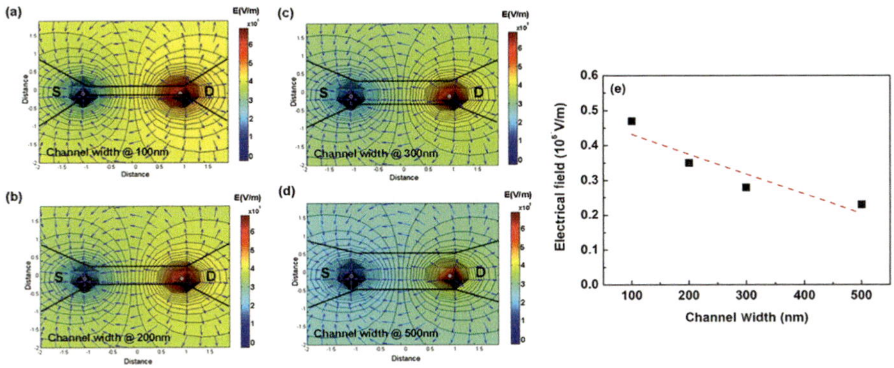

FIGURE 4.11 For PSWs, the channel electric field simulations are performed with channel widths of (a) 100, (b) 200, (c) 300, and (d) 500 nm. The electric field strength of corresponding simulations is linearly fitted in (e) [9] [Open access].

as shown in the current–time course data, show the same result. By tracking the voltammetric current decrease over a constant voltage operation cycle, the stability of the fabricated glucose sensor is confirmed. There is no decrease in the catalytic current (Icat) when the sensor is exposed to 30 voltage cycles between 0 and 0.50 V at a scan rate of 10 mV/s. The continued performance of the NP-based enzyme sensor under constant electrocatalysis even after repeated voltage cycles demonstrates its stability and longevity. The absence of any degradation in the catalytic current further confirms the sensor's reliability for long-term use in glucose detection applications, demonstrating that its electrocatalytic efficiency remains high. To evaluate the lifetime of the enzyme sensor, the team kept it in a 0.1 M phosphate buffer solution at a neutral pH of 4 °C and checked the glucose content every 2–3 days. During the first 30 days, the sensor's current interaction with glucose remained almost constant, and after 2 months, it remained at 70%. Therefore, using an AuNP-modified electrode is a highly effective method for maintaining GOx activity. The storage stability may be attributed to the strong covalent bonding of AuNPs and their exceptional biocompatibility. Figure 4.11 shows the channel electric field intensity simulation results for (a) AuNPs (PSWs) at wavelengths of 100 nm, (b) 200 nm, (c) 300 nm, and (d) 500 nm at a drain-to-source voltage of 5 V. The electric field strengths of PSW wave channels at wavelengths of 100, 200, 300, and 500 nm are approximately 0.47, 0.35, 0.28, and 0.23 mV/m, respectively, as shown in Figure 4.11e. The electric field strengths of PSW wave channels are arranged as follows: for each wavelength, E > E (200 nm) > E (300 nm) > E (500 nm) [4].

4.4 KEY PROPERTIES OF NANOSTRUCTURED MATERIALS

Due to their distinct mechanical, chemical, and physical properties that distinguish them from their bulk counterparts, nanomaterials have received significant attention in recent decades. Their key properties depend largely on their nanoscale size, high surface area-to-volume ratio, and the presence of an interface, all of which combine to produce unique phenomena.

One of the most important of these properties is improved mechanical behavior. Because of grain boundary hardening and limited dislocation motion, the hardness, strength, and wear resistance of nanoscale metals and ceramics often improve. The Hall–Pietsch effect predominates at the nanoscale, with yield strength increasing as grain size decreases. However, beyond a certain point, the so-called reverse Hall–Pietsch effect may emerge, where further grain refinement leads to grain softening.

Furthermore, the optical properties of nanomaterials differ significantly from those of bulk systems. Size-dependent absorption and emission properties result from phenomena such as LSPR in metallic NPs such as gold and silver, or quantum confinement in semiconductor NPs. These properties make nanomaterials extremely useful in biological imaging, sensors, and photonic devices.

Another important feature is their electrical and electronic behavior. In semiconductor nanostructures, the reduced dimensions produce discrete energy levels and tailorable band gaps because of quantum size effects. The high electron mobility and conductivity make CNTs, graphene, and semiconductor nanowires ideal for energy storage and nanoelectronics applications.

In addition, nanomaterials exhibit increased chemical reactivity. Their high surface area-to-volume ratios enhance their performance in energy conversion, heterogeneous catalysis, and photocatalysis, providing more active sites for catalytic reactions. For example, due to their high surface activity and stability, titanium dioxide NPs are widely used as photocatalysts for solar energy harvesting and environmental remediation.

Nanostructures also exhibit remarkable thermal properties. In some cases, enhanced phonon scattering at grain boundaries and interfaces leads to a decrease in thermal conductivity with shrinkage; in other engineered nanostructures, heat transfer can be engineered for thermoelectric applications. Depending on the intended use, this tunability allows the materials to be adapted for thermal insulation or heat dissipation [21].

Moreover, the magnetic properties of nanomaterials behave differently. For example, superparamagnetism is defined as the ability of ferromagnetic NPs to exhibit a magnetization that vanishes in the absence of an external field but responds strongly to it. This property is particularly useful in contrast agents for magnetic resonance imaging (MRI), targeted drug delivery, and data storage technologies.

Enhanced mechanical strength, optical tunability, excellent electrical conductivity, better chemical reactivity, customized thermal management, and unique magnetic responses are just a few examples of the special qualities offered by nanomaterials. These properties, largely attributed to surface phenomena and nanoscale effects, make them highly attractive for a variety of applications in energy, electronics, healthcare, and environmental protection. Monitoring and managing diabetes, one of the most challenging diseases in modern medicine, is one of the most pressing concerns in the biomedical sector. Numerous analytical sensors have been developed that feature properties such as low cost, high sensitivity, and rapid response for continuous glucose measurement.

Based on the detection mechanism and enzyme function, glucose sensors can be broadly divided into enzymatic and non-enzymatic types. Due to their high sensitivity, low detection limits, and good selectivity, enzymatic glucose sensors, especially those utilizing the enzyme glucose oxidase, have been widely used. Because of their large surface area and high electrical conductivity, nanostructured materials, such as zinc oxide and other metal oxides, have been added to glucose bioelectrodes over the past decade to improve sensor performance. However, due to the extreme sensitivity of enzyme activity to changes in pH and temperature, these systems often suffer from poor stability, compromising their long-term reliability. The development of enzyme-free glucose biosensors is currently under increasing focus, with increasing efforts to address these issues. By providing active catalytic sites, enhancing electron transfer, and ensuring robust stability, nanomaterials play a critical role in these designs, opening the door to next-generation biosensing technologies that meet the demands of real-time diabetes monitoring. The use of various metal oxides as a sensing layer in their sensors is the main source of raw materials used in constructing bioelectrodes for non-enzymatic communication with each other. These electrodes are made of a variety of materials, including aluminum nanodots used as a sensing layer on fluorinated tin oxide (FTO) substrates [22]. Many experts are interested in creating nanostructured metal oxide sensors that act as a bridge between the electrode and the enzyme. The band gap of copper is 1.2 eV [23]. This oxide has attracted significant interest for its potential use in temperature superconductors, optical and physical sensors, biosensors, chemical and physical sensors, and catalysts [24]. Some research suggests that the production of copper oxide NPs should be considered. The mechanical, chemical, and physical properties of copper oxide NPs have been studied in the literature [25]. The study of the morphological and structural properties of NPs requires careful tuning of their properties. Copper oxide NPs can be synthesized in several ways, and the electrochemical efficiency of this type of glucose determination is greatly influenced by their shape and morphology [23]. Better results with a wide detection range, low detection limit, and high sensitivity have been obtained using the hydrothermal method to generate copper oxide NPs. Copper produced by direct magnetic sputtering onto a glassy carbon (GC) substrate at a pressure of 4.0×10^{-2} mbar in an argon environment is shown in the SEM images in Figure 4.12. The resulting precipitates have a nodular structure, according to the SEM analysis taken from above. The cross-section shows the characteristic columnar shape. According to Thornton's classification, this shape belongs to the first zone of the theoretical model ($T/Tm = 0.22$ and 30 mTorr). The columnar grains in this zone are defined by open boundaries [22,26]. The graph in Figure 4.12d and e shows that the average diameter of the resulting copper columns under the applied preparation conditions is 121.0 nm $\pm$ 27.2 and their length is 2.52 ± 0.23 μm. Figure 4.12, which shows the link between the diameter dimensions analyzed in the SEM images from above and the column dimensions measured in the cross-section, is the result of a top-down analysis of the copper column dimensions [5].

This material has proven useful in non-enzymatic glucose biosensors. To utilize the non-enzymatic glucose biosensor and compare its catalytic and enzymatic activities, copper NPs are synthesized using three distinct chemical synthesis techniques: glycine [26], hydrazine [27], colloidal solution combustion [28], and solution combustion [29]. Therefore, a variety of characterization studies are used to investigate the optical, morphological, structural, and physical aspects of the synthesized nanocomposite electrode. Using the potentiometric scanning method,

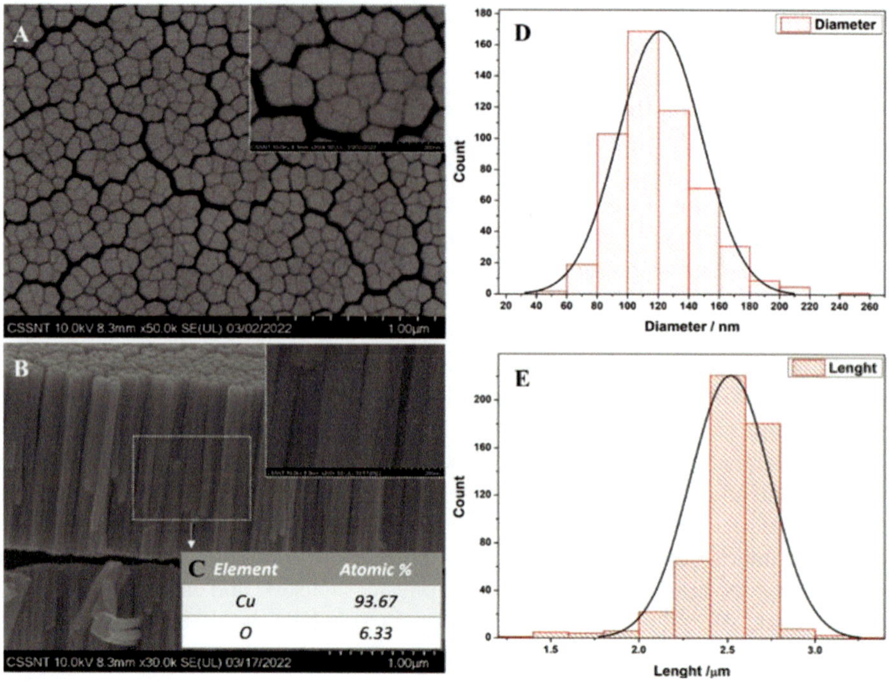

FIGURE 4.12 SEM images of CuNCs: (a) top-view; (b) cross-section; (c) elemental analysis; (d,e) their corresponding diameter and length distribution. [10] [Open access].

Figure 4.12 shows the electrochemical behavior of modified and unmodified copper NPs at a scan rate of 25 mV/s. The anodic peaks (A1, B1, A2, B2) of the cyclic voltammograms show the formation of Cu1+ and Cu^{2+}, while the cathodic peaks (C1, D1, C2, D2) show the transformation of Cu^{2+} to Cu^{1+} and Cu^{1+} to Cu^{0}.

Alkaline conditions are used to synthesize and fabricate copper oxide (CuO) NPs, which have shapes ranging from flower to peony [29]. Recently, copper NPs with microstructures based on nanoflowers and Cu_2O precursors have been synthesized using a simple method [30–34].

4.5 CHARACTERIZATION TECHNIQUES

Understanding and improving the properties of nanomaterials can only be achieved using advanced characterization techniques that provide insights into the structure, morphology, composition, and functional behavior of these materials. Size, shape, surface chemistry, and crystallinity significantly influence the properties of these materials, making it essential to employ a variety of complementary methods.

One of the most widely used methods is XRD, which uses peak broadening analysis to reveal information about average crystallite size, phase identification, and crystal structure. Electron microscopy techniques, such as scanning electron microscopy (SEM) and transmission electron microscopy (TEM), are essential for examining surface morphology, particle size, and internal structure at nanometer resolution. Defects and lattice fringes in nanocrystals can also be observed using high-resolution electron microscopy [35].

In short, characterizing nanomaterials requires a multi-technological approach to comprehensively analyze their shape, structure, composition, and functional responses. By linking characterization data with synthetic methods and applications, researchers can design nanomaterials specifically tailored to technological advancements in energy, electronics, catalysis, and biomedicine.

Furthermore, AFM enables the exploration of mechanical, electrical, and adhesive properties at the nanoscale and enables 3D characterization of surface topography at sub-nanometer resolution. Spectroscopic techniques are essential for examining chemical bonds, vibrational modes, and surface composition. These techniques include FTIR spectroscopy, Raman spectroscopy, XPS, and energy-dispersive X-ray spectroscopy (EDS). While photoluminescence (PL) spectroscopy provides information on defect states and recombination processes, UV–visible spectroscopy is used to examine the optical behavior by examining the absorption, bandgap transitions, and plasmonic properties of NPs. Techniques such as CV, vibrating sample magnetometry (VSM), and electrochemical impedance spectroscopy (EIS) can be used to study electrical and magnetic properties, respectively. Thermogravimetric analysis (TGA) is frequently used to study thermal stability and phase transitions. Figure 4.13 shows a map of the SiNc particle clusters. By overlapping the focused beam on the transparent glass substrate in Figure 4.13b with the real-time topography in Figure 4.13a, the end scanner can image the focal spot. The full width at half maximum (FWHM) of the focal spot is approximately 350 nm. The end scanner in Figure 4.13d accurately images the SiNc clusters after focusing on the spot. Because the PiFM follows the derivative of the optically induced force, it exhibits a better SNR than the real-time topography in Figure 4.13c.

The elongated focal spot in Figure 4.14b is clearly visible in the tip scan image on the clean silicon substrate. The sample scan then shows the graphene layers in Figure 4.14c and d, after the tip is stopped at the mid-position. The gray dashed line segment of the graphene in Figure 4.14d, which is redrawn in Figure 4.14e, provides an estimate of the spatial resolution of the QTF-based PiFM measurement at less than 80 nm. An interesting observation can be made in Figure 4.14d, where the topography measurement in Figure 4.14c and the PiFM image show quite a clear contrast. The circular region in Figure 4.14d, in particular, consists of many patches with uneven patterns, even though the topography appears uniformly thick.

4.6 APPLICATIONS OF NANOSTRUCTURED MATERIALS

With their remarkable mechanical, optical, electrical, thermal, and chemical properties resulting from nanoscale effects, nanomaterials have become a revolutionary class of materials. They are fundamental to the development of both technology and society due to their broad applications in various fields. Nanomaterials, including graphene, CNTs, and semiconductor quantum dots, are used in transistors, light-emitting diodes (LEDs), photodetectors, and solar cells in the electronics and optoelectronics industries [36]. Their high electron mobility, optical response, and

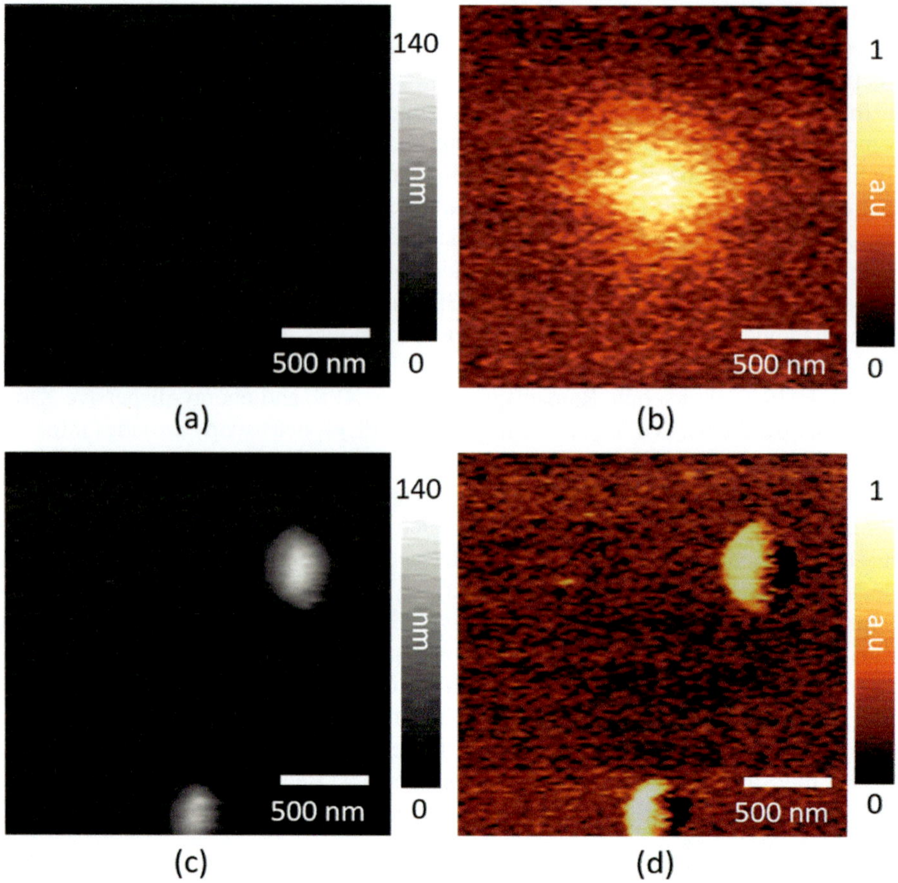

FIGURE 4.13 Mapping of SiNc molecular clusters on a glass substrate using induced force microscopy (PiFM) at 800 nm: To determine the focal spot, the tip scanner uses (a) the topography and (b) a PiFM image of SiNc; after parking the tip in the middle of the spot in (b), the sample scanner uses (c) the topography and (d) a PiFM image of SiNc [11] [Open access].

adjustable band gaps enable the production of faster, more compact, and more efficient devices. Nanomaterials are essential components in batteries, fuel cells, and supercapacitors in the energy generation and storage sector. For example, metal oxide NPs and carbon nanomaterials improve the performance of electrodes by providing larger surface areas, enhanced conductivity, and faster charge and discharge rates. Similarly, hydrogen fuel cells equipped with nanocatalysts offer improved durability and efficiency. Nanostructures are used in the environmental sector for pollution management, air purification, and water purification. For example, nanoscale sorbents and membranes provide effective removal of heavy metals and impurities from water, while titanium dioxide (TiO_2) NPs act as photocatalysts for the degradation of organic pollutants. Various hybrids were produced using $NaBH_4$ ratios ranging from 0% to 100%. XRD and transmission electron microscopy (TEM) investigations revealed

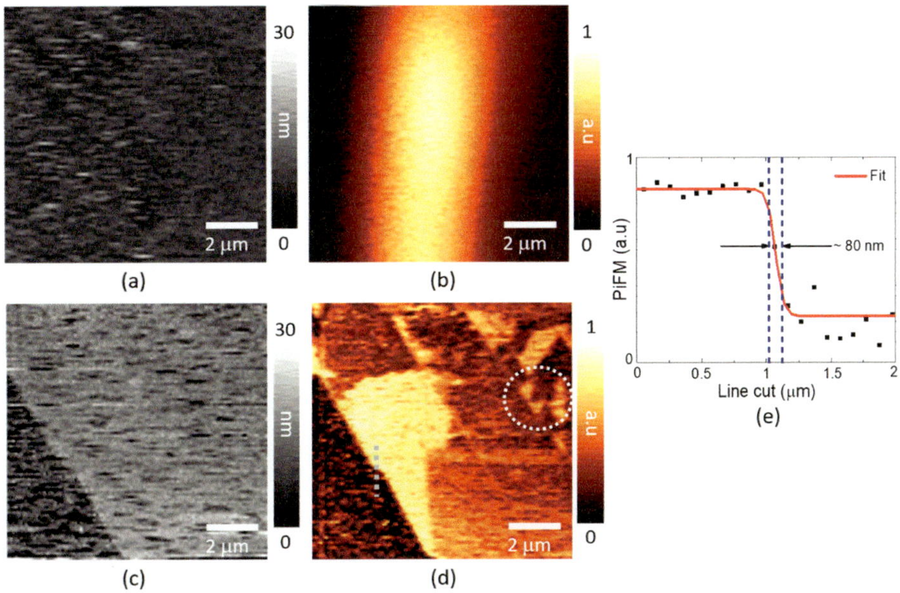

FIGURE 4.14 At 660 nm, PiFM mapping of graphene on a Si substrate: To locate the long focal spot, the tip scanner uses (a) the topography and (b) a PiFM image of graphene; once the tip is parked at the midpoint in Figure 4.5b, the sample scanner uses (c) the topography and (d) a PiFM image of graphene. (e) Estimated spatial resolution of QTF-based PiFM [11] [Open access].

that the oxidation states and sizes of copper nanoparticles (CuNPs) varied depending on the degree of reduction. Higher $NaBH_4$ concentrations were associated with larger NPs (up to 13 nm in size) and a greater presence of Cu(I). Copper(I) was the most common species at lower reducing agent concentrations, with the minimum particle size being less than 6 nm. Furthermore, larger and more agglomerated bio-nanohybrid NPs were produced as the reducing agent was increased (Figure 4.15).

Nanomaterials are revolutionizing diagnostics, imaging, and drug delivery in biomedicine. Lipid- and polymer-based NPs enable targeted and controlled drug release, reducing side effects and enhancing therapeutic outcomes. Magnetic NPs are used as contrast agents in MRI and as carriers in targeted drug delivery systems, while AuNPs and quantum dots are widely used in bioimaging and biosensing due to their optical tunability.

By increasing the strength, durability, and corrosion resistance of structural components, nanostructures are gaining significant importance in mechanical engineering and materials science. Because of their exceptional mechanical performance, nanocomposites containing CNTs or ceramic NPs are suitable for use in the construction, automotive, and aerospace industries. To increase crop yields while reducing their negative environmental impact, the agricultural and food industries also use nanostructures in smart packaging, nanosensors for pathogen detection, and controlled-release fertilizers and pesticides. Nanomaterials are now essential in various fields, including advanced manufacturing, electronics, energy, the

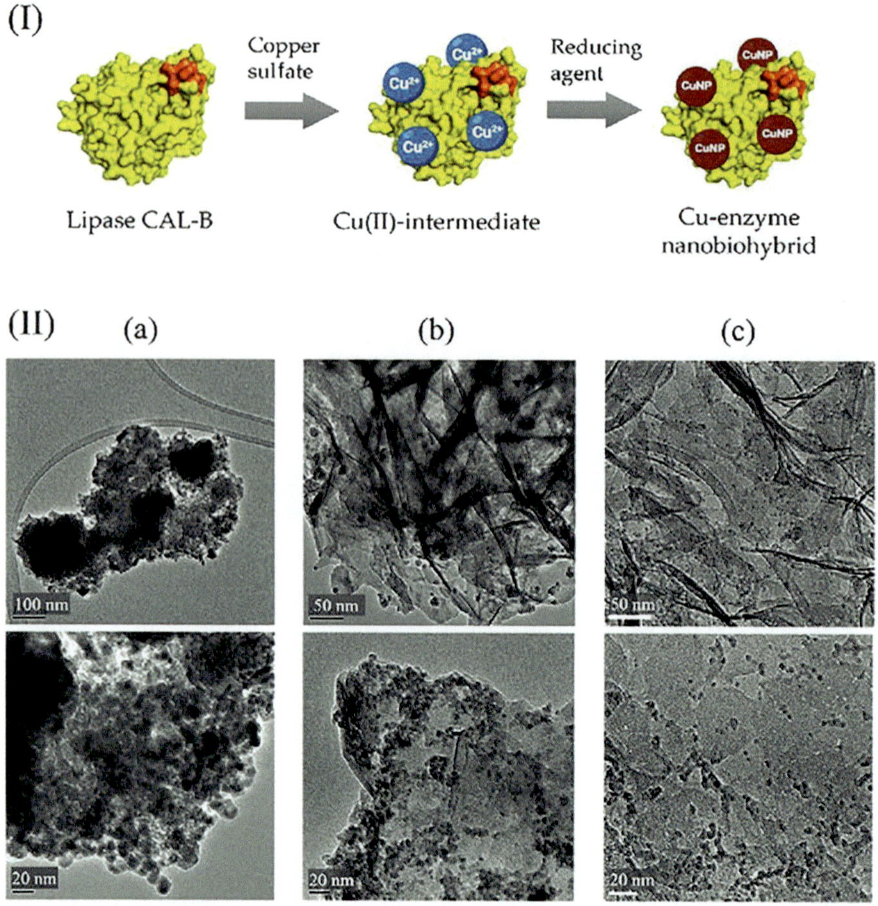

FIGURE 4.15 (I) Schematic of the synthesis of a copper-enzyme biohybrid nanocomposite. (II) TEM images showing the copper–enzyme biohybrid nanocomposites containing (a) 100% reducing agent, (b) 30% reducing agent, and (c) 10% reducing agent [12] [Open access].

environment, and biomedicine. Their adaptability stems from their modifiable nanoscale properties, which continually stimulate developments that address global issues in technology, healthcare, and sustainability [37]. Table 4.1 provides a comprehensive analysis of the scholarly articles, including their standard, local, and global citation values. Global citations reflect the overall worldwide impact of the text, while local citations indicate how frequently the document is cited within a specific dataset.

4.7 CHALLENGES AND FUTURE PROSPECTS

The synthesis, characterization, and use of nanomaterials have seen significant progress, but several challenges hinder their widespread commercialization and long-term reliability. A major obstacle is the scalable and cost-effective synthesis of nanomaterials with

TABLE 4.1
Most Cited Documents in AM and Industry 4.0 and 5.0

DOI	Year	Local Citations	Global Citations	Normalized Local Citations	Normalized Global Citations
[82]	2016	6	105	25.71	4.97
[83]	2016	5	77	21.43	3.64
[84]	2017	1	3	64.00	0.11
[85]	2019	1	2	45.14	0.08
[86]	2019	1	5	45.14	0.20
[87]	2019	1	5	45.14	0.20
[88]	2019	1	34	45.14	1.35
[89]	2021	1	1	145.67	0.06
[90]	2016	1	22	4.29	1.04
[91]	2018	1	7	58.67	0.18
[92]	2018	1	3	58.67	0.08
[93]	2020	1	4	104.75	0.19
[94]	2018	1	2	58.67	0.05
[95]	2019	1	4	45.14	0.16
[96]	2017	1	30	64.00	1.10
[97]	2020	1	7	104.75	0.33
[98]	2019	1	8	45.14	0.32
[99]	2021	1	5	145.67	0.32
[100]	2020	1	6	104.75	0.28
[101]	2020	1	2	104.75	0.09

Source: [40] [Open access].

consistent size, shape, and composition. High-quality nanostructures are often produced using laboratory techniques, but transferring these strategies to industrial-scale manufacturing without compromising quality or performance is a significant obstacle.

Stability and reliability under actual operating conditions are another challenge. Many nanomaterials may change their structure and degrade their performance over time due to their sensitivity to temperature, pH, humidity, and radiation. For example, the degradation of biological components limits the operational stability of enzymatic nanobiosensors, limiting their long-term use.

There are also significant concerns about toxicity and environmental impact. Although the high reactivity of nanostructures is beneficial for applications such as medicine and chemical catalysis, it also raises concerns about their long-term effects on ecosystems and human health, as well as their biocompatibility and safe disposal. Obtaining regulatory approvals and commercialization is increasingly difficult due to the lack of standardized toxicity assessment procedures. Technologically, integrating nanostructures into existing systems and devices presents integration challenges. Reproducible performance, scalability, and compatibility with conventional manufacturing processes remain key challenges.

The topic of nanomaterials holds tremendous potential for prospects. Advances in green synthesis and sustainable nanotechnology are expected to reduce their negative environmental impacts and lower production costs. Future smart devices, energy systems, and biological equipment will be made possible by creating multifunctional and hybrid nanostructures that integrate mechanical, optical, electrical, and catalytic capabilities. Furthermore, the development of artificial intelligence and machine learning in materials science is opening up new opportunities to optimize synthesis pathways, predict performance, and accelerate the discovery of nanomaterials.

Enzyme-free nanobiosensors, targeted nanomedicine, and nanoimplants with enhanced biocompatibility are expected to be widely used in healthcare in the future. Nanostructures are also expected to become essential in the energy and environmental fields, such as next-generation batteries, hydrogen storage systems, and photocatalysis [38,39]. In the future, the scientists will focus their efforts on detecting tiny concentrations of glucose and implantable miniature, subcutaneous and wearable information sensors that can be placed inside the body. Consequently, the researchers are faced with numerous obstacles that may arise shortly. In parallel with the efforts to prepare these sensors for continuous glucose monitoring, there will be a search for new materials that have excellent selectivity, low detection limit, wide detection range and fast response time. Under certain changes, the metal NPs can exhibit glucose concentrations ranging from micromolar to nanomolar [33,34]. The researchers have examined the stability of sensor and influence of interfering species. Several fabricated sensors have demonstrated excellent stability and anti-jamming properties. It is possible to continue the development and testing of this kind of sensor for any danger to the human body. Overall, metal NPs, which can improve existing technologies and reduce the cost of sensors are the way of future for glucose detection [32–36].

According to the recent studies, noble metal NPs hold great promise for the development of advanced label-free biological and chemical sensors. These sensors exploit the special properties of nanostructured noble metals, including their excellent conductivity, high surface-to-volume ratio, LSPR and outstanding light transmittance and reflectivity. These properties collectively enhance the performance of the sensors. The amperometric approach remains the most widely used method for glucose measurement. Alternative measurement capabilities, on the other hand, are provided by LSPR sensors, which have shown remarkable sensitivity to refractive index changes at metal/insulator interface. As can be seen from the above description, metal NPs have a wide range of applications in glucose determination. Many nanomaterials have been studied as glucose sensors as technology has advanced, without the intervention of other species. Compared to bulk metal, metal NPs have a significantly larger surface area, allowing them to offer a larger number of active sites for GOx immobilization and high electron transfer rates. Amperometric sensors are critical for glucose detection. As more as people have diabetes, the researchers are finding this area very attractive [5].

As expected, the glucose-ampere reaction at 0.1 M NaOH, 100 nM is very sensitive with a reaction time of 1 s when glucose is added. The large surface area of CuNP/PEDOT/GO electrode and excellent conductivity may be responsible for the fast diffusion of glucose and ultrafast reaction [35]. The created sensor electrode

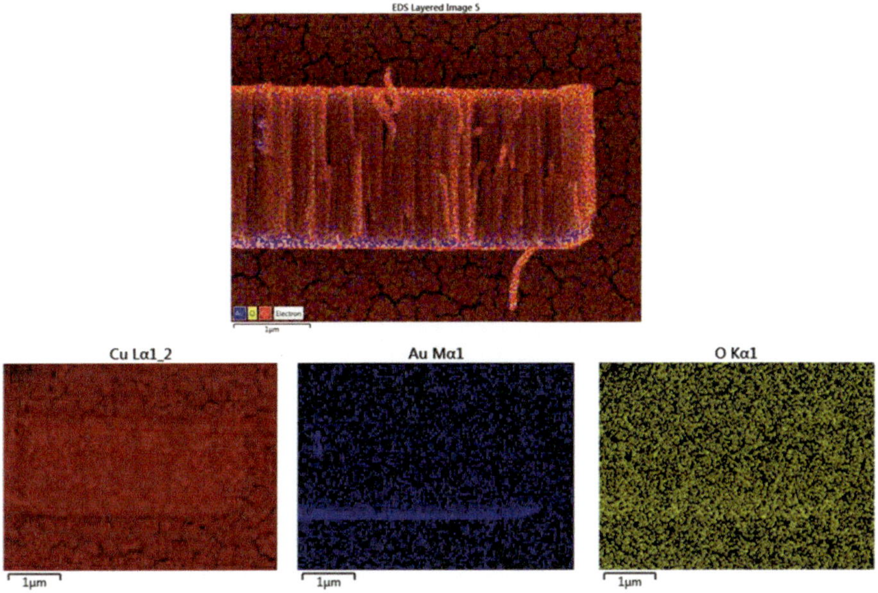

FIGURE 4.16 EDX cross-sectional study of Au-CuNC [5] [Open access].

has demonstrated a linear response to glucose concentration in the range, 0.1–1.3 mM with correlation coefficient, $R^2 = 0.9986$. The sensitivity of sensor is 911.96 μA/mM, and detection limit is 47 nM. After 30 days, the sensor retained almost 92% of its original response, demonstrating a high degree of sensor stability [38]. For electrochemical measurements, a nanometer-thick, ~30 nm gold layer is deposited using electron beam evaporation before the fabrication of Cu-NCs using DC sputtering. This is done to improve the adhesion of Cu-NCs to a GC substrate in alkaline environments. EDX plot shown in Figure 4.16 depicts the presence of the gold layer [5].

In addition, the sensor development capabilities have been expanded by techniques such as attenuated total reflection and fiber-based glucose sensors. This chapter provides a comprehensive summary that includes enzymatic and non-enzymatic sensors based on copper, gold and silver NPs. This covers the current and future development of noble metal NP-based sensors in various application areas, reflecting the current interests and developments in this field

PROBLEMS

1. What is the main reason for the difference in the properties of bulk materials and nanomaterials?
2. How do the mechanical properties of materials change with nanoscale size?
3. What is the role of the surface area-to-volume ratio in applications involving nanomaterials?

4. What are the most important characterization methods for studying nanomaterials?
5. Give two key examples of how nanomaterials are used in energy or medicine.

REFERENCES

1. Harish, V., Ansari, M. M., Tewari, D., Gaur, M., Yadav, A. B., García-Betancourt, M. L., ... & Barhoum, A. (2022). Nanoparticle and nanostructure synthesis and controlled growth methods. *Nanomaterials*, 12(18), 3226.
2. Slattery, A. D., Shearer, C. J., Shapter, J. G., Blanch, A. J., Quinton, J. S., & Gibson, C. T. (2018). Improved application of carbon nanotube atomic force microscopy probes using peakforce tapping mode. *Nanomaterials*, 8(10), 807.
3. Kim, D. S., Yang, X., Lee, J. H., Yoo, H. Y., Park, C., Kim, S. W., & Lee, J. (2022). Development of GO/Co/chitosan-based nano-biosensor for real-time detection of D-glucose. *Biosensors*, 12(7), 464..
4. Guo, S., et al. (2021). Fast response GaAs photodetector based on constructing electron transmission channel. *Crystals*, 11(10), 1160.
5. Chiang, H. C., Wang, Y., Zhang, Q., & Levon, K. (2019). Optimization of the electrodeposition of gold nanoparticles for the application of highly sensitive, label-free biosensor. *Biosensors*, 9(2), 50.
6. Lin, J. J., Hsu, P. Y., Wu, Y. L., & Jhuang, J. J. (2011). Characteristics of polysilicon wire glucose sensors with a surface modified by silica nanoparticles/γ-APTES nanocomposite. *Sensors*, 11(3), 2796–2808.
7. Madden, J., et al. (2021). On-chip glucose detection based on glucose oxidase immobilized on a platinum-modified, gold microband electrode. *Biosensors*, 11(8), 249.
8. Potejanasak, P. (2021). A study on fabrication process of gold microdisk arrays by the direct imprinting method using a PET film mold. *Crystals*, 11(12), 1452.
9. Lin, J. J., Hsu, P. Y., Wu, Y. L., & Jhuang, J. J. (2011). Characteristics of polysilicon wire glucose sensors with a surface modified by silica nanoparticles/γ-APTES nanocomposite. *Sensors*, 11(3), 2796–2808.
10. State, S., Enache, L. B., Potorac, P., Prodana, M., & Enachescu, M. (2022). Synthesis of copper nanostructures for non-enzymatic glucose sensors via direct-current magnetron sputtering. *Nanomaterials*, 12(23), 4144.
11. Jahng, J., Kwon, H., & Lee, E. S. (2019). Photo-induced force microscopy by using quartz tuning-fork sensor. *Sensors*, 19(7), 1530.
12. Ortega-Nieto, C., Losada-Garcia, N., Prodan, D., Furtos, G., & Palomo, J. M. (2023). Recent advances on the design and applications of antimicrobial nanomaterials. *Nanomaterials*, 13(17), 2406.
13. Huang, L., et al. (2001). Engineered collagen–PEO nanofibers and fabrics. *J. Biomater. Sci. Polym. Ed.*, 12(9), 979–993.
14. Geng, X., Kwon, O.-H., & Jang, J. (2005). Electrospinning of chitosan dissolved in concentrated acetic acid solution. *Biomaterials*, 26(27), 5427–5432.
15. Um, I. C., et al. (2004). Electrospinning and electro-blowing of hyaluronic acid. *Biomacromolecules*, 5(4), 1428–1436.
16. Jin, H.-J., et al. (2002). Electrospinning Bombyx Mori silk with poly(ethylene oxide). *Biomacromolecules*, 3(6), 1233–1239.
17. Lee, T. D., & Ebong, A. U. (2017). A review of thin film solar cell technologies and challenges. *Renew. Sustain. Energy Rev.*, 70, 1286–1297.

18. Siampour, H., Abbasian, S., & Moshaii, A. (2020). Copper columnar nanostructures fabricated by glancing angle deposition as a robust and scalable method for high sensitive non-enzymatic glucose detection. *Appl. Surf. Sci.*, 518, 146182.
19. Augustin, A., Udupa, K. R., & Bhat, K. U. (2019). Characterization of DC magnetron sputtered copper thin film on aluminium touch surface. *Trans. Indian Inst. Met.*, 72, 1683–1685.
20. Borisevich, V., Senchenkov, S., & Titov, D. (2016). Magnetron sputtering of copper on thermosensitive polymer materials of the gas centrifuge rotors. *J. Phys. Conf. Ser.*, 751, 012013.
21. Hajimammadov, R., et al. (2018). Random networks of core-shell-like Cu-Cu_2O/CuO nanowires as surface plasmon resonance-enhanced sensors. *Sci. Rep.*, 8, 4708.
22. Abermann, R., & Koch, R. (1980). In situ determination of the structure of thin metal films by internal stress measurements: Structure dependence of silver and copper films on oxygen pressure during deposition. *Thin Solid Film.*, 66, 217–232.
23. Li, X., Wang, Y., Yin, C., & Yin, Z. (2020). Copper nanowires in recent electronic applications: Progress and perspectives. *J. Mater. Chem. C*, 8, 849–872.
24. Bhagat, M., et al. (2021). Review—Multifunctional copper nanoparticles: Synthesis and applications. *ECS J. Solid State Sci. Technol.*, 10, 063011.
25. Liu, X., et al. (2021). One-pot synthesis of uniform Cu nanowires and their enhanced non-enzymatic glucose sensor performance. *J. Mater. Sci.*, 56, 5520–5531.
26. Hassaninejad-Darzi, K., & Yousefi, F. (2015). Electrocatalytic oxidation of glucose on the modified carbon paste electrode with sodalite nanozeolite for fuel cell. *Iran. J. Hydrog. Fuel Cell*, 1, 47–58.
27. Medhi, A., & Mohanta, D. (2022). Deciphering highly sensitive non-enzymatic glucose sensor based on nanoscale CuO/PEDOT-MoS2 electrodes in chronoamperometry. *ECS Adv.*, 1, 046504.
28. Sattarahmady, N., & Heli, H. (2012). A non-enzymatic amperometric sensor for glucose based on cobalt oxide nanoparticles. *J. Exp. Nanosci.*, 7, 529–546.
29. Viswanathan, P., et al. (2019). Polydopamine-wrapped Cu/Cu(II) nano-heterostructures: An efficient electrocatalyst for non-enzymatic glucose detection. *Colloids Surf. A Physicochem. Eng. Asp.*, 580, 123689.
30. Khedekar, V. V., & Bhanage, B. M. (2016). Simple electrochemical synthesis of cuprous oxide nanoparticles and their application as a non-enzymatic glucose sensor. *J. Electrochem. Soc.*, 163, B248–B251.
31. Alam, M. M., & Howlader, M. M. R. (2021). Nonenzymatic electrochemical sensors via Cu native oxides (CuNOx) for sweat glucose monitoring. *Sens. Bio-Sens. Res.*, 34, 100453.
32. Reitz, E., et al. (2008). CuO nanospheres based nonenzymatic glucose sensor. *Electroanalysis*, 20, 2482–2486.
33. Kim, D. S., et al. (2022). Development of GO/Co/chitosan-based nano-biosensor for real-time detection of D-glucose. *Biosensors*, 12(7), 464.
34. Madden, J., et al. (2021). On-chip glucose detection based on glucose oxidase immobilized on a platinum-modified, gold microband electrode. *Biosensors*, 11(8), 249.
35. Guo, S., et al. (2021). Fast response GaAs photodetector based on constructing electron transmission channel. *Crystals*, 11(10), 1160.
36. Kim, J., Jang, Y. J., & Jang, Y. H. (2023). Electrodeposition of stable noble-metal-free Co-P electrocatalysts for hydrogen evolution reaction. *Materials*, 16(2), 593.
37. Stoica, I., Barzic, A. I., Ursu, C., Stoian, G., Hitruc, E. G., & Sava, I. (2023). Atomic force microscopy probing and analysis of polyimide supramolecular systems for sensor devices. *Sensors*, 23(9), 4489.

38. Potejanasak, P. (2021). A study on fabrication process of gold microdisk arrays by the direct imprinting method using a PET film mold. *Crystals*, 11(12), 1452.
39. Yonas, S., Gicha, B. B., Adhikari, S., Sabir, F. K., Tran, V. T., Nwaji, N., ... & Tufa, L. T. (2024). Electric-field-assisted synthesis of Cu/MoS2 nanostructures for efficient hydrogen evolution reaction. *Micromachines*, 15(4), 495.
40. Dehghan, S., Sattarpanah Karganroudi, S., Echchakoui, S., & Barka, N. (2025). The integration of additive manufacturing into industry 4.0 and industry 5.0: A bibliometric analysis (trends, opportunities, and challenges). *Machines*, 13(1), 62.

5 Metallic, Ceramic, and Polymeric Nanomaterials

5.1 INTRODUCTION TO NANOMATERIALS

The structural properties of materials at the nanoscale, typically ranging from 1 to 100 nm, define a unique class of materials known as nanomaterials. Materials at this scale exhibit distinct physical, chemical, and biological properties that distinguish them from their larger counterparts. These unique properties are primarily due to their high surface-to-volume ratio, which enhances surface-related phenomena such as reactivity and adsorption, and quantum confinement effects that modify electronic, optical, and magnetic behaviors [1–3].

Nanoparticles, one-dimensional (1D) nanowires and nanotubes, 2D thin films and nanosheets, and 3D nanolattices are examples of nanomaterials, which are typically classified according to their content, size, and shape. They can also be composite, polymeric, ceramic, or metallic, each with unique properties.

Two approaches are used in the synthesis of nanomaterials: top-down and bottom-up. Top-down techniques that reduce bulk materials to nanoscale dimensions include lithography, milling, and etching. Bottom-up techniques, atom by atom or molecule by molecule, which include chemical vapor deposition (CVD), sol–gel processes, and self-assembly, produce nanostructures. Each approach has advantages and disadvantages in terms of cost, scalability, and precision.

Nanomaterials are used in a wide range of industries, such as electronics, energy, healthcare, and environmental technology. For example, polymeric nanomaterials are being studied for drug delivery systems and lightweight nanocomposites, ceramic nanomaterials improve the thermal and mechanical stability of coatings and devices, and metallic nanoparticles are used in biological imaging and catalysis. Despite their enormous potential, issues remain related to long-term stability, toxicity, environmental impact, and large-scale production. To safely and successfully integrate nanoparticles into contemporary technology, these concerns must be addressed through sustainable synthesis, regulation, and interdisciplinary studies. Coating is the application of one layer of material over another, known as a substrate. Its primary function is to protect materials from deterioration caused by external factors by acting as a mediator between the substrate and the environment. Additionally, coatings offer aesthetic appeal [2].

DOI: 10.1201/9781003758556-5

Coatings that contain at least one dimension, with particle sizes ranging from 1 to 1000 nm, are called nanocoatings. Because these coatings are more durable and long-lasting than conventional coatings, they are more resistant to corrosion. Additional substrate properties, such as antibacterial properties, stain and wrinkle resistance, hydrophobic and hydrophilic behavior, ultraviolet (UV) protection, and antistatic properties, enhance the substrate's bulk properties [3].

By creating nature-inspired soft nanocoatings, it is hypothesized that additional capabilities, such as electronic or anti-infective applications, can be incorporated into composite nanomaterials. This focuses on two techniques for creating protein- and wax-based nanocoatings using metalloproteins with antiviral and antibacterial properties, which are reticular proteins [1]. Vapor-phase processes include CVD, laser

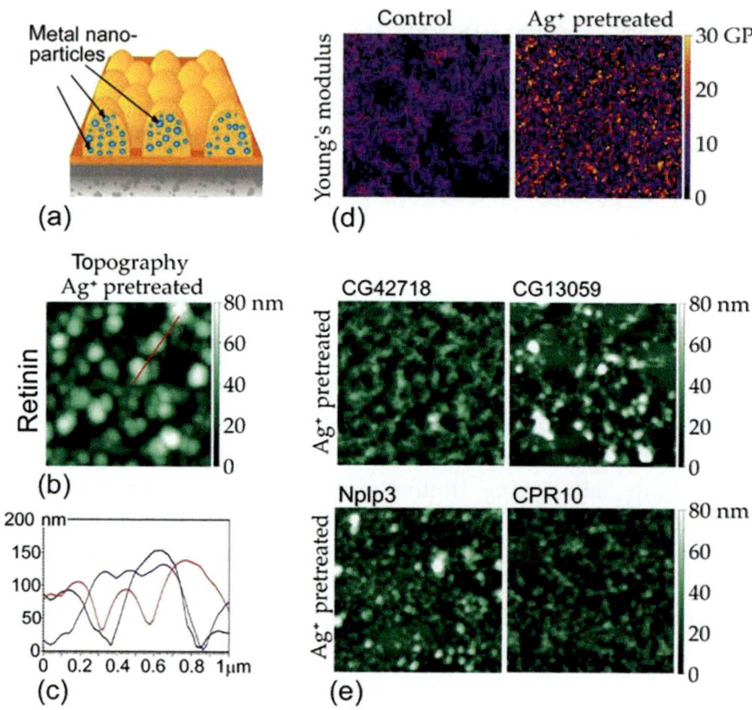

FIGURE 5.1 Arming with metal nanoparticles before self-assembly (a) produces a prominent structure of retinoin-based nanocoatings, which incorporate silver nanoparticles (b). The cross-section used to plot the heights is labeled with the red line. (c) Representative heights of the nanostructures are shown through cross-sections taken through the Ag^+-pretreated nanocoatings; three distinct cross-sections are shown in different colors, with black and blue representing other images and red representing the image in (b). The Young's modulus of the Ag^+-pretreated nanocoatings is midway between those of metal- and protein-based surfaces. The structure of the silver-containing nanocoatings is based on the proteins CG42728, CG13059, Nplp3, and CPR10, which are similar to retinoin. The color scales next to the images represent Young's modulus and surface height for a 2 × 2 μm scan in (b, d, e) [1] [Open access].

ablation, vapor condensation, plasma arc, and flame welding [4–6]. Furthermore, it is hypothesized that our nature-inspired formulation of soft nanocoatings can be used as a basis for incorporating a wide range of additional capabilities into composite nanomaterials, such as electronic or anti-infective applications. The efficiency of this method, which encapsulates metal nanoparticles in retinoin-based nanolayers before they self-assemble, is demonstrated. Furthermore, we argued that as the metals will form bioconjugates and be capped with protein, this strategy may be more environmentally friendly (Figure 5.1a). In this method, we chose copper and silver as metals. The topology of the retinin-based nanocoatings formed from silver oxides ($AgNO_3$) shows that, under these conditions, retinin retains its self-assembly ability (Figure 5.1b). The resulting nanostructures are more pronounced compared to the absence of metals (see Figure 5.1b and c) and are characterized by a characteristic size ranging from 50 to 300 nm in diameter and up to 150 nm in height (Figure 5.1b and c).

Nanocoatings are used in industrial pulleys, turbines, hydraulic shafts, boiler tubes, pumps, landing gear, and aircraft engines to prevent corrosion and erosion problems. They are also used as decorations on watches, pencils, cars, and cosmetics. Nanocoatings on banknotes also prevent counterfeiting. This chapter discusses tannins in detail. In addition, efforts have been made to integrate the various processing technologies used in their production. Impact coatings, also known as basecoats, are characterized by layers of one material placed on top of another. They are often applied to materials to protect them from damage caused by external impacts. They act as an intermediary between the environment and their substrate and are also a decorative element. Liquid-phase technologies include sol–gel, hydrothermal, electrolysis, microemulsion, and sedimentation. Industrial cylinders, turbines, hydraulic shafts, boiler tubes, pumps, landing gear, and aircraft engines use nanocoatings to prevent corrosion and erosion problems

5.2 METALLIC NANOMATERIALS

Due to their outstanding mechanical, electrical, optical, and catalytic properties, metallic nanomaterials are among the most widely researched types of nanomaterials. The behavior of metals such as copper, nickel, platinum, silver, and gold differs significantly at the nanoscale compared to their bulk state. The main reasons for these differences are their high surface-to-volume ratio and quantum confinement effects, which alter electronic structures and increase surface reactivity.

Several techniques, such as chemical reduction, electrochemical deposition, laser ablation, and physical vapor deposition, can be used to synthesize metallic nanomaterials. The effectiveness of any technique in specific applications depends largely on its ability to control particle size, shape, and surface functionality. For example, colloidal synthesis is commonly used to create stable suspensions. Metallic nanomaterials exhibit excellent electrical conductivity, high surface plasmon resonance (SPR), and increased catalytic activity. These nanoparticles are highly valuable in industries including electronics, energy storage, and environmental cleanup due to their properties. Platinum and palladium nanoparticles are effective catalysts in fuel cells, while gold and silver nanoparticles are used in biomedicine for biosensing, imaging, and targeted drug delivery.

Despite their promising potential, metal nanoparticles face challenges in large-scale production, cost-effectiveness, stability, and environmental safety. Green synthesis techniques and biocompatible surface modifications are essential due to concerns about cytotoxicity and environmental impacts. Future research is expected to focus more on multifunctional systems and hybrid nanostructures that combine metallic nanoparticles with advanced technology, with a particular focus on environmentally friendly production techniques. Starting with the fundamentals of nanotechnology, this field of study examines several methods for producing and modifying nanofilms, with an emphasis on the influence of size effects at the nanoscale. This chapter covers several techniques for creating nanofilms, including comprehensive justifications supported by numerous examples. The discussion is divided into sections covering important topics such as the sol–gel process, size-dependent properties, nanoparticles, thin film production, nanotechnology fundamentals, and the numerous uses of nanocoatings [3–5]. In addition, the method of encapsulating metallic nanoparticles in nano-reticulated coatings before their self-assembly is investigated, providing information on new approaches that could increase the utility and versatility of nanomaterials. Due to their high impact strength, hardness, and low porosity relative to the substrate, nanocoatings offer superior corrosion resistance compared to conventional coatings [1,6]. The properties of the substrate are also influenced by its antibacterial properties, stain and wrinkle resistance, hydrophobicity, hydrophilicity, UV protection, and antistatic properties [7]. Examples of materials that can be coated with nanoparticles include nano-titanium dioxide, nano-titanium dioxide, nickel and nickel arsenic oxide, nano-aluminum and magnesium hydroxide nano-layered double-layered nanostructures (NANO-ALDs), nickel and molybdenum nano-alloys, silver nanoparticles, silver-based nanomaterials, zinc nanoparticles, copper nanoparticles, and carbon nanotubes [8]. Furthermore, this approach is proposed to be more environmentally friendly because the metals will encapsulate themselves with protein and form biocomplexes. A lateral force microscope (LFM) image and an atomic force microscope (AFM)-coupled topographic image of a freshly cleaved molybdenum disulfide (MoS_2) sample are shown in Figure 5.2a; both images show a smooth surface [5]. However, the simultaneously recorded C-AFM map (Figure 5.2b) shows inhomogeneous lateral current injection across the nanocontact between the MoS_2 surface and the conducting tip (made of boron-doped diamond). The surface density of these structures, depending on the sample and location, range between 1010 and 1011 cm². According to Figure 5.2c, the scanning tunneling microscope (STM) maps revealed the presence of distinct circular depression-like structures with the MoS_2 lattice superimposed on them.

5.3 CERAMIC NANOMATERIALS

Ceramic nanoparticles are a prominent family of nanomaterials characterized by their exceptional mechanical strength, resistance to corrosion, wear, and chemical degradation, as well as their exceptional thermal stability. Ceramic materials, such as titania (TiO_2), zirconia (ZrO_2), silica (SiO_2), and alumina (Al_2O_3), exhibit unique properties at the nanoscale compared to their raw material counterparts. These properties include enhanced hardness, higher fracture toughness, superior ionic conductivity,

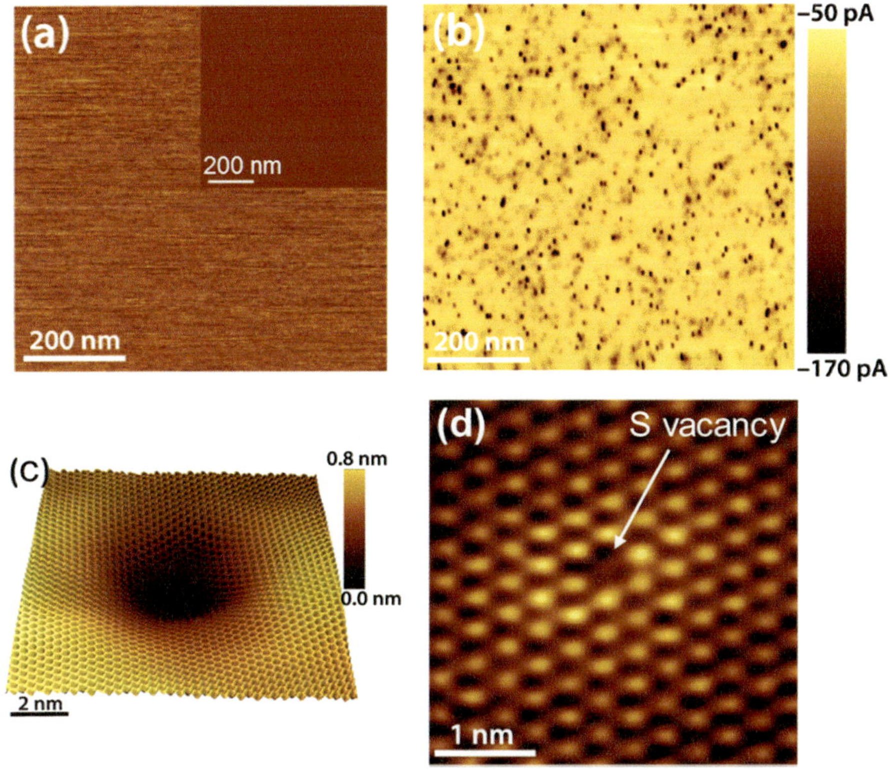

FIGURE 5.2 (a) Topographic (inset) and lateral force microscope (LFM) images of the MoS_2 surface. (b) Simultaneous C-AFM image. (c) Scanning tunneling microscope (STM) image showing a subsurface defect corresponding to one of the high-current regions in panel (b). (d) Sulfur void in the upper sulfur layer as seen in the STM image [2] [Open access].

and unique optical behaviors, including photocatalytic activity and luminescence. These properties arise from quantum confinement effects, increased surface area, and reduced grain size.

Ceramic nanoparticles are manufactured through a variety of processes, including sol–gel processing, hydrothermal synthesis, CVD, combustion techniques, and mechanical milling. Each technology has its own advantages in tuning particle size, shape, and crystallinity critical factors for improving performance. There are numerous industrial applications for ceramic nanoparticles. Energy-related uses of nanoceramics include solid oxide fuel cells, lithium-ion batteries, and photocatalysts for hydrogen production and environmental cleanup. These materials are used in electronics as dielectrics, optical coatings, and sensors. Due to their excellent biocompatibility, they can also be utilized in various biomedical applications, including bone tissue engineering, drug delivery systems, and dental implants. Additionally, ceramic nanocoatings enhance the corrosion and heat resistance of aircraft and automotive components.

Despite these advantages, controlling microstructural consistency, achieving affordable, large-scale production, and ensuring long-term stability under harsh operating conditions remain challenging tasks. Nanocoatings have been increasingly used in biotextile applications in recent years to improve their performance, such as biofiltration materials, medical clothing, textile implants, cell culture substrates, and other medical textiles [2–5]. Thin films, nanocapsules, and nanoparticles deposited on textile surfaces can be used to fabricate these nanocoatings; their typical diameters range from a few tens to hundreds of nanometers. This finding supports the idea that metallic nanoparticles, such as silver, can be incorporated into lattice nanostructures. Estimating the size of silver nanoparticles in these systems involves considering the regions with the highest Young's modulus, which is approximately 100 nm, which is smaller than the typical size of adjacent nanostructures, which is approximately 300 nm. Energy-distributed X-ray spectroscopy (EDS) and scanning electron microscopy (SEM) are used to further analyze the materials to confirm these results, as shown in Figure 5.3 [1].

Many nanomaterials, including metal nanoparticles, colloidal nanoparticle suspensions or solutions, electrospun nanofibers, nanocoated systems with controlled conductivity (nanocapsules), plasma spray coatings, and cold plasma polymerized coatings, are of interest to the textile industry. For example, polymeric nanofibers are

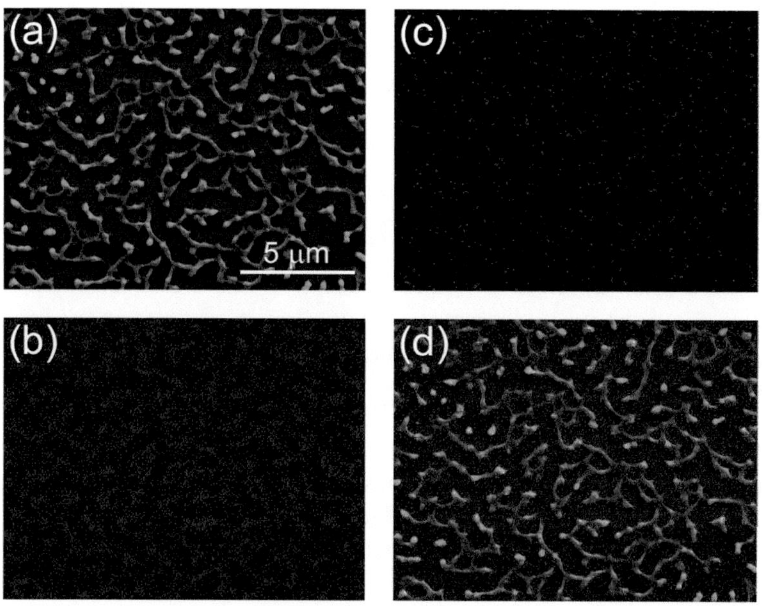

FIGURE 5.3 SEM and energy dispersive X-ray spectroscopy (a) of carbon (b) nanocoatings based on retinal nanoparticles encapsulating silver nanoparticles (c) and silver (d) Panels (a–c) that have been combined. Keep in mind that the overlap of carbon signals with the nanostructures also contains silver dots. Observe the increased scale (15 × 10 μm) in contrast to the earlier SEM pictures [1] [Open access].

deposited on films, foams, and nonwovens by electrospinning [1,2] to form a thin, flexible, and undetectable layer. Commercial applications include medical devices, textiles, and multipurpose filters (antibacterial, antiviral, and antiparticulate).

5.3.1 Synthesis and Fabrication Methods

Given the significant impact of the preparation process on the functional, chemical, and physical properties of nanostructures, the synthesis and fabrication of nanomaterials constitute a fundamental aspect of nanoscience and nanotechnology. These methods can be broadly divided into "bottom-up" and "top-down" approaches. Top-down methods use processes including mechanical milling, lithography, etching, and laser ablation to reduce bulk materials to the nanoscale. While these methods are useful for creating nanostructures with precise dimensions, they often have drawbacks in terms of scalability and surface defects.

Bottom-up methods, in contrast, rely on the formation of nanostructures through the assembly of atoms or molecules. Many techniques are used in this category, including sol–gel processing, hydrothermal and pyrolysis synthesis, CVD, and self-assembly.

Due to its ease of use, low processing temperatures, and ability to produce highly purified materials with tailored porosity, the sol–gel method is the most popular technique for producing oxide-based ceramic nanomaterials. CVD and physical vapor deposition (PVD) are known for producing thin films with exceptional homogeneity, while hydrothermal techniques are particularly useful for synthesizing crystalline nanoparticles under mild conditions. In recent years, biosynthetic and eco-friendly approaches have focused on finding sustainable, cost-effective, and environmentally acceptable methods for producing metal and polymer nanoparticles from plant extracts and microbial systems.

The intended use, the required physical and chemical properties, and the overall feasibility of the project significantly influence the choice of synthesis and manufacturing technology. Recent developments in scalable processes and hybrid approaches are increasingly important for addressing these issues and achieving industrial applications. Surface engineering studies also demonstrate how growth conditions affect the quality of nanostructures. As demonstrated by an AFM study, surfaces produced on sapphire substrates with an inaccurate orientation (misorientation) under controlled conditions, for example, generally exhibit lower root mean square (RMS) roughness values over small scan areas (2×2 μm) compared to samples with an A-orientation (Figure 5.4).

However, heterogeneously oriented (M) samples also exhibit large-scale waviness, resulting in higher roughness values when evaluated across wider areas (e.g., 8–9 nm for 1C at 465 μm × 350 μm versus 2–3 nm for 2C), consistent with previous reports [4,5]. CVD remains one of the most flexible thin-film growth fabrication techniques. It enables the controlled deposition of elementary and complex semiconductors, crystalline or amorphous materials with various aspect ratios, and metal alloys.

CVD produces non-volatile solids that are atomically deposited onto suitable substrates by reacting volatile precursors of the target material with other gases, as illustrated in Figure 5.5. Because of this feature, CVD is a highly flexible method for creating superior nanofilms with precise control over composition, thickness, and homogeneity.

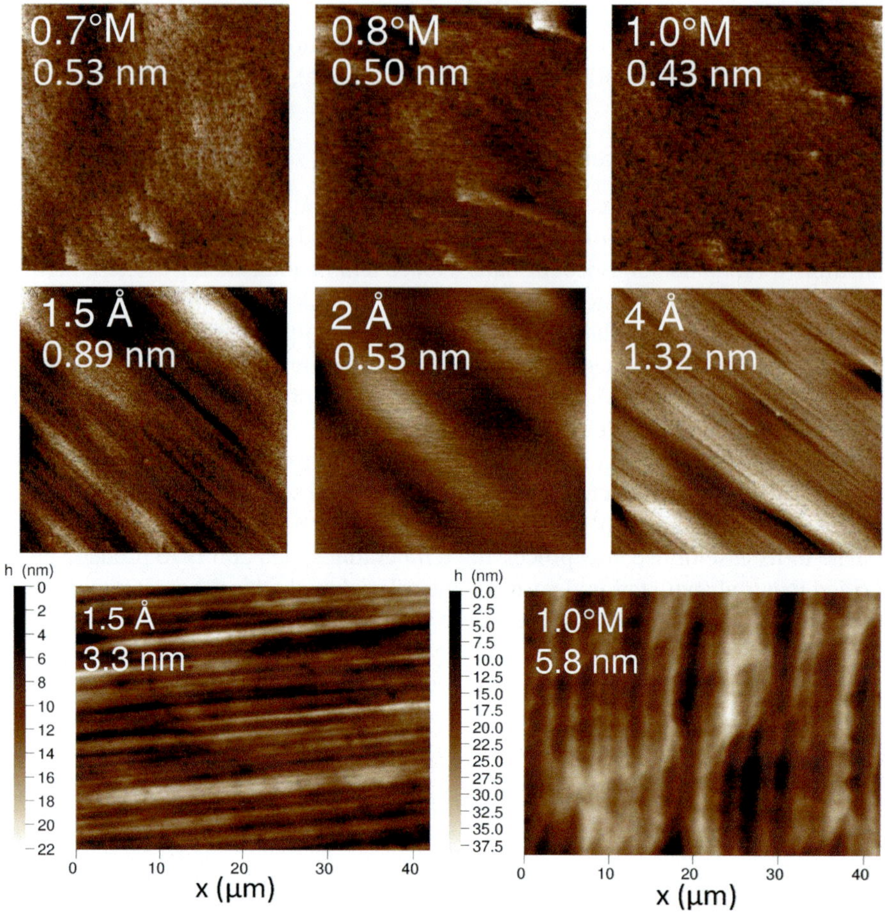

FIGURE 5.4 Atomic force microscope (AFM) images, each measuring 2 μm × 2 μm, of gallium nitride growth on sapphire at different orientational biases. The graph displays orientational variations and square roughness values. The Z scale was set to 4.5 nm for all images, except for those where the substrate tilt angle used in the experiment was 4 Å or 10 nm. For flat wafers, a 45° angle is typically used for testing. Samples with a 1.5° bias, and growth characterization values (1°) are shown in the bottom row. The sample with a 1° bias has a higher square roughness and more ripples around the step boundaries [2] [Open access].

In this experiment, the effect of growth conditions, particularly ammonia flow, on the surface properties of gallium nitride (GaN) was investigated. A sapphire substrate was studied at 1,150 °C with a modest growth rate of 0.45 nm/s, and another at 1,180 °C with a much higher Trimethylgallium (TMGa) flow rate. These experiments aimed to obtain smoother surfaces, particularly with reduced misorientation. Each run had a different ammonia flow, but all other parameters remained the same.

At a 0.3 M ammonia flow, the GaN surfaces showed noticeable changes that could be seen even with the naked eye. Two clear trends emerged from the results: high

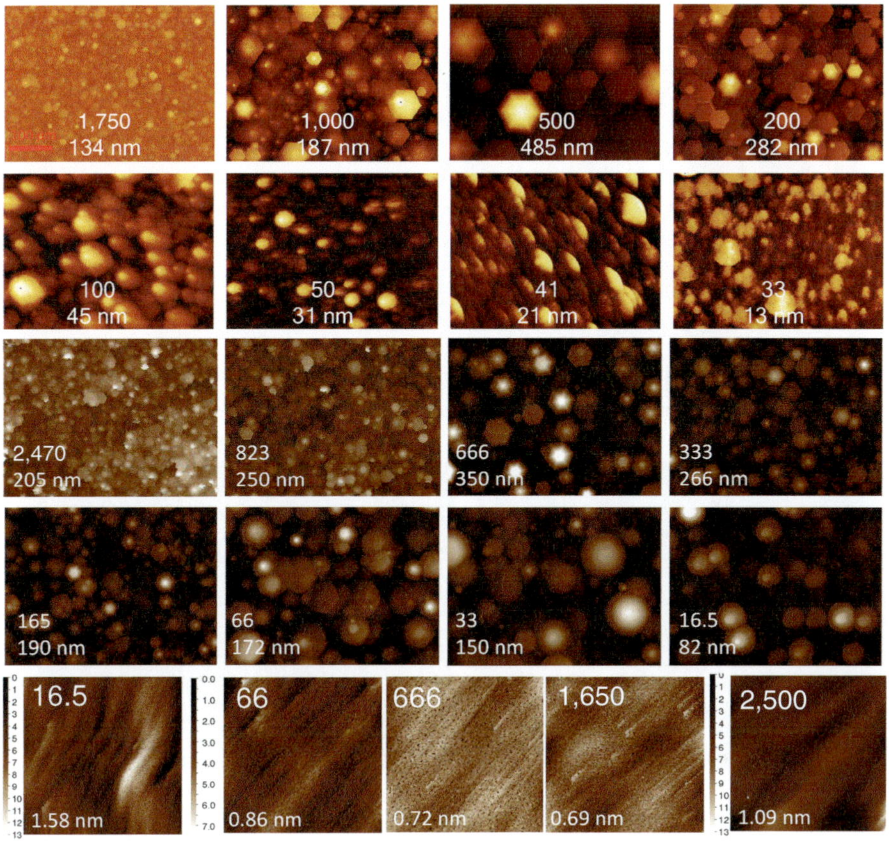

FIGURE 5.5 GaN arcs were produced under different conditions: micrographs and 3D white-light surface maps. The first two rows are surface maps of 0.3 M GaN arcs, covering an area of 465 μm × 350 μm, produced for 30 min at 1,180 °C with a constant TMGa flow and decreasing NH_3 flow. The first value represents the V/III ratio, while the second value represents the average RMS roughness under these conditions. The middle two rows are surface maps of 0.2 M GaN arcs, covering the same area (465 μm × 350 μm), produced for 45 min at 1,150 °C with a constant TMGa flow and decreasing NH_3 flow. Again, the V/III ratio is shown first, followed by the average RMS roughness. The bottom row: AEM images of 2 μm × 2 μm sample grown for 45 min at 1,150 °C on 2° misoriented sapphire at very low, medium, and maximum NH_3 flux levels [2] [Open access].

V/III ratios (above 500) caused the high-hexagonal linewidth (HHL) to shrink and become shallower. This indicates that ammonia flow, especially under different growth temperatures and flow conditions, significantly affects gallium nitride surface morphology and crystallinity. These results provide important guidance for optimizing growth conditions to produce smoother and more stable gallium nitride surfaces.

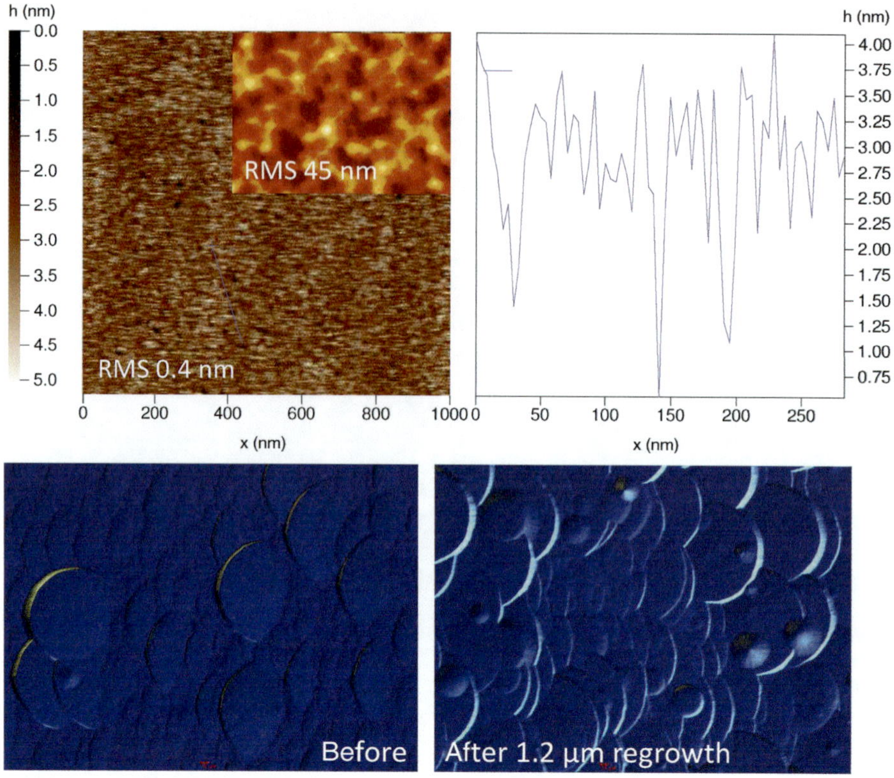

FIGURE 5.6 After two growth stages in V/III 500 and 12 at 1,150 and 0.2 °C, the top row shows an atomic mass micrograph of the surface. A 3D image of the surface with dimensions of 930 × 700 µm is included in the supplement. Bottom row: Nomarski V/III 1,000 images taken at 1 µm scale before and after regrowth at 1,180 °C [2] [Open access].

Even for N-polarized gallium nitride, the best sample in this series is located deep within the absorption-limited growth region, at the edge of stability. (Ga-polarized gallium nitride layers absorb all their energy within minutes.) The 13-nm-wide mean-square roughness region may still be suitable for devices, although it is three to four times larger than the typical 4–6 nm gallium nitride surface roughness. Furthermore, there is some evidence of atomic gradients on the surface of this sample. This surface is an N-polarized gallium nitride surface with a slight bias in orientation, and only the edges of the gradients are visible. Surfaces, such as those shown in Figure 5.6, are typically featureless, containing pores about 1 nm deep [2].

The chemical power of CVD processes is manifested in the combination of chemicals and precursors that can be used to deposit a specific film. Vapor supersaturation influences the rate of film formation, while CVD and PVD rely on substrate temperature to control the film growth rate. These two factors work together to influence grain size, shape, texture, and the degree of overlap. Low gas supersaturation and high substrate temperatures promote the formation of single-crystal films on substrates. Polycrystalline, nanocrystalline, and even amorphous films have been formed as a result of low substrate temperatures and high gas supersaturation. The five main processes in the vapor phase approach are flame synthesis, laser ablation, CVD, PVD, and plasma arc (Figure 5.6) [2–4].

5.3.1.1 Physical Vapor Deposition

PVD is an evaporative coating technique that involves the transfer of materials at the atomic level. The steps involved in this process are summarized below. First, the material is physically converted into a vapor (either a gas plasma or a high-temperature vacuum); then the vapor moves from the source to the substrate in a low-pressure region; finally, the vapor condenses on the substrate, forming a thin film. The resulting films are extremely rough (roughness > 10 nm) and contain many closely packed particles, as illustrated in several previous examples [3–7] (Figures 5.7a–c). This is true regardless of the deposition period. The predominance of adsorbed molecular aggregate in this case indicates the presence of aggregated particles [3].

PVD is typically used to deposit films with thicknesses ranging from a few nanometers to thousands of nanometers. It can also be used to create layered coatings, graded deposits, and free-standing structures, especially thick deposits [9]. PVD covers a wide range of deposition processes, including reactive evaporation, ion deposition, electron beam or hot bath evaporation, and other deposition techniques. PVD includes ion beam and plasma spray-based processes. To evaluate the flexibility of the chemical vapor deposition (FCPVD) technique, several organic molecules (Alq_3, NPD, and PCBM) were investigated; all showed similar results (Figure 5.8). In contrast to the extremely rough surfaces (RMS roughness greater than 10 nm) of single-layer chemical vapor deposition (SCPVD) substrate films, which contained a large number of aggregated particles, FCPVD-fabricated films have ultra-smooth surfaces (RMS roughness less than 0.4 nm) free of any noticeable particles or defects (Figure 5.8). Furthermore, FCPVD technology has been applied to a number of substrates, including silica, quartz, mica, and indium oxide (Figure 5.8). These results suggest that FCPVD technology can be used to produce ultra-smooth films on a range of organic molecules and substrates by inhibiting crystallization [3].

These results indicate that rough films are crystalline, while smooth films are amorphous (Figure 5.9). q_{xy} and q_z are the in-plane and out-of-plane scattering vectors, respectively. Strong, anisotropic scattering peaks can be seen in the SCPVD-prepared rubrene film, which is a mixture of orthorhombic and triclinic phases. The large, aggregated particles can form thermodynamically favorable crystalline forms (Figure 5.9a) [3]. However, the FCPVD-prepared rubrene film exhibits broad peaks with anisotropy at $q \sim 0.68$ Å^{-1}, indicating the presence of an amorphous phase with favorable kinetic properties (Figure 5.9b) [3]. While SCPVD-fabricated films exhibit distinct and anisotropic crystalline phases, as evidenced by the scattering peaks (Figure 5.9c–g), FCPVD-fabricated Alq_3, NPD, and PCBM films exhibit nucleation inhibition comparable to that of rubrene (Figure 5.9d–h). Thermodynamically favorable triclinic and α-phases have been reported for deposited Alq_3 and NPD films, respectively [6,7]. However, FCPVD-fabricated Alq_3, NPD, and PCBM films exhibit significant scattering peaks, indicating that the amorphous phase is kinetically advantageous (Figure 5.9d–h) [3]. These results suggest that rapid substrate cooling

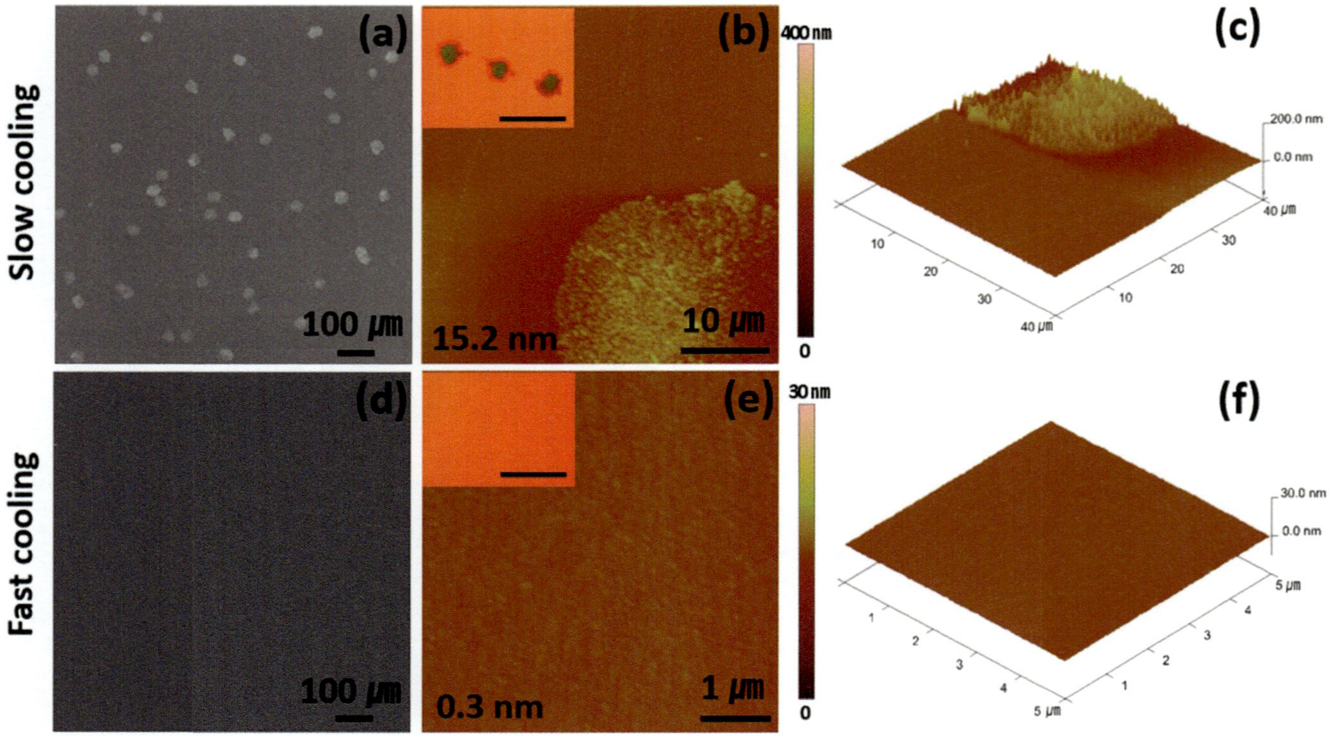

FIGURE 5.7 Scanning electron microscope (SEM) and atomic force microscope (AFM) images of rubrene films as a function of cooling rate. The top row images (a–c) and bottom row images (d–f) were produced using SCPVD and FCPVD, respectively, at a deposition time of 100 s. SEM images of each film (scale bar: 50 μm) are shown as insets [3] [Open access].

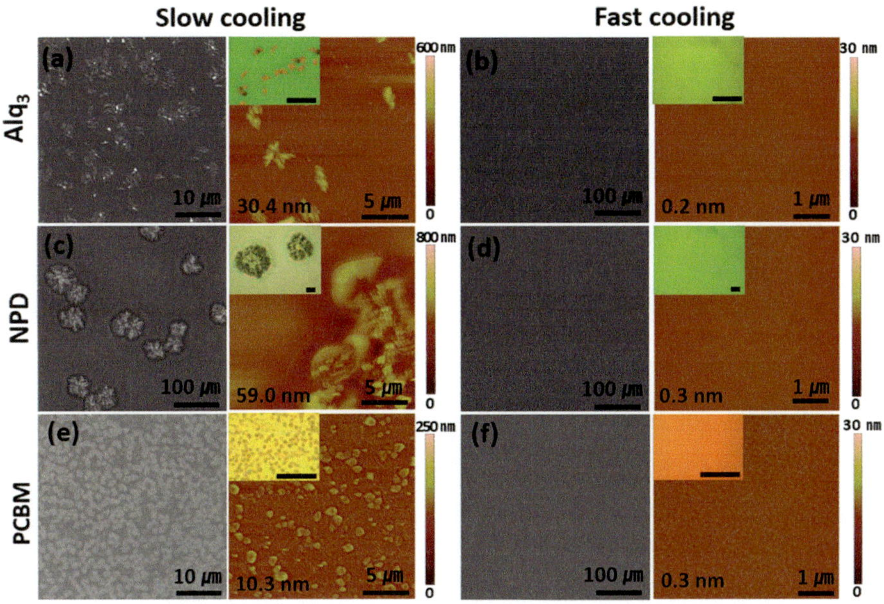

FIGURE 5.8 Differentiation of organic films as a function of substrate cooling rate. Atomic energy and scanning electron microscopy (SEM) images below show, as a function of cooling rate, Alq3 films (a, b), NPD films (c, d), and PCBM films (e, f). The films on the right (b, d, f) and left (a, c, d) sides are produced using SCPVD and FCPVD processes. To each film, 10-μm optical microscope images are attached [3] [Open access].

effectively inhibits the nucleation of adsorbed molecules, facilitating the formation of ultra-smooth organic thin films [3].

Arc source deposition, whether filtered or not, can also be referred to as PVD. This process is classified into two categories: ablation and vaporization. Vaporization refers to the heating process used to form thin layers, while spraying uses gaseous ions (plasma) to remove atoms or molecules from a solid material. Many specialized techniques have been added to both approaches [1–3]. When applying titanium nitride to a steel substrate using PVD, results were found to be better with nanoparticles than with micron-sized particles.

5.3.1.2 Chemical Vapor Deposition

The membrane composition, surface morphology, thickness, and crystal structure must be controlled during the construction of CVD reactors. Horizontal and vertical systems can be distinguished by the direction of gas flow to the substrate carrier (receiver). Special reactors are designed for deposition in low pressure and in air. Radiation is used to heat the substrate carrier using conductive radio frequencies. Reactors with hot or cold walls can heat internally. Therefore, field-emission scanning electron microscope (FESEM) characterizations show that phase composition changes are visible in the layer morphology. The film grown at 800 °C, shown in Figure 5.10a, exhibits the

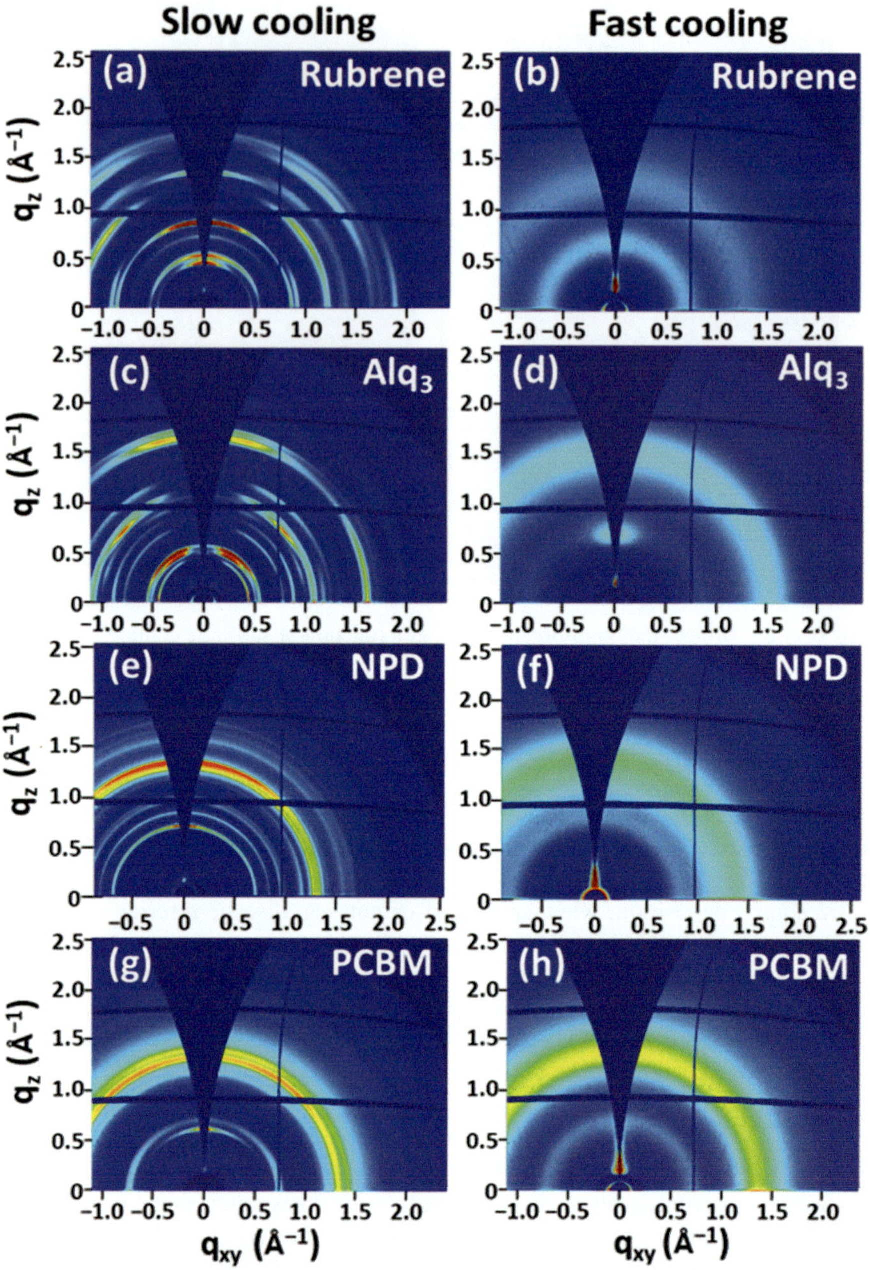

FIGURE 5.9 Characterization of organic thin film structures. GIWAXS patterns of SCPVD (left column) and FCPVD (right column) films, including (a, b) Rubrene films, (c, d) Alq3 films, (e, f) NPD films, and (g, h) PCBM films [3] [Open access].

FIGURE 5.10 Scanning electron microscope (SEM) images of Fe–Zn thin films based on Ba–Ce, which are more suitable for extremely high temperatures: 800, 850, 900, and 950 °C for liquid magnesium (100). SEM images of Fe–Zn show that this is the hottest element on the planet [4] [Open access].

formation of a nanolayer with evenly distributed grains, hundreds of nanometers in size, between rod- or plate-shaped grains, a few microns in size.

The inset in Figure 5.10a shows the film thickness, which is 875 ± 25 nm. The surface of the sample obtained at 850 °C, shown in Figure 5.10b, is characterized by uneven grains. Numerous micron-sized triangular grains are present. In particular, these triangular helical nanoplatelets exhibit a helical development [3], which is consistent with the high reflectivity of cerium dioxide, with the same pattern observed at 28.85 °C and the growth of cerium dioxide. The thickness of the cross-section (inset in Figure 5.10b), which exhibits this rather peculiar shape, can be estimated to be between 700 and 930 nm. For the sample obtained at 900 °C, Figure 5.10c depicts smaller and more uniformly distributed triangular grains [4].

EDX analysis is used to determine the purity and composition of $BaCeO_3$ thin films. The spectrum shows characteristic peaks of oxygen, cerium, and barium. The results are shown in Figure 5.11a. Specifically, the Mα lines of Ce show peaks at 0.88 keV in the range of 4.7–5.9 keV, the Kα lines of O show peaks at 0.53 keV, and the L lines of Ba show peaks in the range of 4.4–5.5 keV. Quantitative EDX analysis proves the accuracy of the chemical readings for pure $BaCeO_3$ films with a Ba to Ce ratio of nearly 1:1 over the entire surface [4].

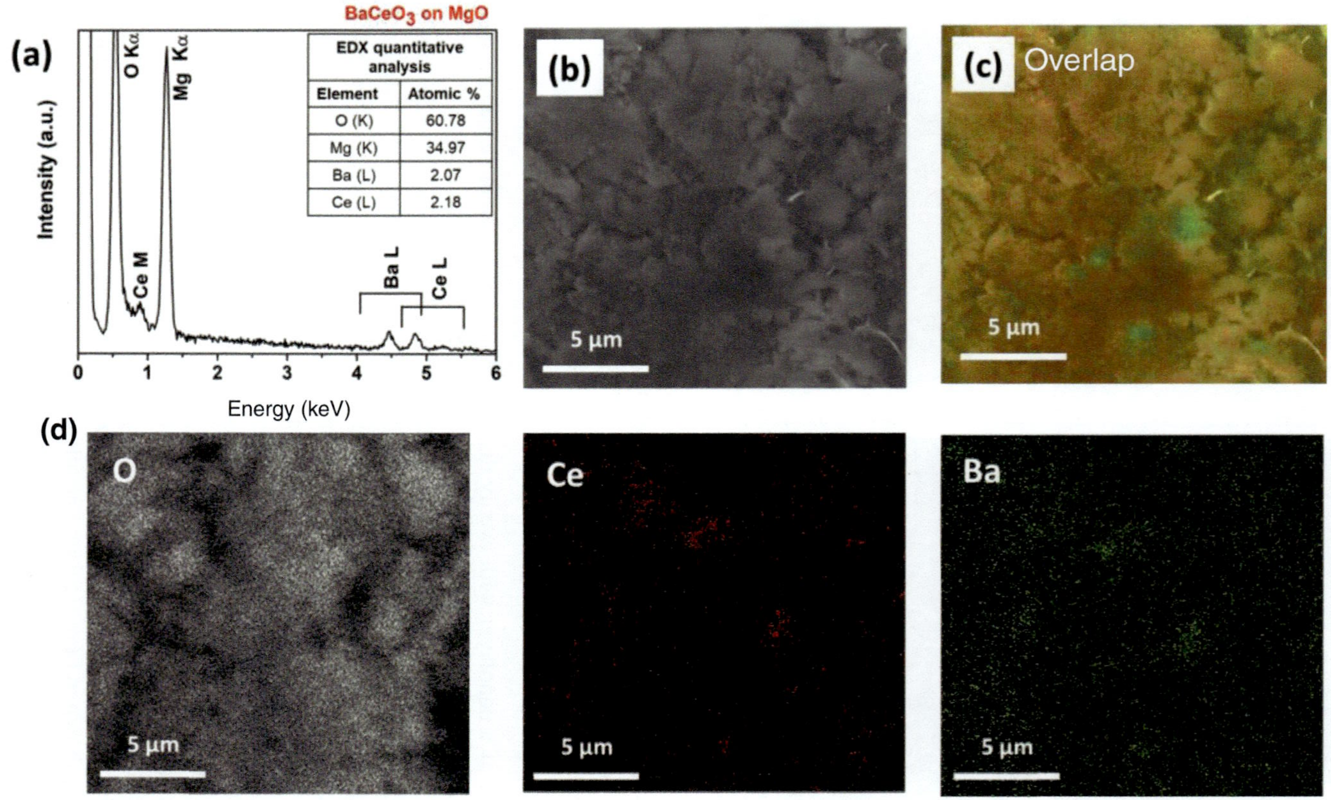

FIGURE 5.11 (a) EDX spectra, (b) relative FE-SEM image, (c) interference map image, and (d) EDX maps of O, Ce, and Ba elements show the $BaCeO_3$ film deposited on MgO at 950 °C [4] [Open access].

The purpose of rotating substrate carriers is to improve flow dynamics and film homogeneity. Reactors that enable simultaneous deposition on multiple substrates, including barrel reactors, pancake reactors, and rotating disk reactors, have been designed to produce petroleum vapor deposition on a large scale. When the substrate is exposed to one or more volatile precursors via vapor deposition, the reaction or annihilation of these precursors leads to the formation of specific deposits on the substrate surface (Figure 5.11) [6]. The success of the process depends largely on the system conditions. The temperature ranges between 800 and 2,000 °C during this process. Coating thicknesses up to 20 μm with excellent adhesion are achieved [7,8]. Characterization techniques such as SEM and XRD have concluded that the pore surface characteristics changed as a result of CVD nanocoating.

5.3.1.3 Plasma Arc

A plasma torch is used to heat, melt, and cut electrically conductive materials into distinctive patterns and designs. This manufacturing process is called plasma cutting, also known as fusion plasma cutting or artistic plasma cutting. Waterjet, laser, and oxyfuel cutting are alternatives to plasma cutting. The main difference here is the size of the features. While relief engraving encompasses all surface features at various scales, roughness focuses on slight deviations from a perfectly smooth surface. For example, the surface roughness of polytetrafluoroethylene (PTFE) samples increases with varying exposure times to the radiofrequency plasma torch, as shown in the AFM images in Figure 5.12a–d [8]. Roughness is the degree of irregularity of a smooth surface, regardless of any regularity or directional structure. On the other hand, regularity or organized structural elements are referred to as texture. Relief is a more general term that includes all surface elements and their features, while texture refers to specific directional structures or properties. Figure 5.12e–g shows teardrop, ripple, and honeycomb structures shown in AFM or SEM images, respectively [5].

Plasma cutting can cut metals with thicknesses ranging from 0.5 to 180 mm, including copper, aluminum, steel alloys, and structural steel. Plasma cutting is not suitable for non-conductive materials such as rubber and plastic. Different forms of plasma cutting include water injection, plasma arc cutting, secondary plasma arc cutting, and standard (or conventional) plasma arc cutting. This process, known as plasma cutting, uses a hot, accelerated plasma jet to cut electrically conductive materials. This jet adheres to the material, transfers energy, and melts it when an electrically conductive gas such as oxygen, nitrogen, argon, or a mixture of gases ionizes, generating a plasma arc. This is the precise mechanism behind plasma cutting. A special state of matter called plasma arises when a gas is heated to the point where atoms lose their electrons, leaving behind a mixture of free electrons and ions (atoms that have gained or lost electrons). The atoms lose their electrons in a process known as ionization [10].

The most common states of matter are solid, liquid, and gas. However, plasma is often referred to as the fourth state of matter because it can exist both inside and outside galaxies (e.g., in the centers of stars and lightning). Technically, plasma is the most common state of matter in terms of mass and volume [11]. Collisions can

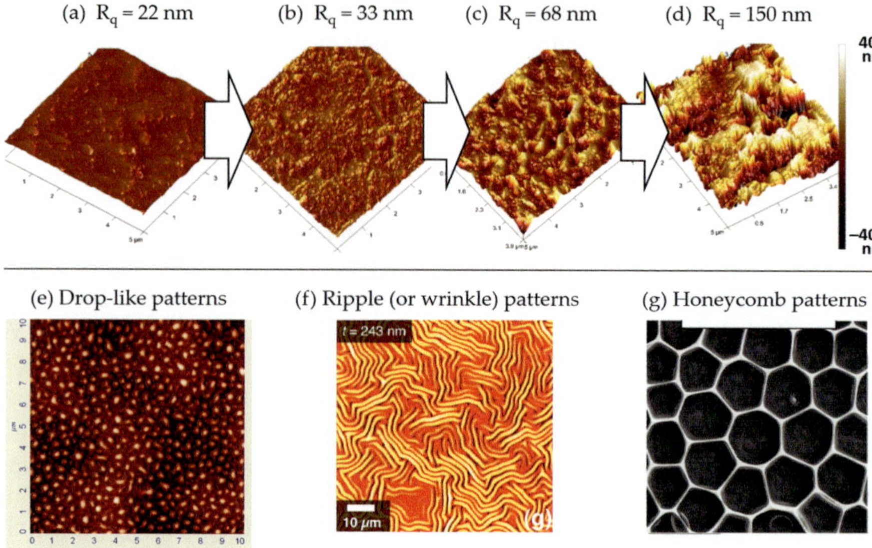

FIGURE 5.12 Atomic energy microscopy (AFM) images of PTFE surfaces exposed to a radiofrequency plasma torch for varying exposure times—0, 2, 26, and 50 min—are shown in panels (a–e). Rq stands for root mean square roughness. (e) AFM image of plasma-treated polypropylene showing a droplet-like pattern. (f) AFM image showing undulation of plasma-treated polydimethylsiloxane (PDMS). (g) Scanning electron microscope (SEM) image showing the honeycomb structure of plasma-treated FEP. (All rights reserved. American Chemical Society, 2009 [5] [Open access].)

split molecules into their constituent atoms, which can then be isolated from their outer shell electrons to form ions. As additional atoms lose their electrons, the gas reaches a critical point where the number of atoms accepting free electrons balances the number of atoms losing them [12]. Coatings with a high critical percolation coefficient (CPSP) have shown lower corrosion resistance compared to substrates with a low critical percolation coefficient [13].

5.3.1.4 Flame Synthesis

Flame spray pyrolysis (FSP), an industrially scalable technology for producing nanomaterials, produces a variety of metal nanoparticles or metal oxides with specific physical and chemical properties [14]. FSP relies on a complex process architecture, involving atomization of the starting materials, their combustion, and the synthesis of nanoparticles [15]. This enables rapid, single-step synthesis, avoiding the post-production modifications that other methods may sometimes require. Therefore, FSP is gaining increasing importance in the development of nanodevices that allow the in situ insertion of nanoparticles into complex functional matrices [16]. Therefore, FSP not only revolutionizes the production of nanomaterials but also paves the way for the creation of next-generation nanodevices because of recent developments

[17]. Consequently, this report provides a revised overview of the latest advances in nanodevice development, novel material synthesis, reactor projects, and innovative processes in FSP technology.

Figure 5.13 shows high-resolution transmission electron microscope (TEM) images of carbon films grown on a copper foil at different growth times (copper foil over a flame, ethanol flow rate 2 mL/min). Figure 5.13a and b show images of a TEM grid without a grown sample. According to these figures, each TEM grid is supported by a separate carbon layer, and there is a gap at the center of each grid that is not physically connected. Consequently, carbon films are produced at the center of the grids during synthesis experiments [18].

5.3.2 Method of Liquid Phase

Liquid-phase synthesis is another technique for producing silicate-based phosphorus compounds. Compared to the high-pressure process, which yields only 20% conversion, this method can yield approximately 90% more. In liquid-phase synthesis, solvents are widely used as the reaction medium. The solvent spectrum can cause significant differences in reaction rate and order. This process is known as the solvation effect, and the free energy of solvation can be quantitatively analyzed. Compared with gas-phase synthesis techniques, much less attention has been paid to understanding the processes that determine the appearance during liquid-phase synthesis. One example is the change in the cyclic nucleotide-gated potassium (K^+) channel (SthK), which is specifically catalyzed by the presence of cyclic guanosine monophosphate (cGMP). Some bacteria exhibit this phenomenon and are used as a model for investigating cyclic nucleotide-gated channel processes. When additional cGMP is added using HS-AFM, the channel state changes from an active/open state to a quiescent/closed state. Starting with an initial concentration of 0.1 mM cAMP, cGMP is added in increasing doses until the final concentration reaches 7 mM. The surface roughness and agglomeration of the 2D crystals are clearly altered in the image. Figure 5.14a shows how this transition begins with molecules outside the crystal patch and then gradually progresses from the left (0.1 mM cAMP) to the right (7 mM cGMP) toward the center [19].

Green phosphorus (Ba_{1-x}, Sr_{1-x})SiO_4:2xEu^{2+}) is produced using the liquid-phase precursor technique, where x = 0.03, 0.05, 0.1, and 0.15 [7]. The emission intensity of phosphorus produced using the liquid-phase precursor is reportedly significantly higher than that of phosphorus produced using other synthesis techniques. Aluminum, boric acid, or metal can easily decompose SiO_2, which is the most commonly used film. When forming a thin film, highly acidic solutions are also produced in addition to basic solutions.

5.3.2.1 Electrochemical Method

The study of chemical reactions involving charge separation, usually in liquid media such as solutions, is known as electrochemistry. Charge separation is often associated with charge transfer, which can occur homogeneously in solutions between different chemical compounds or heterogeneously across the surface of electrodes. Therefore, it clearly has a wide range of applications, which will be discussed later.

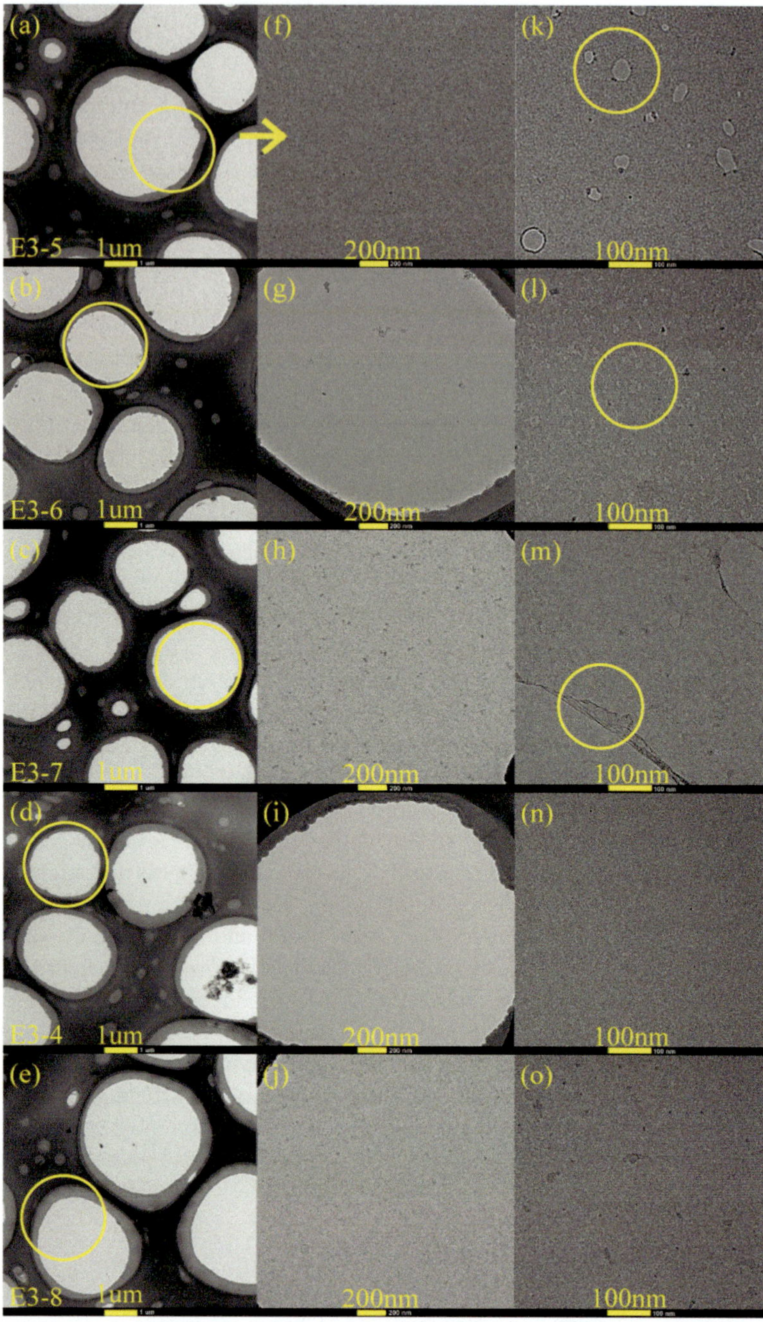

FIGURE 5.13 TEM images of carbon films with different growth time at ethanol flow rate of 2 mL/min and height of 14.0 mm: (a,f) 10 s (E3-5); (b,g) 30 s (E3-6); (c,h) 2 min (E3-7); (d,i) 5 min (E3-4); (e,j) 10 min (E3-8). HRTEM images of carbon films with different growth time at ethanol flow rate of 2 mL/min and height of 14.0 mm: (k) 10 s; (l) 30 s; (m) 2 min; (n) 5 min; (o) 10 min. [6] [Open access].

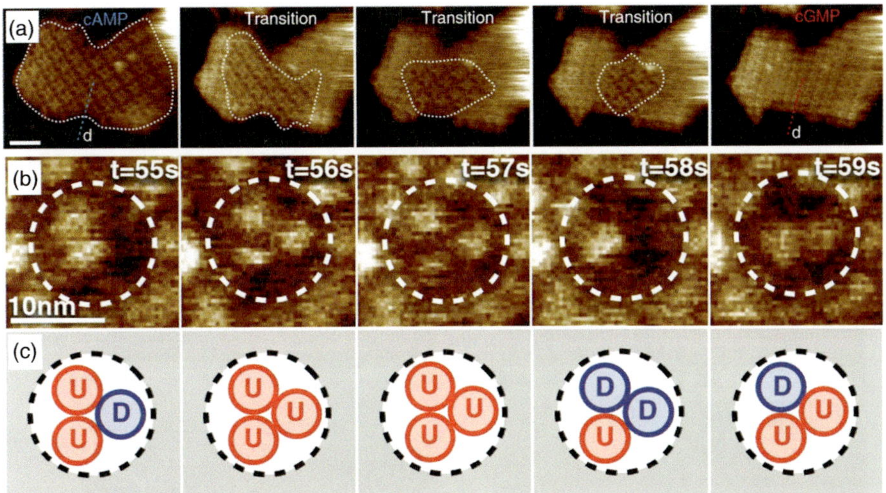

FIGURE 5.14 (a) Real-time high-resolution HS-AFM imaging showed ligand-stimulated dynamic conformational changes in SthK from 0.1 mM initial cAMP to 7 mM final cGMP. Conformational modifications in SthK channels are indicated by the dotted line. Scale is 30 nm. (b) High-resolution HS-AFM images showing conformational changes in GltPh trimers in sequential frames. (c) Schematic representation of the up (U) or down (D) state of each protomer in the AFM images shown in (b) [10] [Open access].

Half-reactions often occur at different electrodes immersed in a solution within the same cell or in separate cell compartments, and are spatially separated at the electrode surfaces. Conductive paths connecting the electrodes enable the completion of the electrical circuit and the transfer of charge both internally and externally.

When the total number of Gibbs energy changes at both electrodes is negative, batteries and fuel cells can capture the released electrical energy. To balance the positive Gibbs energy and stimulate the electrode processes that convert molecules (electrolysis), external electrical energy can be applied if the difference is positive. Therefore, if the cell design allows, the products of the two electrode reactions can be separated, just as in industrial electrolysis processes [12].

To maintain electrical neutrality, at least two half-reactions involving charge transfer must occur simultaneously and in opposite directions: oxidation (loss of electrons or increase in oxidation state) and reduction (gain of electrons or decrease in oxidation state). No pitting corrosion was observed with increasing body-centered cubic (BCC) phase volume fraction and aluminum content in $Al_{0.7}CoCrFeNi$ alloy.

However, as shown in Figure 5.15, localized corrosion at the dendritic/interdendritic boundaries and selective dissociation of the (Al, Ni)-rich BCC structural phase were observed. Thus, when the microstructure changed from a single-phase solid solution to a multiphase one, the researcher believed that the decomposition of the passivation coating shifted from pitting corrosion to dissociation at the phase boundaries, which reduced the corrosion resistance [9].

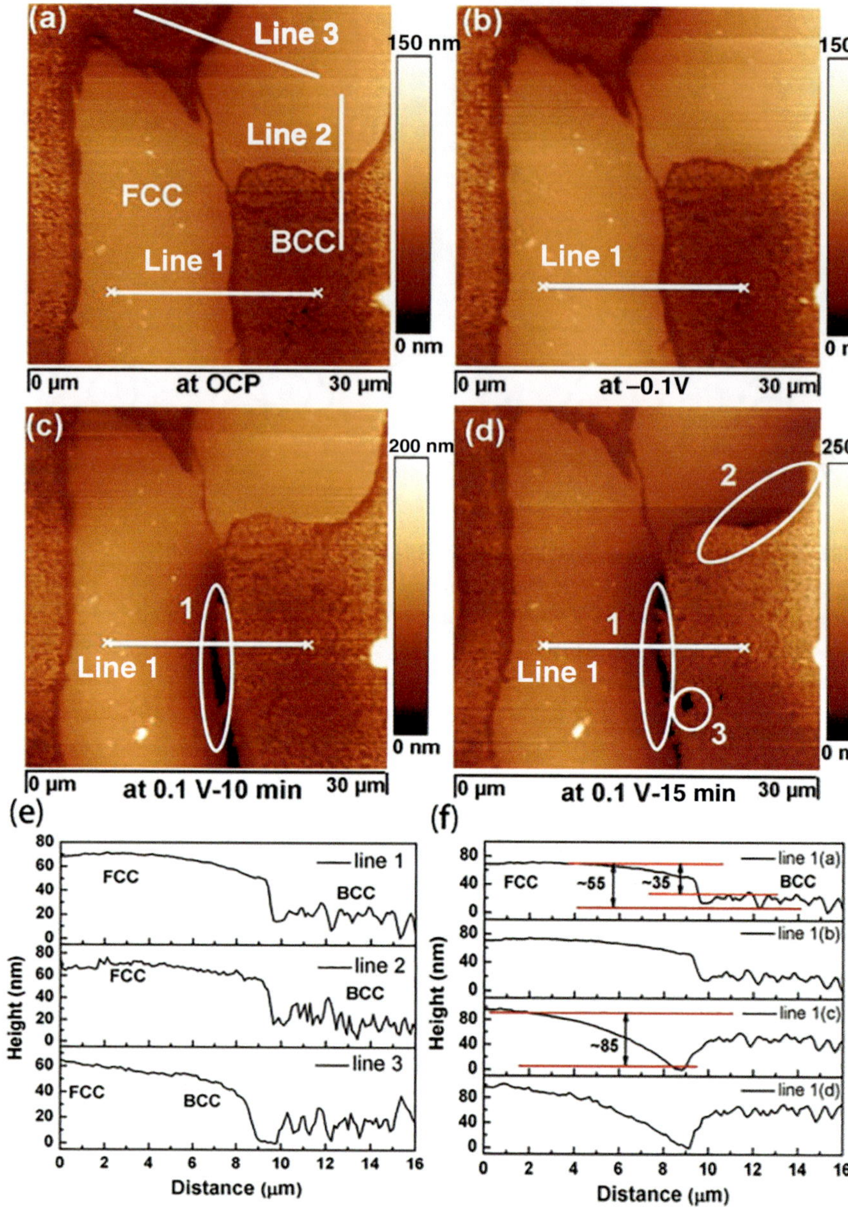

FIGURE 5.15 In situ EC-AFM topographic images of $Al_{0.7}CoCrFeNi$ HEA at (a) open circuit potential (OCP), (b) 0.1 V, (c) 0.1 V, and (d) during a 15-min soak at 0.1 V compared to SCE after a 10-min soak in 3.5 wt% NaCl solution. (e) Trilinear profiles cross the phase boundary shown in (a). (f) Vertical profiles along surface line 1 in (a–d) [11] [Open access].

5.3.2.2 Precipitation

Chemical precipitation is the most widely used technique for removing dissolved (ionic) metals from solutions, such as wastewater from processes containing hazardous metals. Ionic metals are converted to an insoluble (particulate) form via a chemical reaction between the precipitating reagent and soluble metal compounds. The resulting particulates are removed from the solution by sedimentation and/or filtration. Neutralization, precipitation, coagulation/flocculation, particle–liquid separation, and dewatering are typically essential processes in this process.

Factors that influence the effectiveness of a chemical precipitation process include the type and quantity of ionic metals in solution, the precipitant used, reaction conditions (especially the pH of the solution), and the presence of other chemicals that may impede precipitation. Hydroxide precipitation, sometimes known as pH precipitation, is the most widely used chemical precipitation technique and produces metal hydroxides using precipitants such as calcium hydroxide (lime) or sodium hydroxide (caustic). The hydroxides of each dissolved metal must precipitate at a pH between 7.5 and 11.0 for chromium and cadmium, respectively. Metal hydroxides become more soluble at both high and low pH levels due to their amphoteric nature. Therefore, depending on the optimum pH for precipitation of this metal, the metal may dissolve or return to solution. As most wastewater from operations contains mixed metals, precipitating these different metals as hydroxides can be a complex process [20].

5.3.2.3 Sol–Gel

The sol–gel process is a common wet chemistry technique for applying nanocomposite membranes. Material sources initially produced by this method include primary metal oxides and alkoxides, metal nanoparticles, tetraethoxysilane for a silica matrix, stabilizers, catalysts, and porosity additives. The sol then undergoes hydrolysis and polycondensation, ultimately forming a gel network. Discrete colloidal particles or connected polymer networks are examples of the basic structure or shape of a solid phase [21].

After the sol–gel feedstock is coated on a substrate, the compounds undergo thermal decomposition and dissociation, organic matter and water are extracted from the resulting network, and crystals form and grow. This process transforms the sol–gel feedstock into a nanocomposite membrane. The thermal decomposition behavior of the sol–gel feedstock significantly affects the crystal size and porosity of the membrane [22]. On the other hand, ash consists of a continuous liquid phase and a dispersed solid phase. Ash differs from a non-colloidal liquid in that it contains solid particles dispersed throughout the liquid [23].

The complex textural properties of the enzymes were examined using atomic force microscopy (AFM). AFM images show complex enzyme molecules of natural horseradish peroxidase (HRP) and lignin peroxidase (LiP) deposited on the surface (Figure 5.16a and b). The presence of silica agglomerates on the surface of immobilized magnetite indicates that the enzymes are encapsulated (Figure 5.16c). Figure 5.16e_and f show pure magnetite nanoparticles and silica-coated magnetite nanoparticles, respectively [23]. Centrifugation of a sol can precipitate dispersed particles in a liquid. This is not the case for a non-colloidal liquid solution, such as a salt dissolved in water [24]. Figure 5.16 summarizes the different sol–gel techniques.

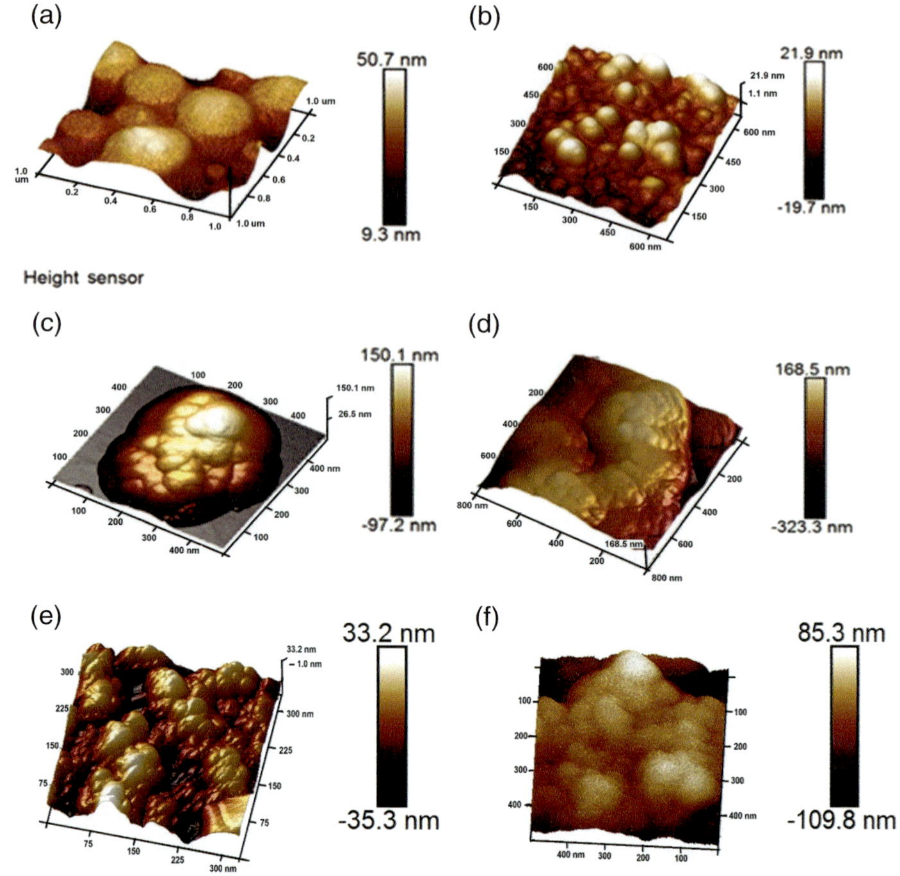

FIGURE 5.16 AFM images of native magnetite (e), silicon-coated magnetite (f), sol–gel-coated HRP (c) and LiP (d) complexes, and native HRP (a) and LiP (b) enzymes [14] [Open access].

Figure 5.16a shows sol–gel formation. Figure 5.16 illustrates the chemical steps involved in sol–gel formation, which include hydrolysis, condensation, coagulation, aging, drying, and pressing. The precursor is then activated by adding an acid (e.g., hydrochloric acid) or a base (e.g., potassium hydroxide). The activated precursors then coalesce to form a network. The network grows and ages with temperature and time until it reaches the volume of the container. The viscosity of the liquid then increases dramatically until coagulation occurs [24].

5.4 POLYMERIC NANOMATERIALS

A rapidly evolving class of materials known as polymer nanomaterials combines the physical and chemical properties of the nanoscale with the adaptability of polymers. These materials, typically described as polymer-based structures with dimensions of nanometers (1–100 nm), have attracted significant interest due to their biocompatibility, adaptable properties, and versatility in diverse industries, such as electronics, energy, medicine, and environmental remediation.

Several techniques, including emulsion polymerization, self-assembly, template-assisted processes, and advanced green chemistry methods, can be used to manufacture polymer nanomaterials. The performance and usability of the resulting nanomaterials are determined by the degree of control each process provides over particle size, shape, and surface functionality.

Polymer nanomaterials can be structurally classified as nanoparticles, nanofibers, nanogels, nanocapsules, or nanocomposites. Their high surface area-to-volume ratio, enhanced mechanical properties, and distinctive optical and electrical behaviors are all results of their nanoscale size. Their capabilities can be further enhanced by incorporating ligands, biomolecules, or inorganic nanoparticles into their surfaces, enabling them to act as catalysts, biosensors, or tailor-made drug delivery vehicles [25].

Biocompatibility and biodegradability are among the most significant benefits of polymeric nanoparticles, especially when manufactured from natural polymers such as polylactic acid, chitosan, or alginate. This makes them highly promising for biological applications, including tissue engineering, controlled drug release, and imaging. In the energy sector, polymeric nanocomposites are used to create durable, flexible, and lightweight electrodes for batteries and supercapacitors.

Despite their tremendous potential, the widespread manufacture and application of polymeric nanoparticles still face several hurdles. Careful attention must be paid to issues related to cost-effectiveness, scalability, reproducibility, and long-term environmental safety. Furthermore, a comprehensive assessment of the toxicity and lifecycle effects of these materials is required, as regulatory frameworks governing their use in consumer and biomedical products are still evolving.

In short, polymeric nanomaterials offer unprecedented opportunities for innovation in diverse fields, given their pivotal position in polymer research and nanotechnology. Future developments in materials science, healthcare, and sustainable technologies are expected to be significantly influenced by ongoing improvements in synthesis processes, characterization techniques, and application-oriented design [21–26].

5.5 COMPARATIVE ANALYSIS OF METALLIC, CERAMIC, AND POLYMERIC NANOMATERIALS

The three main classes of nanomaterials—metallic, ceramic, and polymeric—are characterized by unique compositions, properties, and potential uses. These three classes are compared to illustrate their complementary advantages and disadvantages, which form the basis for selecting appropriate nanomaterials.

1. *Metallic Nanomaterials:* Metallic nanomaterials, including gold, silver, copper, and nickel nanoparticles, exhibit high electrical and thermal

conductivity, strong mechanical strength, and distinctive optical properties resulting from surface plasmon resonance. These properties make them ideal for use in biological imaging, sensing, electronics, and catalysis. They can be synthesized, allowing precise control of size and shape, by chemical reduction, laser ablation, or electrochemical techniques. However, these materials often face challenges, including limited biodegradability, high cost (for noble metals), and susceptibility to oxidation.

2. *Ceramic Nanomaterials:* Ceramic nanomaterials, composed of oxides, carbides, and nitrides, exhibit exceptional thermal stability, hardness, chemical inertness, and corrosion resistance. These properties make them essential in energy storage devices, protective coatings, high-temperature applications, and chemical catalysis. Their dielectric and magnetic properties also make them particularly useful in electrical and magnetic applications. However, in some cases, the brittle nature of ceramics may limit their mechanical performance. Furthermore, their complex synthesis procedures, which include vapor deposition and sol–gel processing methods, may hinder large-scale production.
3. *Polymeric Nanomaterials:* Polymeric nanoparticles exhibit unique properties due to their tailorable chemical activity, low density, flexibility, and biocompatibility. These materials have broad applications in drug delivery, tissue engineering, sensors, and environmental cleaning, and can be manufactured in the form of nanofibers, nanogels, nanocapsules, or nanocomposites. The attractiveness of these materials for use in biomedicine is enhanced by the presence of natural and biodegradable polymers (such as polylactic acid and chitosan). However, compared to their metallic and ceramic counterparts, polymeric nanomaterials often have lower mechanical strength and thermal stability, limiting their use in harsh environments. Hybridization with metals or ceramics can improve their performance and produce multifunctional nanocomposites [26].

5.5.1 Comparative Perspectives

Because of their exceptional electrical conductivity and catalytic efficiency, metallic nanoparticles are critical for electronics and energy applications. Due to their strength and toughness, ceramic nanoparticles are widely used in structural and high-temperature applications. Conversely, polymeric nanoparticles are characterized by biocompatibility, processability, and flexibility—all essential properties for medical and environmental applications. An emerging trend is to integrate diverse materials into hybrid nanostructures, aiming to mitigate their unique limitations while combining their complementary advantages.

Comparing polymeric, ceramic, and metallic nanoparticles confirms that no single class can consistently meet all functional needs. Rather, matching their intrinsic properties with their intended uses determines their best application. Future developments are likely to focus on sustainable synthesis techniques, hybrid nanocomposites, and a better understanding of their biological and environmental interactions [26]. Table 5.1

TABLE 5.1
Source's Production in AM and Industry 4.0 and 5.0

Sources	Articles
Journal of Intelligent Manufacturing	118
Lecture Notes in Mechanical Engineering	56
International Journal of Advanced Manufacturing Technology	49
Procedia CIRP	33
AM	32
IEEE Access	30
Proceedings of the International Conference on Physical Micromechanics Materials with Multilevel Hierarchical Structure and Intelligent Manufacturing Technology	30
IFAC Papers Online	28
Sensors	28
Lecture Notes in Networks and Systems	23
30th International Conference on Flexible Automation and Intelligent Manufacturing (FAIM, 2021)	22
Rapid Prototyping Journal	20
Polymers	19
Applied Sciences—Basel	18
Journal of Manufacturing Systems	18
Materials Today: Proceedings	18
Materials	17
Sustainability	17
Applied Sciences (Switzerland)	16
IOP Conference Series: Materials Science and Engineering	16

Source: [27] [Open access].

lists sources of academic publications in the field of technology and manufacturing, along with the number of articles published in each source. This data enables researchers to identify the most prominent and active journals and conferences in smart manufacturing and advanced manufacturing technologies.

5.6 MICROEMULSION

The physical, chemical, biological, and engineering sciences have given rise to the field of nanotechnology. These fields are renowned for their innovative approaches to manipulating individual atoms and molecules. In the context of nanotechnology, a nanoparticle is a small-scale object that behaves as a self-contained unit, possessing transport properties and potential applications. Nanoparticles represent one of the most challenging areas of contemporary scientific and technological research, involving remarkable changes in their optical, magnetic, catalytic, electrical, and drug delivery properties, as well as enhanced physical properties such as mechanical strength, thermal stability, and chemical inertness.

For the development of novel nanodevices with diverse applications in the physical, biological, biomedical, and pharmaceutical fields, simple or complex nanoparticles are essential building blocks. Due to their applications in Raman scattering, catalysis, photonics, electronics, labeling, imaging, and sensing, nanoparticles and nanomaterials have long captivated scientists. Significant differences are often observed when comparing the optical properties of nanoparticles and single crystals with identical chemical compositions. Because of the numerous surface atoms and the quantum confinement of electronic states in nanocrystals, the properties of nanomaterials differ from those of their macroscopic phases.

The surface morphology of different bulk-heterojunction (BHJ) films is analyzed by tap-mode AFM. The Au/PEDOT: PSS film layers are peeled off to expose the BHJ layers (PTB7:EH-IDTBR and PTB7:ITIC) after J–V measurements. Then, background electrolyte (BGE) solution is used to remove the acceptor materials from the BHJ layers. AFM images of BHJ films after BGE washing has revealed obvious differences in the molecular assembly of PTB7, in contrast to AFM images of BHJ films without BGE washing (Figure 5.17a–d). PTB7 has created distinct granular

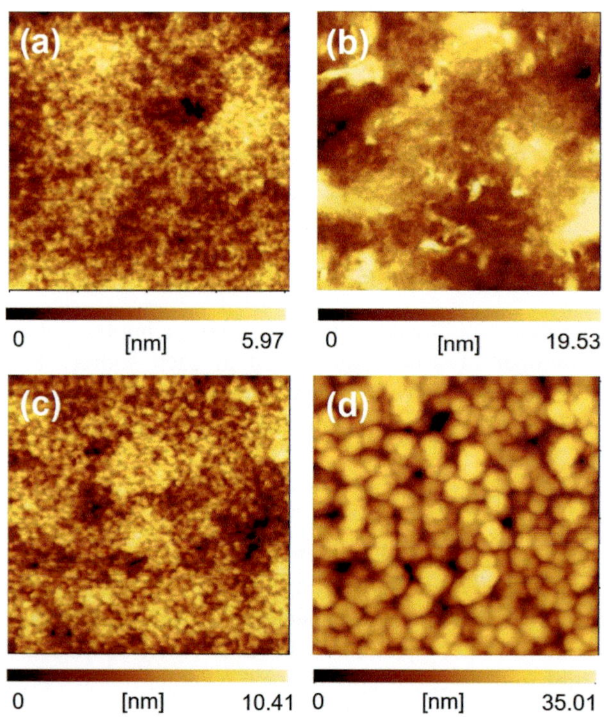

FIGURE 5.17 AFM (2 μm × 2 μm) before (a, b) and after (c) BGE washing. PTB7/ITIC films (b, d) and PTB7/EH-IDTBR films (a, c) [17] [Open access].

aggregates in ITIC-containing BHJ films, while it has created an amorphous-like texture in the EH-IDTBR-containing BHJ films [17].

The properties of matter at this size differ significantly from those of bulk materials and single atoms or molecules. A variety of physical and chemical techniques, including thermal and radiation procedures, have been employed to produce these organic and inorganic nanoparticles. The microemulsion process, which have several microemulsifiers, is one of several preparation techniques that can regulate various particle properties, such as size, geometry, shape, homogeneity, and surface area. Microemulsion processes are used to produce colloidal minerals, colloidal silver chloride, nanocrystalline oxides of Fe_2O_3, TiO_2, and Al_2O_3, and high-temperature oxides of $YBa_2Cu_3O_7$. In addition to summarizing recent studies on the synthesis of organic and inorganic nanoparticles using this technique, this work aims to provide a clear overview of the use of microemulsions to produce nanoparticles and control their particle size and shape [9,12,19,20].

5.7 ADVANCEMENTS IN NANOCOATING TECHNOLOGY

Nanotechnology is one of the fastest-growing fields, primarily used to enhance performance, prevent corrosion, mitigate erosion, and prevent water damage across all disciplines, including mechanical, electrical, electronic, and computer applications. As a result, it is ubiquitous. We also aim to utilize nanotechnology in turning and other machining processes as their use becomes increasingly widespread. In this case, we will be turning hard materials that are difficult to cut after coating them with nanoparticles using conventional techniques. As you can see, we will first examine nanotechnology, the materials we will use, our work environment, and the types of tools we will employ before delving into this project. A nanometer is roughly half the size of a DNA molecule, a thousand times smaller than a red blood cell, and a hundred thousand times smaller than the diameter of a human hair. The ability to control or manipulate matter at the atomic level, develop technologies at the atomic, molecular, and macromolecular levels using a length scale of approximately 1–100 nm in any dimension, and create and apply structures, devices, and systems with unique properties and functions due to their small size are all aspects of nanotechnology. Nanotechnology is the study of molecular engineering. It is an umbrella term for a range of techniques, tools, and processes that involve manipulating materials at the smallest scale, from 1 to 100 nm^2. Two factors make it difficult to apply the laws of classical physics and chemistry at this minute scale. First, the electrical properties of very small particles may differ radically from their larger counterparts. Second, because surface atoms are often the most reactive, the properties of the material change, leading to a marked increase in the surface area-to-volume ratio [21–26].

5.8 APPLICATION OF NANOCOATINGS

Typically, solid, single- or multi-phase structures known as nanocoatings—which are at least 100 nm thick—are applied directly to a surface to impart specific properties or functions to the substrate material. At the nanoscale, different materials exhibit distinct properties. General properties, such as hydrophobicity, electrical conductivity, magnetic permeability, melting point, and reactivity, are greatly influenced by the particle size of the material. The properties of nanomaterials are governed by intermolecular forces, including hydrogen bonds, van der Waals forces, and electrostatic forces. These intermolecular forces operate on length scales ranging from a few angstroms to several nanometers.

The properties of a material change dramatically as its dimensions approach the nanoscale, or the equivalent of the intermolecular force distances. Therefore, by improving the physical and chemical properties of nanocoatings composed of nanoparticles or multiple nanolayers, numerous new applications can be explored. Corrosion-resistant coatings, self-cleaning (non-stick) surfaces, corrosion prevention, and thermal protection are among the main industrial uses of nanocoatings. Because they enable modifications to aircraft engine components, cabins, and airframes to improve performance, reduce operating costs, and enhance fuel efficiency, these applications are crucial for the aerospace industry [22].

It is said that the efficiency of converting solar radiation into thermal energy depends on the type of solar absorber used in the collector panel. An ideal solar absorber is characterized by a high absorption coefficient and low emissivity. It has been discovered that selective nickel-based coatings can be electrochemically deposited to provide good optical performance and reasonable cost for solar absorbers [27]. However, due to the low thermal stability of chemical vapor deposition and environmental concerns, scientists are turning to physical vapor deposition technology for coating medium- and high-temperature solar photovoltaic panels. One of the physical vapor deposition techniques that has been effectively applied to receiver tubes and flat-plate solar collectors is spray coating [28].

The use of nanoparticles to produce high-quality coatings with enhanced performance characteristics is revolutionary, as demonstrated in the research, "Nanocoatings: Design, Development, and Application." Porous nanocoatings, characterized by extremely thin layers of nanoparticles, offer numerous advantages over conventional coatings, including improved surface properties, increased abrasion and corrosion resistance, and antibacterial properties.

Design: To successfully design nanocoatings, appropriate nanomaterials and nanostructures must be selected to achieve specific performance objectives. Understanding how particle size and shape affect the overall coating performance is crucial.

Development: The development phase aims to establish efficient production processes that consistently produce high-quality nanocoatings. Key considerations include quality control, coating stability, and the optimization of preparation procedures.

Applications: Numerous industries, including electronics, energy, construction, and medicine, have demonstrated the effectiveness of nanoporous coatings. These coatings extend the lifespan of materials and improve product performance.

5.9 CHALLENGES AND FUTURE PERSPECTIVES

With their unique size-dependent properties and broad applications in electronics, environmental technologies, energy, and medicine, nanomaterials have become revolutionary materials. Prospects point to promising approaches to overcome these barriers, but despite significant development, several challenges remain that prevent their widespread application [22].

1. Challenges

   Reproducibility and Scalability: While many nanomaterial synthesis techniques offer superior control at the laboratory level, they face challenges when used in industrial production. Scalability and consistency are limited by problems that include unpredictability of batch-by-batch results, high energy consumption, and processing complexity.

   Environmental and Toxicity Issues: The small size and strong reactivity of nanomaterials raise concerns about how they interact with biological systems and ecosystems. These issues require more comprehensive toxicological assessments, as their potential cytotoxicity, bioaccumulation, and long-term environmental persistence remain poorly understood.

   Durability and Stability: Under operating conditions, nanomaterials are susceptible to oxidation, damage, or agglomeration. Consequently, their functionality deteriorates over time in applications such as biomedical devices, sensors, and catalysis. There is an urgent need to develop methods to improve their mechanical and chemical stability.

   High Production Costs: The relatively high cost of advanced nanomaterials is attributed to energy-intensive processes, sophisticated production techniques, and reliance on expensive raw materials. Compared to conventional materials, this reduces their commercial competitiveness.

   Gaps in Regulation and Standardization: Nanotechnology is expanding faster than the development of safety regulations and standards. Public acceptance and industrial adoption are hindered by the lack of clear requirements for production, labeling, risk assessment, and waste disposal.

2. *Future Prospects*

   Sustainability and Green Synthesis: The development of scalable, economical, and environmentally friendly synthesis techniques such as plant-based and bioinspired methods will be essential for the development of nanomaterials while minimizing their negative environmental impact.

   Multifunctional and Hybrid Systems: By combining metallic, ceramic, and polymeric nanoparticles to create hybrid nanocomposites, their complementary properties can be leveraged to create multipurpose platforms for various energy storage, catalysis, and medical applications.

   Improved Modeling and Characterization: Newly developed in-situ characterization and computational modeling techniques will enable a deeper understanding of the mechanisms of nanomaterial development, stability, and interactions. This will accelerate the creation of customized materials with reliable performance.

   Personalized and Biomedical Applications: Nanomaterials have potential applications in the fields of therapeutic diagnostics, regenerative medicine, and targeted drug delivery. Enhancing biocompatibility, reducing toxicity, and integrating nanomaterials into personalized healthcare solutions will be key objectives for future developments.

Recycling and the Circular Economy: The sustainable use of nanomaterials will be ensured by allowing their safe disposal, recycling, and reuse within the framework of a circular economy. This aligns with international goals of resource efficiency and environmental responsibility.

The future of nanomaterials depends on solving current problems through interdisciplinary studies, environmentally friendly engineering, and strict safety regulations. Nanomaterials are expected to play a pivotal role in developing next-generation technologies and promoting sustainable development by addressing concerns related to scalability, toxicity, and cost, while exploring multi-purpose and environmentally friendly designs [28].

PROBLEMS

1. How do metallic, ceramic, and polymeric nanomaterials differ from each other in terms of structure and function?
2. What makes metallic nanoparticles popular in electronics and catalytic applications?
3. What properties allow ceramic nanoparticles to be used in protective coatings and high-temperature applications?
4. What is the role of polymeric nanoparticles in biomedical applications such as tissue engineering and drug delivery?
5. What are the main advantages and disadvantages of each class of nanomaterials, and how can hybrid nanocomposites overcome them?

REFERENCES

1. Kryuchkov, M., Adamcik, J., & Katanaev, V. L. (2022). Bactericidal and antiviral bionic metalized nanocoatings. *Nanomaterials*, 12(11), 1868.
2. Giannazzo, F., Schilirò, E., Greco, G., & Roccaforte, F. (2020). Conductive atomic force microscopy of semiconducting transition metal dichalcogenides and heterostructures. *Nanomaterials*, 10(4), 803.
3. Yoon, Y., Lee, J., Lee, S., Kim, S., & Choi, H. C. (2021). Ultrasmooth organic films via efficient aggregation suppression by a low-vacuum physical vapor deposition. *Materials*, 14(23), 7247.
4. Pellegrino, A. L., Lo Presti, F., & Malandrino, G. (2023). Metalorganic chemical vapor deposition approach to the synthesis of Perovskite BaCeO3 and BaCe0.8Y0.2O_3 thin films. *Molecules*, 28(8), 3303.
5. Dufour, T. (2023). From basics to frontiers: A comprehensive review of plasma-modified and plasma-synthesized polymer films. *Polymers*, 15(17), 3607.
6. Zhu, J., Li, F., Liu, G., Liu, D., Li, Q., & Kan, E. (2018). Synthesis of amorphous carbon film in ethanol inverse diffusion flames. *Nanomaterials*, 8(9), 656.
7. Blumenstein, A., Simon, P., & Ihlemann, J. (2023). High-resolution laser interference ablation and amorphization of silicon. *Nanomaterials*, 13(15), 2240.

8. Ruan, Y., Miyagi, A., Wang, X., Chami, M., Boudker, O., Scheuring, S. (2017). Direct visualization of glutamate transporter elevator mechanism by high-speed AFM. *Proc. Natl. Acad. Sci. USA*, 114, 1584–1588.
9. McNeilly, O., Mann, R., Hamidian, M., & Gunawan, C. (2021). Emerging concern for silver nanoparticle resistance in *Acinetobacter baumannii* and other bacteria. *Front. Microbiol.*, 12, 652863.
10. Rotondi, S. M. C., Ailuno, G., Mattioli, S. L., Pesce, A., Cavalleri, O., & Canepa, P. (2023). Morphological investigation of protein crystals by atomic force microscopy. *Crystals*, 13(7), 1149.
11. Chen, H., Qin, Z., He, M., Liu, Y., & Wu, Z. (2020). Application of electrochemical atomic force microscopy (EC-AFM) in the corrosion study of metallic materials. *Materials*, 13(3), 668.
12. Urnukhsaikhan, E., Bold, B.-E., Gunbileg, A., Sukhbaatar, N., & Mishig-Ochir, T. (2021). Antibacterial activity and characteristics of silver nanoparticles biosynthesized from Carduus crispus. *Sci. Rep.*, 11, 21047.
13. Hövelmann, J., Putnis, C. V., & Benning, L. G. (2018). Metal sequestration through coupled dissolution–precipitation at the brucite–water interface. *Minerals*, 8(8), 346.
14. Pylypchuk, I. V., Daniel, G., Kessler, V. G., & Seisenbaeva, G. A. (2020). Removal of diclofenac, paracetamol, and carbamazepine from model aqueous solutions by magnetic sol–gel encapsulated horseradish peroxidase and lignin peroxidase composites. *Nanomaterials*, 10(2), 282.
15. Xie, R., Li, D., Hou, B., Wang, J., Jia, L., & Sun, Y. (2011). Solvothermally derived Co_3O_4@m-SiO_2 nano-composites for Fischer–Tropsch synthesis. *Catal. Commun.*, 12, 380–383.
16. Huang, G., Kang, W., Geng, Q., Xing, B., Liu, Q., Jia, J., & Zhang, C. (2018). One-step green hydrothermal synthesis of few-layer graphene oxide from humic acid. *Nanomaterials*, 8(4), 215.
17. Nakano, M., Takahara, A., Genda, K., Shahiduzzaman, M., Karakawa, M., Taima, T., & Takahashi, K. (2021). Selective extraction of nonfullerene acceptors from bulk-heterojunction layer in organic solar cells for detailed analysis of microstructure. *Materials*, 14(9), 2107.
18. Ghosh Chaudhuri, R., & Paria, S. (2012). Core/shell nanoparticles: Classes, properties, synthesis mechanisms, characterization, and applications. *Chem. Rev.*, 112, 2373–2433.
19. Arendsen Linda, P., Thakar, R., & Sultan Abdul, H. (2019). The use of copper as an antimicrobial agent in health care, including obstetrics and gynecology. *Clin. Microbiol. Rev.*, 32, e00125–18.
20. Román-Kustas, J., Hoffman, J. B., Reed, J. H., Gonsalves, A. E., Oh, J., Li, L., ... & Alleyne, M. (2020). Molecular and topographical organization: Influence on cicada wing wettability and bactericidal properties. *Adv. Mater. Interfaces*, 7(10), 2000112..
21. Imani, S. M., Ladouceur, L., Marshall, T., Maclachlan, R., Soleymani, L., & Didar, T. F. (2020). Antimicrobial nanomaterials and coatings: Current mechanisms and future perspectives to control the spread of viruses including SARS-CoV-2. *ACS Nano*, 14, 12341–12369.
22. Li, W., Thian, E. S., Wang, M., Wang, Z., & Ren, L. (2021). Surface design for antibacterial materials: From fundamentals to advanced strategies. *Adv. Sci.*, 8, 2100368.
23. Cloutier, M., Mantovani, D., & Rosei, F. (2015). Antibacterial coatings: Challenges, perspectives, and opportunities. *Trends Biotechnol.*, 33, 637–652.
24. Escobar, A., Muzzio, N., & Moya, S. E. (2021). Antibacterial layer-by-layer coatings for medical implants. *Pharmaceutics*, 13, 16.

25. Mas-Moruno, C., Su, B., & Dalby, M. J. (2019). Multifunctional coatings and nanotopographies: Toward cell instructive and antibacterial implants. *Adv. Healthc. Mater.*, 8, 1801103.
26. Staneva, A. D., Dimitrov, D. K., Gospodinova, D. N., & Vladkova, T. G. (2021). Antibiofouling activity of graphene materials and graphene-based antimicrobial coatings. *Microorganisms*, 9, 1839.
27. Dehghan, S., Sattarpanah Karganroudi, S., Echchakoui, S., & Barka, N. (2025). The integration of additive manufacturing into industry 4.0 and industry 5.0: A bibliometric analysis (trends, opportunities, and challenges). *Machines*, 13(1), 62.
28. Kesal, D., Christau, S., Krause, P., Möller, T., & Von Klitzing, R. (2016). Uptake of pH-sensitive gold nanoparticles in strong polyelectrolyte brushes. *Polymers*, 8(4), 134.

6 Industrial Applications of Nanomaterials

6.1 INTRODUCTION

With their remarkable structural, mechanical, electrical, and optical properties at the nanoscale, nanomaterials have emerged as a major driving force behind technological advancements in contemporary industry. They can overcome the drawbacks of conventional materials and open new avenues for industrial solutions because of their high surface-to-volume ratio, customizable physical and chemical behavior, and atomic-level engineering capabilities. Nanomaterials are crucial for miniaturization, accelerating data processing, and enhancing energy efficiency in the semiconductor and electronics industries. These factors are critical for the ongoing development of computing, smart devices, and next-generation communication systems. In the energy industry, their incorporation into solar cells, fuel cells, batteries, and supercapacitors has significantly improved conversion efficiency, storage capacity, and operational lifespan, playing a crucial role in the transition to sustainable and renewable energy supplies. Advanced water purification membranes, nano-adsorbents, and photocatalysts for wastewater treatment, as well as air purification and pollution control technologies that contribute to reducing greenhouse gas emissions and harmful pollutants, are just a few examples of the environmental applications of nanomaterials. Lightweight nanocomposites and high-performance coatings are increasingly being used in the automotive and aerospace sectors. These materials not only increase the resilience of structures to harsh weather conditions and extend their lifespan but also contribute to reducing carbon emissions and fuel consumption. Nanomaterials improve the mechanical strength, thermal insulation, fire resistance, and self-cleaning properties of building materials and protective coatings in the construction sector, extending their lifespan and reducing maintenance costs. Biomedicine and pharmaceuticals are among the fastest-growing application areas, where nanoparticles are being used in regenerative medicine, advanced imaging technologies, targeted drug delivery, and biosensors to enable highly specialized treatments with better patient outcomes and fewer side effects. Nanocatalysts, which increase reaction rates, enhance selectivity, and reduce energy consumption, have been of great benefit to the chemical industry. This has made large-scale chemical production more economical and environmentally friendly. Despite the rapid growth of these applications, challenges persist related to large-scale and economical production,

DOI: 10.1201/9781003758556-6

long-term stability, and the need for stringent safety and regulatory frameworks to address the potential health and environmental risks associated with the industrial use of nanomaterials. The future of industrial nanomaterials is expected to focus on sustainable synthesis methods, hybrid nanostructures with multifunctional properties, and scalable technologies that balance innovation and responsibility to ensure nanotechnology continues to impact industries in technologically advanced and socially acceptable ways. In this regard, hydrogels provide a suitable substrate for incorporating nanoparticles as microspheres or cross-linking agents, preserving the inherent contact properties of the hydrogel network while enhancing its functional performance. Most nanoparticles and composite materials exhibit a wide range of electrochemical, mechanical, optical, thermal, dielectric, acoustic, magnetic, and electrocatalytic properties, expanding their potential and enabling applications in biotechnology fields such as controlled drug delivery, separation devices, catalysis, antibacterial activity studies, and biofilm inhibition. Furthermore, by blending nanoparticles with various polymers and clay minerals, interconnected nanocomposites with well-ordered multilayer structures can be created. Despite their frequent occurrence in partially exfoliated forms, these polymer nanocomposites maintain their exceptional structural and functional advantages .Composite nanomaterials, a special class of materials made by adding nanoparticles to a solid matrix, significantly improve the mechanical, electrical, optical, and thermal properties of the base material. These materials have attracted increasing attention. Nanoparticles hold great promise for a variety of advanced applications due to their synergistic effects, particularly in the fabrication of sensors, medical devices, and other technologies that require high performance, multifunctionality, and reliability [1–4].

In the field of sensors, composite nanomaterials have been used to produce highly sensitive, selective, and responsive sensors. Signal detection and response time can be improved by incorporating nanoparticles, such as metal oxides, graphene, or carbon nanotubes, into a polymer matrix. These sensors show promising potential for improving the performance, functionality, and compatibility of implants and biodiagnostic devices in the medical device sector by detecting a wide range of chemical, biological, and environmental analytes at low concentrations. The use of composite nanomaterials in prosthetics, tissue engineering, and drug delivery systems is currently being investigated. Modifying the strength, flexibility, and conductivity of composite nanomaterials could enhance drug delivery, make medical devices more tissue-like, and provide better integration with the body. Furthermore, because composite nanomaterials are customizable to meet specific needs, they can be used to manufacture biosensors, wound dressings, and implant coatings that prevent infection or promote healing [1–3].

Sensors and medical devices based on composite nanomaterials enable smarter, more effective, and more personalized medical solutions by integrating contemporary nanotechnology with traditional materials science. Their application in therapeutics, medical diagnostics, and health monitoring systems could revolutionize current medicine by providing faster, more accurate, and more cost-effective solutions. The recent integration of nanoparticles with conventional materials has led to the production of composite nanomaterials, which combine several remarkable features not found in conventional materials. Nanocomposites, which are created by combining nanoparticles in a matrix of metals, polymers, or ceramics, have exceptional mechanical, electrical, thermal, optical, and chemical properties that make them ideal for a variety of advanced applications. The results of these drugs have been particularly encouraging. Nanoparticles in sensors have revolutionized the detection of many biological, chemical, and environmental variables. By increasing their surface area, conductivity, and functionality, nanoparticles such as graphene, carbon nanotubes, quantum dots, and metal oxide nanoparticles improve the detection capabilities of sensors.

Nanoparticles can exhibit improved selectivity, sensitivity, and response time in sensors, even at low analyte concentrations. For example, gas sensors based on nanocomposite technology can detect ammonia, carbon dioxide, and nitrogen dioxide, which are very useful for environmental and industrial monitoring. Biosensors are another important application of nanocomposite technology, where the enhanced surface area and functionality of nanoparticles enable the detection of specific biomarkers such as DNA, proteins, or glucose [4–6].

6.2 ELECTRONICS AND SEMICONDUCTORS

The future of electronics and semiconductors will be significantly influenced by nanomaterials, as the ongoing demand for faster, smaller, and more energy-efficient devices drives innovation at the nanoscale. Their outstanding optical, thermal, and electrical properties enable the production of components that surpass traditional materials like silicon in terms of performance and physical limitations. Because of their exceptional material transport, flexibility, and ability to operate at extremely small dimensions, two-dimensional nanomaterials, such as graphene, carbon nanotubes, and others, are potential alternatives to silicon-based technologies [1–3].

These materials enable the construction of transparent electronics, flexible integrated circuits, and nanoscale transistors—all essential components for wearable technologies and next-generation computing. Another prominent example is quantum dots, which offer size-dependent tuning of electrical and optical properties. This makes them ideal for use in highly efficient solar cells, photodetectors, quantum dot light-emitting diodes (QLEDs), and advanced display technologies. Similarly, semiconductor nanowires and metal oxide nanoparticles have been widely used in memory devices, thin-film transistors, transparent conductive films, and nanosensors, offering enhanced stability, sensitivity, and versatility. The incorporation of nanomaterials into microelectronics has also made it possible to create nanoconnections, insulating layers, and thermal management systems—all essential for maintaining performance as device dimensions continue to decrease. Furthermore, the revolutionary impact of nanomaterials on data transmission speed and energy efficiency is demonstrated by their application in optoelectronic devices such as lasers, photonic circuits, and optical communication systems. Despite these advances, several hurdles remain to be overcome, such as large-scale production, controlling costs, and ensuring the long-term reliability of nanomaterial-based devices.

However, nanomaterials are likely to remain of paramount importance in the electronics and semiconductor sectors, driving the transition to faster, smarter, and greener technologies, given rapid advances in synthesis methods, device design, and large-scale manufacturing. Today, clay-based polymer nanocomposites are becoming attractive options due to their low cost, tunable performance, and ease of production. To create well-dispersed nanostructures within a polymer matrix, these nanocomposites are typically fabricated by melt intercalation, solution exfoliation, and in situ polymerization. Each of these processes starts with organically modified clays, or sometimes virgin clays. These hybrid materials are ideal for electronic applications because they combine the flexibility and processability of polymers with the mechanical robustness and barrier properties of clay minerals.

One of the most important applications in this field is the development of flexible, wearable pressure sensors that convert mechanical stimuli or external deformations into quantifiable electrical pulses. The four main transduction methods used by these sensors—piezoresistive, capacitive, piezoelectric, and triboelectric—offer unique advantages in sensitivity, response time, and compatibility with future wearable electronics. The innovative sensor designs, coupled with advanced nanomaterials, demonstrate how nanotechnology is revolutionizing the development of multi-purpose, long-lasting, and environmentally friendly electronic devices.

When an external force is applied to a piezoresistive pressure sensor, the associated electrical signal changes (Figure 6.1a). These sensors require little signal processing and feature a simple device design that is easy to fabricate. However, they are limited in some applications. Capacitive pressure sensors change the capacitance of the device by changing the area of the dielectric layer upon application of an external mechanical force, as shown in Figure 6.1b. When piezoelectric pressure sensors generate forces that produce unequal charges within the material, charge polarization occurs (Figure 6.1c). Tribological pressure sensors can produce electrostatic charges as a result of the cyclic mechanical energy of positive and negative electrode contact and separation, depending on the friction material and the properties of the different friction electrodes (Figure 6.1d) [1–4].

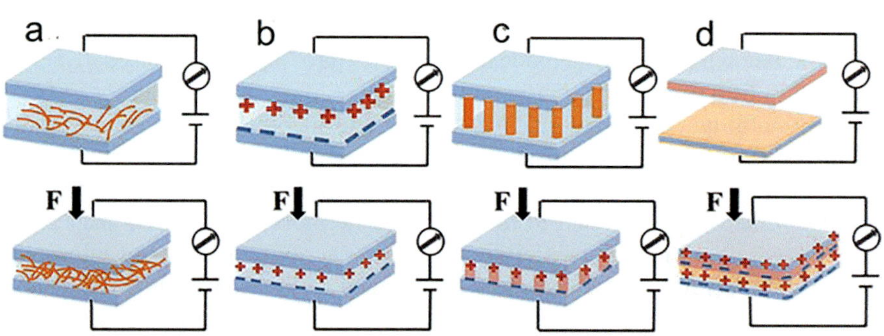

FIGURE 6.1 Schematic representations of four common transduction mechanisms: (a) piezoelectric, (b) capacitive, (c) piezoelectric, (d) triboelectric [1] [Open access].

Clay nanocomposites exhibit enhanced physical properties, including barrier properties, fire and corrosion resistance, thermal and chemical stability, and enhanced flexibility and toughness. Polymer nanocomposites are made from a variety of polymers. Understanding the nanomechanical properties of adaptable matrix materials or polymers, including polylactic acid (PLA), is critical, especially when used in a variety of sensors or devices. Along with topographic and morphological studies, these properties are examined using the advanced AM-FM technique of amplitude-frequency modulation atomic force microscopy (AFM). By closely examining surface properties at the nanoscale, this method sheds light on the mechanical behavior of polymer surfaces.

Scanning electron microscopy (SEM) was used to examine thin films fabricated by coating poly(3-hexylthiophene) (P3HT) and a styrene-butadiene-styrene (SBS) suspension by spinning it with anisole and chloroform. The resulting P3HT nanofibers exhibit a fingerprint pattern indicative of microphase separation of SBS in an anisole and chloroform solution, enabling the formation of nanocomposites with tailored properties. The resulting P3HT nanofibers exhibit a fingerprint pattern indicative of microphase separation of SBS, which can be utilized to fabricate nanocomposites with tailored properties for applications such as sensors or medical devices.

This study combines several approaches, such as AM–FM, AFM, and SFM, to investigate the morphology and nanomechanical performance of composite nanomaterials. This demonstrates how the unique structure of these polymers can expand their use in specific applications. This result indicates that the presence of SBS polymers in the suspension does not affect the synthesis of P3HT nanofibers. Furthermore, due to the microscale rupture at the P3HT nanofiber interface, a significant proportion of the P3HT nanofibers visible in the SFM image are embedded in the matrix (Figure 6.2) [1].

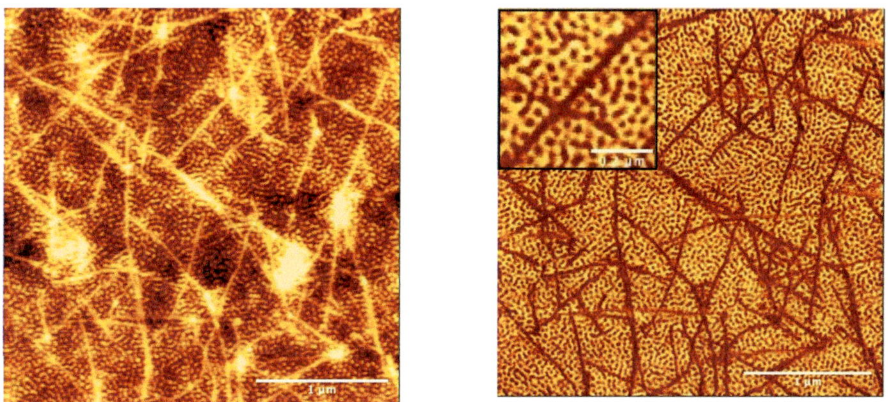

(a) (b)

FIGURE 6.2 (a) SFM topography and (b) phase pictures of the thin film produced by spin-casting a suspension of SBS (0.5 wt%) and poly(3-hexylthiophene) (P3HT) (0.05 wt%) in a 70:30 (v/v)) chloroform and anisole (3 × 3 μm). SBS's microphase separation and P3HT's nanofiber production are seen in both pictures. The phase picture's magnified image is displayed in the upper left inset of (b) [2] [Open access].

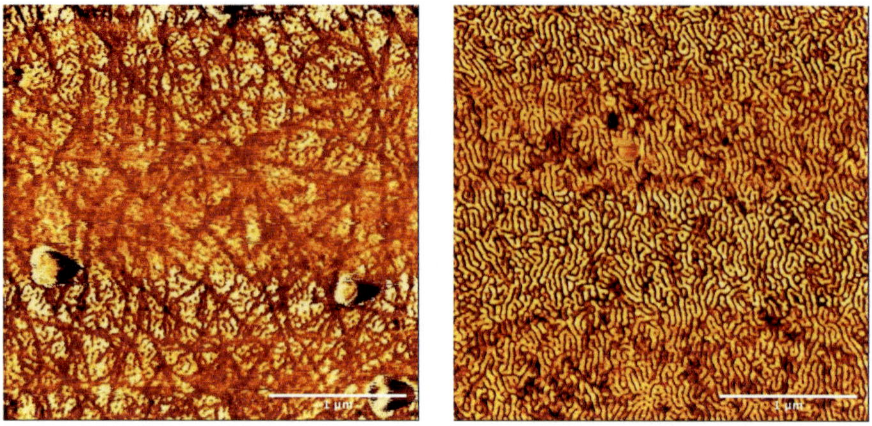

(a) (b)

FIGURE 6.3 SFM phase image of thin film obtained by spin-casting a solution of P3HT (0.05 wt%) and SBS (0.5 wt%) in a 70:30 (v/v) mixture of chloroform and anisole (3 × 3 μm) in (a) unstretched and (b) stretched sections after elongation treatment [2] [Open access].

These polymers are widely used in low-cost applications due to their simple production and relatively low mechanical requirements. Fabricating nanocomposites is one way to increase the value of these products. As the P3HT content increases, the conductivity also increases, but the mechanical strength appears to decrease. Therefore, the conductivity of a free film containing P3HT (0.05%) and SBS (5%), a ratio that produces a free film with sufficient mechanical strength, is examined after stretching. SFM is performed after stretching the film twice. SFM phase images of the stretched and unstretched sections are shown in Figure 6.3. In the unstretched section, P3HT nanofibers are regularly visible, but they are completely absent in the stretched section [5].

6.3 ENERGY GENERATION AND STORAGE

With the growing global demand for efficient, sustainable, and renewable energy systems, nanomaterials are gaining increasing importance as a revolutionary force in the energy sector. Their exceptional surface-to-volume ratio, tunable porosity, quantum confinement effects, and enhanced catalytic activity are just a few examples of their distinctive structural and functional properties that enable significant advances in energy generation and storage technologies, offering solutions not possible using conventional materials. For example, in the field of solar energy, nanoscale semiconductors, such as metal oxide nanoparticles, quantum dots, and perovskites, have revolutionized the design of photovoltaic devices by improving light absorption, allowing variable band gaps, and enabling efficient charge separation and transport. These properties enable the creation of low-cost, highly efficient solar cells that perform better than conventional silicon-based systems and hold tremendous potential for large-scale renewable energy production.

The kinetics of important electrochemical reactions, particularly oxygen reduction, have been significantly improved in the fuel cell field using nanocatalysts based on platinum, palladium, gold, and transition metal oxides. This has led to increased efficiency and durability while reducing the overall catalyst load. Similarly, nanomaterials have improved hydrogen technologies by facilitating safe storage using graphene systems, hydrides, and porous nanoframeworks, as well as efficient generation through photocatalytic water splitting. The realization of hydrogen as a reliable and clean energy source depends on these developments. Nanotechnology has significantly contributed to the development of energy storage systems. Nanostructured electrode materials, such as silicon nanowires, graphene, carbon nanotubes, and metal oxide nanocomposites, improve mechanical stability, conductivity, and ion diffusion rates in lithium-ion batteries, resulting in higher energy density, faster charging, and longer lifetime.

Beyond lithium-ion systems, nanomaterials open new horizons for next-generation technologies, such as solid-state batteries, sodium-ion batteries, and lithium-sulfur batteries. In these systems, nanoscale design mitigates problems of dendritic growth, volume expansion, and poor stability. Concurrently, supercapacitors have achieved high energy density, fast charge-discharge cycles, and long lifetimes using conductive polymers, metal-organic frameworks, and nanocarbons. This has helped them bridge the performance gap between batteries and conventional capacitors. Despite these advances, challenges remain, including affordable, large-scale synthesis, long-term stability in demanding operating environments, and safety and environmental concerns that hinder the widespread use of nanomaterials in the energy sector.

Developing scalable and sustainable manufacturing techniques, along with comprehensive material lifecycle analyses, will be essential to address these issues. Nevertheless, the integration of nanomaterials into energy generation and storage technologies is expected to remain a key factor in the global transition to more sustainable, efficient, and cleaner energy systems. The future of energy is expected to be significantly influenced by nanotechnology, which combines industrial scalability and nanoscale innovation to create reliable, efficient, and environmentally friendly systems. Beyond the energy industry, nanoparticles show promising potential in the production of advanced polymer nanocomposites, which are increasingly viewed as preferred alternatives to conventional polymers due to their improved mechanical, functional, and physical properties. Studies comparing polymer nanocomposites with conventional polymers show that the addition of nanoparticles significantly enhances performance while maintaining the polymers' lightweight and processable properties. For example, AFM was used to characterize poly(methyl methacrylate) (PMMA) doped with titanium dioxide (TiO_2) nanoparticles or halloysite nanotubes (HNTs) at concentrations of 1 and 3 wt%. The results demonstrate a significant improvement in surface homogeneity and morphology.

The addition of clay, typically at levels between 2% and 8%, to nanopolymer composites has been shown to significantly improve their mechanical strength, fatigue resistance, and dimensional thermal stability, while reducing their permeability to gases and liquids. Importantly, these improvements occur without compromising the optical clarity of the pure polymer, qualifying these composites for applications requiring transparency. Furthermore, nanocomposites exhibit greater biodegradability, conductivity, and fire resistance—attributes that expand their industrial applications to include coatings, packaging, biomedical devices, and environmentally friendly materials. These results demonstrate how the addition of nanoparticles to polymer matrices not only overcomes the drawbacks of conventional polymers but also produces multipurpose materials that combine sustainability, durability, and enhanced performance, highlighting the critical role of nanotechnology in a variety of industrial sectors. Examples of such fillers include silicon carbide, aluminum nitride, and boron nitride. The reaction time can be accelerated by using materials with high thermal conductivity, which can be particularly useful in dynamic temperature or heat detection [6]. Thermal conductivity determines the amount and speed of heat extraction from the contacting object, by mechanically assembling a thermosensitive nanocomposite of chitosan and CuMnO, ranging from 21 to 80 °C.

6.4 ENVIRONMENTAL APPLICATIONS

In the fields of pollution management, air purification, and water treatment, nanomaterials have become highly effective tools for addressing pressing environmental issues. They effectively interact with pollutants at the molecular level because of their unique surface chemistry, strong reactivity, and large specific surface area, providing more sustainable and effective treatments than conventional technologies. Nanomaterials, such as carbon nanotubes, graphene oxide, TiO_2, and silver nanoparticles (AgNPs), have been used in water treatment to remove organic pollutants, heavy metals, dyes, and microbiological contamination. While photocatalytic nanoparticles, particularly those based on TiO_2, help degrade persistent organic pollutants when exposed to sunlight, nanoscale membranes and adsorbents offer higher selectivity and adsorption capacity. These methods not only improve water quality but also help create economical and energy-efficient treatment systems.

Nanomaterials are essential for removing fine particulate matter, volatile organic compounds, and harmful gases from the air. While carbon nanomaterials are very effective for gas adsorption and filtration applications, metal oxide nanoparticles, such as zinc oxide (ZnO), TiO_2, and copper oxide (CuO), exhibit excellent catalytic activity for the degradation of hazardous chemicals. Furthermore, photocatalytic nanocoatings can effectively degrade air pollutants when applied to building materials and urban infrastructure, improving the quality of life in congested areas.

Nanotechnology also offers interesting approaches in environmental monitoring and soil remediation. Hazardous chemicals, pesticides, and petroleum hydrocarbons can be stabilized or degraded by nanoparticles, and pollutants can be detected at very low concentrations in real time using nanosensors. The efficiency of waste-to-energy technologies increases, and greenhouse gas emissions are reduced when industrial processes utilize nanocatalysts, resulting in the production of nanomaterials. Nanomaterials are also being used in renewable energy-based ecosystems, where they are integrated into solar-powered desalination units, photovoltaic reactors, and self-cleaning coatings. By reducing reliance on fossil fuels, these developments not only improve environmental protection but also support global sustainability goals.

Nanomaterials have tremendous potential, but their use in the environment is limited by issues of toxicity, long-term stability, and the environmental risks posed by the release of nanoparticles. Environmentally friendly synthesis techniques, careful risk assessment, and regulations that ensure safe use without endangering the environment or public health are all essential to address these issues. However, the integration of nanotechnology into environmental protection strategies highlights the vital role nanomaterials will play in building a sustainable future by providing a pioneering method for producing cleaner air, better water, and healthier ecosystems. Beyond its environmental uses, nanotechnology has achieved significant progress in the field of polymer nanocomposites, particularly in improving their mechanical performance. The addition of organoclay to polymer matrices has been shown to significantly increase their strength, toughness, and fatigue resistance. This improvement is primarily due to the high stiffness of the polymer matrix and the strong affinity between the polymer chains and organoclay lamellae, which together increase structural integrity and load transfer efficiency. In conventional glass fiber-reinforced composites, repetitive deformation and poor fatigue behavior are typically attributed to stress concentrations reduced by stronger edge connections within the nanomaterial network. Consequently, polymer–organic clay nanocomposites exhibit improved mechanical properties, such as increased stiffness, enhanced load-bearing capacity under cyclic loads, and increased dimensional stability. These advantages demonstrate how nanoscale design can overcome the long-standing drawbacks of conventional composite materials, expanding their application in structural and engineering fields.

Due to its high stiffness, which increases its rigidity and resistance to deformation, this composite is more durable and able to withstand mechanical loads. However, it allows the clay particles to be evenly distributed throughout the polymer matrix. When these components are combined, the material's ability to withstand mechanical loads is enhanced, opening up a wide range of complex applications for polymer nanocomposites. Figure 6.4 shows relatively low-magnification images of a polystyrene (PS) and SBS blend. According to research, the islands are believed to exhibit the SBS phase, while PS is believed to form the matrix [7]. It is usually impossible to deduce a specific mechanical characteristic from the phase difference, as the phase depends on a wide range of physical properties. As mentioned earlier, the phase image (Figure 6.4c) can be used to differentiate between different polymers intuitively,

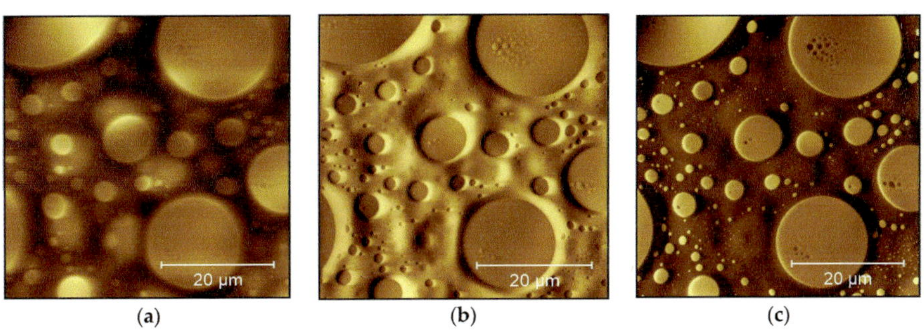

FIGURE 6.4 AFM images of (a) a combination of polystyrene and styrene–butadiene–styrene block copolymer (PS/SBS) (b) topography (height range 2.4 μm) amplitude; and (c) phase image (phase range 137°) [4] [Open access].

rather than relying on numbers. Some studies suggest that the darker phase regions (i.e., the smaller or more passive phase) contain the softer material [20–22], while others suggest the opposite [8]. The darker (smaller) phase in this case represents the stiffer material, as the SBS thermoplastic elastomer is softer than the adjacent brittle PS matrix. Note that the amplitude image (Figure 6.4b) highlights the edges, making them appear more pronounced, while only the topographic image (Figure 6.4a) shows the actual height differences [9].

The morphology of these samples was determined because the amount of nanofillers in the above formulations is well above the leakage current threshold [10,11]. This provides a useful criterion for assessing how well the nanofillers are dispersed in the resin. The entire surface of the epoxy nanocomposites is examined to provide reliable results. The carbon nanofillers can be seen and distinguished from the surrounding epoxy matrix when used for AFM examination [12]. Remarkably, at higher magnification, PS residues can also be observed within the SBS islands, as shown in Figure 6.2. In this case, the topographical variations (Figure 6.5a) are much smaller, the amplitude image (Figure 6.5b) provides a clearer view of the distribution of the islands within, and the phase image (Figure 6.5c) again shows clear distinctions.

After this comprehensive testing of the phase imaging technique, ABS was examined as an additional sample of known composition. The results are shown in Figure 6.6, which was taken with an edge length of 1 μm and at a higher resolution than the previous images. Although it is impossible to clearly distinguish the islands and matrix from the amplitude (Figure 6.6b), and the height range in Figure 6.6a is only 16 nm, the amplitude exhibits a strong contrast (phase range, 90°) that enables the two different polymers to be distinguished.

A specific nanocomposite material is reviewed, explaining its exceptional mechanical properties, such as 40% tensile strength, 60% flexural strength, 68% elongation, and 126% flexural strength, in terms of strength and elastic modulus. It then cites research using rapid thermal processing (RTP) that demonstrated similar properties in nylon-6 clay-based nanocomposites. According to this section, the final nanostructure and larger aspect ratio of the clay layers, which contribute to the material's strength,

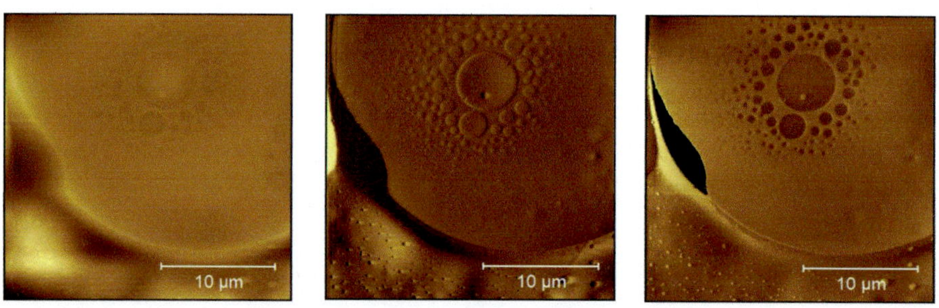

(a) (b) (c)

FIGURE 6.5 HR-AFM images of polystyrene and styrene–butadiene–styrene (PS/SBS): (a) topography (height range: 1.0 μm); (b) amplitude; phase image (phase range: 55°; (c) except for the black area that is not accurately measured) [4] [Open access].

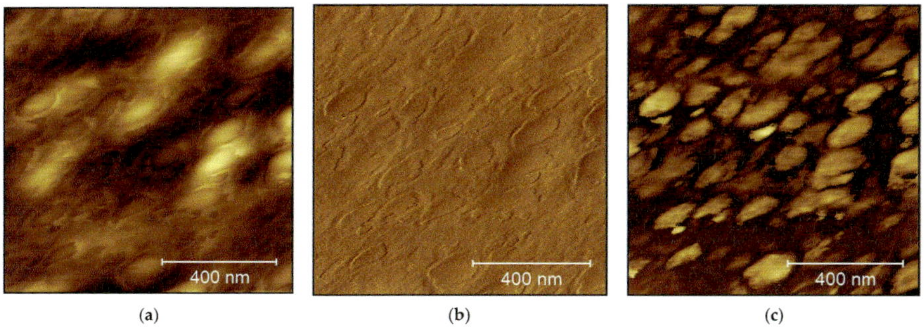

FIGURE 6.6 AFM images of acrylonitrile butadiene styrene (ABS) surface: (a) topography (height range: 16 nm); (b) amplitude; (c) phase image (phase range: 90°) [4] [Open access].

are likely the reason for the increased elastic modulus (stiffness). The development of exfoliated (disintegrated) nanostructures, such as amine-heat-treated montmorillonite (MMT) epoxy nanocomposites, has also seen significant progress, further enhancing the material's properties. Flexible temperature sensors are then discussed, highlighting the need for efficient operation on complex or uneven surfaces. Therefore, thin materials with high stretchability and a low elastic modulus are ideal. Polymeric materials such as polyimide (PI), polydimethylsiloxane (PDMS), polyethersulfone (PES), polyethylene terephthalate (PET), and polyurethane (PU) are recommended for these sensors due to their flexibility. This links the enhanced mechanical properties of nanocomposites to their potential applications in flexible and stretchable sensors that operate on a variety of surfaces.

6.5 AUTOMOTIVE AND AEROSPACE

Because of their ongoing search for stronger, lighter, and more flexible materials capable of withstanding harsh operating conditions, the automotive and aerospace sectors were among the first to utilize nanotechnology. Nanomaterials are highly useful in the design of next-generation cars and aircraft, offering special combinations of mechanical strength, thermal stability, and corrosion resistance—properties often unmatched by conventional materials. One of nanotechnology's most significant contributions to these fields is the development of lightweight nanocomposites, which combine carbon nanotubes, graphene, and nanoclay in metal or polymer matrices to produce components with exceptional strength-to-weight ratios. Reducing the weight of cars and aircraft directly contributes to increased fuel economy, reduced carbon emissions, and lower operating costs—all of which are vital to addressing global sustainability issues.

Nanomaterials are also widely used in surface treatments and protective coatings. Engine parts, turbine blades, and structural elements exposed to harsh conditions can last longer because of the exceptional resistance of ceramic and metal nanocoatings to corrosion, oxidation, and abrasion. Oxide-based self-cleaning and anti-icing nanocoatings also improve safety and reduce maintenance requirements, especially in aircraft applications where reliability under extreme weather conditions is critical. Furthermore, thermally insulating nanocoatings increase engine resistance to heat, enabling higher operating temperatures and greater energy efficiency. Fuel efficiency and energy storage are two exciting areas.

To increase energy density, speed up charging, and extend cycle life, automotive manufacturers are increasingly using nanomaterials in batteries and supercapacitors for electric vehicles. Similarly, nanocatalysts improve reaction kinetics and durability in fuel cell technologies, facilitating the transition to cleaner propulsion systems. By creating lighter, more efficient composite materials for aircraft fuselages, wings, and internal structures, nanomaterials in aerospace engineering help reduce the overall weight of aircraft while maintaining or even improving their structural integrity. Nanotechnology also enables the development of advanced monitoring and sensing systems. Real-time information on stress, temperature, and wear can be obtained via nanosensors embedded in automotive and aircraft components, facilitating predictive maintenance and enhancing operational safety. These "smart materials" represent a step toward the development of autonomous cars and aircraft capable of adapting to changing operating conditions.

Despite their numerous benefits, challenges remain regarding large-scale production, controlling costs, and ensuring the long-term reliability of nanomaterial components in harsh environments. Nevertheless, the use of nanomaterials in the aerospace and automotive sectors is changing performance standards and accelerating the transition to greener, safer, and more efficient transportation systems. Typically, larger amounts of volatile chemicals are formed due to the thermoplastic nature of nanocomposites [13]. The thermal stability of nanocomposites made of a polymer matrix and several organic layers has been improved [14]. For example, the crosslinking of nanocomposites made of Marmur [15] can be explained by increased thermal stability. Figures 6.7 and 6.8 show high-resolution microscopic images of the fracture surfaces (height and amplitude profiles) of heat-cured epoxy resin samples containing TBD + 3% EG, TBD + 0.64% multi-walled carbon nanotubes (MWCNTs), and TBD + 1% CNFs. In addition, transmission electron microscopy (TEM) images of heat-cured carbon nanofillers, expanded graphite (EG), MWCNTs, and micro- or nanocellulose (CNFs) were obtained. When added to epoxy resin, TEM images provide a reliable way to visualize the morphological properties of nanoparticles and help distinguish them from each other. Notably, TEM images of a carbon nanofiber sample (see Figure 6.7) [3] show the morphological changes in the carbon nanofibers resulting from heat treatment. Figure 6.8: HR-AFM micrographs of the MWCNT fracture surface.

6.6 CONSTRUCTION AND COATINGS

Nanomaterials have revolutionized the construction and advanced coatings industry, offering solutions that improve sustainability, utility, and durability in ways that traditional materials cannot. Nanoparticles, which enhance mechanical, thermal, and chemical performance, have greatly benefited the construction industry, which previously relied on bulk materials such as cement, steel, and polymers. For example, adding nanoparticles of TiO_2, carbon nanotubes, and silica nanoparticles to cementitious

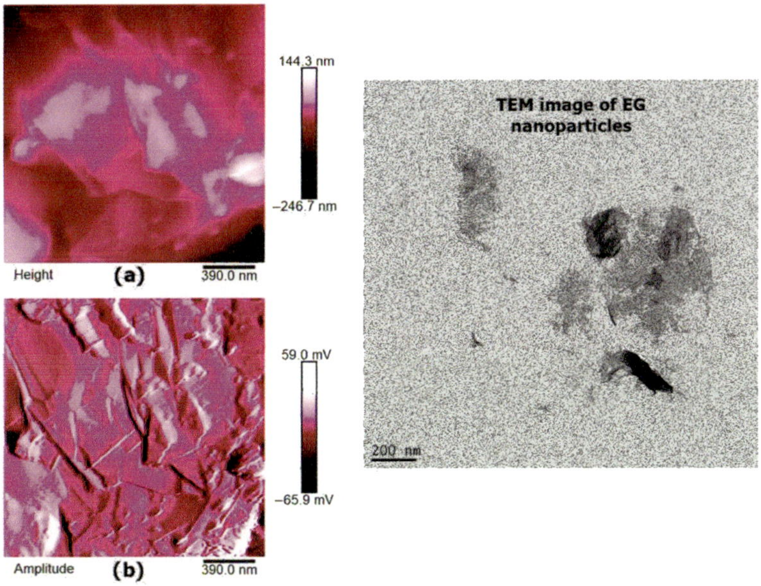

FIGURE 6.7 TEM of EG nanoparticles (see right) and high-resolution atomic electron micrograph of the fracture surface of TBD+3%EG sample in tapping mode (see left: (a) height profile, (b) amplitude profile) [3] [Open access].

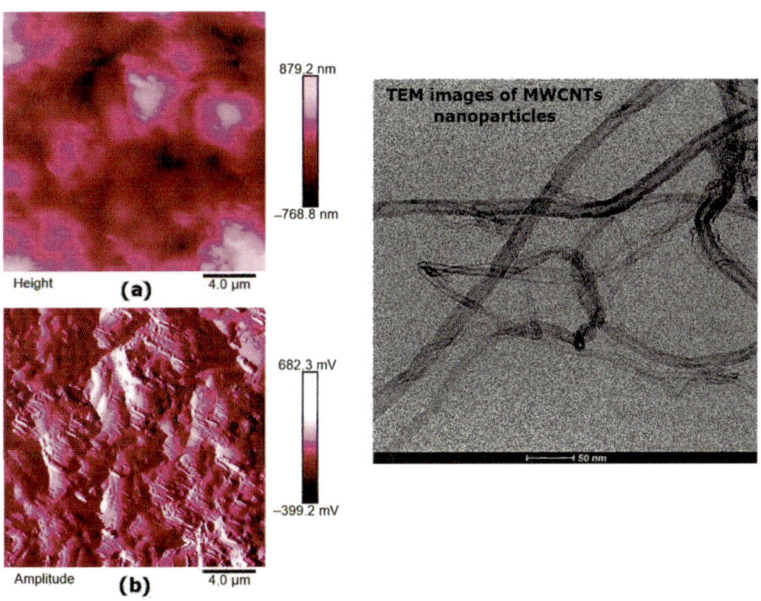

FIGURE 6.8 HR-AFM micrographs of fracture surface of MWCNT sample in tapping mode: (a) height profile, (b) amplitude profile and (right) transmission electron micrograph of nanoparticles made from MWCNTs [3] [Open access].

materials has been shown to increase compressive strength, reduce porosity, and accelerate hydration reactions, resulting in stronger concrete with a longer service life. Furthermore, nanoparticles increase resistance to freeze–thaw cycles, cracking, and shrinkage, reducing maintenance costs and extending the life of infrastructure.

Nanotechnology has revolutionized the protective and beneficial surfaces of coatings. Self-cleaning coatings made from ZnO or TiO_2 nanoparticles utilize photocatalytic capabilities to decompose organic contaminants and prevent dirt buildup and biological growth. By making surfaces water-repellent, hydrophobic nanocoatings enhance resistance to weathering and corrosion and prevent moisture damage. To reduce the risk of contamination, antimicrobial nanocoatings—often containing copper or AgNPs—are increasingly used in public buildings, hospitals, and food processing plants. Furthermore, flame-resistant nanocoatings containing nanoclay or alumina nanoparticles significantly enhance fire resistance, essential for the security of both residential and commercial buildings. In addition to providing protection, nanocoatings support sustainability and energy efficiency. Reflective nanoparticles, or nanogels, used in thermal insulation coatings help control building temperatures and reduce energy consumption for heating and cooling. By controlling light transmittance, transparent nanocoatings applied to glass surfaces enable smart windows to adapt to changing environmental conditions, supporting eco-friendly building designs. Nanomaterials in architectural aesthetics enable the development of coatings with enhanced UV protection, color fastness, and scratch resistance, ensuring that buildings maintain their appearance and functionality over time.

Standard applications are not the only avenue for collaboration between nanotechnology and building materials. New studies show promising signs of self-healing coatings and concretes infused with nanocapsules that release repair chemicals in response to microcracks, providing pioneering approaches for infrastructure maintenance. Similarly, real-time monitoring of structural health can be achieved using nanosensors embedded in building materials. This enhances safety management by enabling early detection of stress or damage.

Despite these encouraging developments, barriers to the widespread use of nanomaterials in coatings and construction remain, such as the need for comprehensive safety assessments of human and environmental exposure, integration with existing manufacturing processes, and production costs. However, the application of nanotechnology in surface engineering and construction is expected to expand rapidly, driving the industry toward more resilient, smarter, and sustainable built environments. Polymer nanocomposites have demonstrated better barrier properties against gases such as oxygen, nitrogen, carbon dioxide, water vapor, and hydrocarbons compared to conventional cement and coating systems. These properties are critical for improving the long-term performance of building and protective materials [16]. Material properties have been further enhanced by incorporating natural biomolecules into nanocomposite structures. For example, fish protein extracts from *Clupeonella cultriventris caspia* have been successfully used as antioxidant supplements.

Increasing the concentration of these protein extracts was found to improve the reinforcing effects by increasing the fiber diameter of electrospun mats (Figure 6.9). Furthermore, compared to mats made solely of polyvinyl alcohol (PVA), the addition of chitosan and protein hydrolysates to nanofiber mats significantly alters the surface topography, raising the surface roughness index. This improvement is attributed to increased viscosity of the spinning solution, which promotes the development of stronger and more structured fibers [17]. These changes enhance the mechanical properties of the resulting nanofibers, as well as their functional properties, such as durability, biocompatibility, and gas barrier effectiveness. These developments demonstrate how organic-inorganic hybrid nanocomposites can be used with conventional building materials to provide smart coatings, protective barriers, and multi-purpose surfaces that support sustainability goals.

Interest in laminate-reinforced polymer nanocomposites has grown. Several studies have shown that the clay laminate aspect ratio significantly reduces its permeability, with high aspect ratios significantly enhancing the gas barrier effectiveness. The best gas barrier properties are observed in polymer nanocomposites containing 2% organoclay and having a fully exfoliated relative permeability coefficient of PI. This interest can be explained by the remarkable properties of the polymer matrices low processing temperatures, low weight, low cost, versatility, and ease of fabrication as well as the significant improvement in their mechanical and physical properties resulting from the addition of a small percentage of polymer. For pure polymers (top row) and synthesized nanopolymers (bottom row, D = 4.5 nm, φ = 2.31%), local crystallization lines are displayed in the center of the simulation region at approximately 15 ns (higher crystallization with nanoparticle addition) and 56 ns (lower crystallization with nanoparticle addition). The red sections (colors available online) contain crystallizations greater than 90% (see also images in the supplementary material showing sections above the simulation region). At the end of crystallization, the cross-section boxes display the corresponding molecular orientations of the pure and synthesized polymers; the molecules are shown in different colors for clarity. Figure 6.10 [18] shows movies of molecular crystallization of the pure and synthesized polymers and is available in the supplementary material.

6.7 BIOMEDICAL AND PHARMACEUTICAL APPLICATIONS

With their unique physical and chemical properties, nanoscale size, and ability to precisely interact with biological systems, nanomaterials have revolutionized the biomedical and pharmaceutical industries. Their multifunctional properties, variable surface chemistry, and high surface area-to-volume ratio enable the creation of advanced therapeutic, diagnostic, and regenerative platforms—a feasibility unattainable with conventional materials. To deliver therapeutics directly to target tissues or cells, nanoparticles, such as liposomes, polymeric nanocarriers, metal nanoparticles, and dendrimers, have been extensively studied. Ultimately, this focused delivery increases the safety and efficacy of therapeutics by reducing systemic

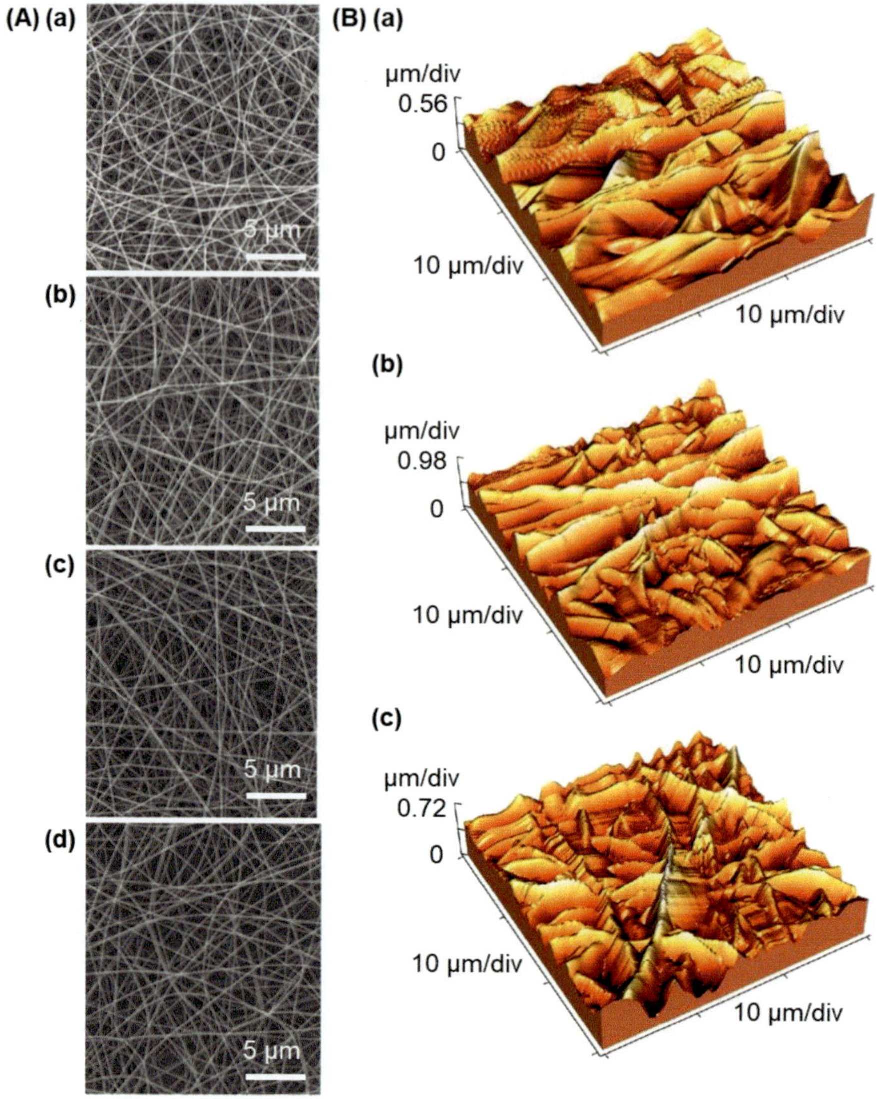

FIGURE 6.9 (A) SEM images of electrospun polyvinyl acetate chitosan nanofibrous mats with four different protein extract concentrations: (a) 0 mg/mL, (b) 1 mg/mL, (c) 3 mg/mL, and (d) 5 mg/mL. (B) 3D AFM images of electrospun nanofibrous mats made from (a) polyvinyl acetate alone, (b) polyvinyl acetate mixed with chitosan, and (c) polyvinyl acetate and chitosan mixed with protein solutions [6] [Open access].

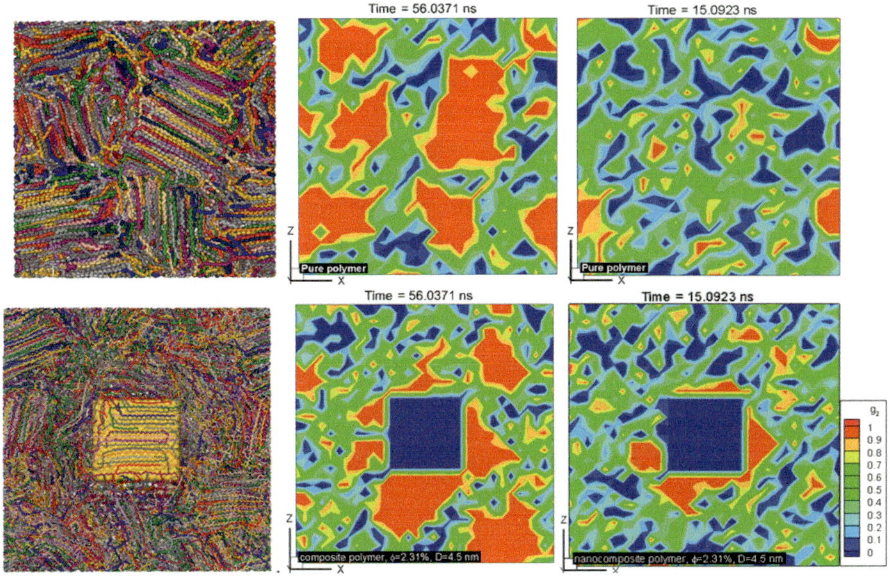

FIGURE 6.10 Molecular crystallinity of neat and synthesized polymers can be seen in supplementary material [30] [Open access].

toxicity, increasing bioavailability, and enabling controlled release. Furthermore, nanoparticles are essential for diagnostic imaging. To improve accuracy, sensitivity, and real-time tracking of disease progression, quantum dots, gold nanoparticles, and superparamagnetic iron oxide nanoparticles (SPIONs) have been added to imaging modalities such as fluorescence, computed tomography, and magnetic resonance imaging (MRI). These particles are excellent contrast agents due to their optical and magnetic properties, enabling earlier and more accurate diagnosis. Similarly, nanosensors integrated into lab-on-a-chip devices enable rapid recognition of infections, biomolecules, or genetic markers, opening the door to point-of-care diagnostics and personalized therapy.

Regenerative medicine and tissue engineering are two critical application areas. By mimicking the natural extracellular matrix, nanostructures, hydrogels, and nanofibers can promote cell adhesion, proliferation, and differentiation. Osteoconductivity can also be enhanced by adding nanoparticles such as bioactive glass or hydroxyapatite to polymer matrices. By providing antibacterial activity, maintaining moisture balance, and releasing therapeutic chemicals in a controlled manner, nanomaterials also aid wound healing.

In addition, nanotechnology has been incorporated into pharmaceutical formulations to enhance the solubility, stability, and targeted delivery of poorly soluble drugs. For example, hydrophobic drugs can be encapsulated in lipid nanocarriers and polymer nanocomposites to improve their absorption and dispersion in the body. Precision medicine is seeing new applications because of advances in smart nanomaterials, such as stimuli-responsive systems that release drugs in response to changes in pH, temperature, or enzyme activity. Despite these advances, issues remain regarding the toxicity, biocompatibility, and long-term safety of nanomaterials in clinical settings. To ensure that laboratory discoveries can be safely transformed into practical treatments, regulatory structures must change. However, the application of nanotechnology in biomedicine and pharmaceuticals continues to expand rapidly, providing a pioneering approach for drug delivery, diagnostics, regenerative medicine, and sustainable healthcare. In addition to these biological uses, research into nanomaterials with special electrical and ionic properties has gained increasing importance, particularly in the fabrication of implantable medical devices and bioelectronics. The protection of interlayers of water and its movement within their structures enables remarkable ionic conductivity, although clay layers are typically considered electrical insulators. The performance of natural clay resources can be significantly altered by the interposition of neutral molecules or functional nanostructures, which regulate ion mobility, electrical conductivity, and dielectric behavior. Typically, natural clay resources exhibit only ionic conductivity.

For example, compared to untreated clay, crown ether-modified clay exhibits several orders of magnitude higher ionic conductivity. This simple and effective method underscores the potential for improving ion transport and engineering permeation pathways in fine clay materials [19]. These changes directly impact biomedical technologies, as drug release systems, biosensors, and electrochemical devices that interact with biological tissues rely on controlled ionic transport. Researchers are developing multifunctional nanocomposites that combine ionic conductivity and mechanical reinforcement by blending clay nanostructures with biocompatible polymers or matrices. In addition to improving the functionality of medical devices, these hybrid systems open up new possibilities in tissue engineering, electrochemical scaffolds, and smart therapeutic platforms. The chemical composition of the membrane surfaces was studied at the microscopic spatial level using energy-dispersive X-ray (EDX). The EDX images in Figure 6.11 show the distribution of C, S, Ca, and Ti elements on the surface of the O1 nanoparticle sample. Specifically, S (see Figure 6.11c) is present as $BaSO_4$, while Ca (see Figure 6.11d) and Ti (see Figure 6.11e) are present as $CaCO_3$ and TiO_2, respectively.

The future development of conductivity can be predicted. At a maximum temperature of 600 K, polyethylene oxide (PEO) and ion-conducting clay nanocomposites exhibit stronger ionic conductivity than clay, preventing the transport of anions (negatively charged clay layers) and the interaction of complex cations. Similar to the nanocomposite, PEO (40%) dissolved directly in Li-MMT exhibits maximum conductivity, ranging from 105 to 104 S/cm in a direction parallel to the clay [20].

6.8 CATALYSIS AND CHEMICAL INDUSTRY

Based on their applications in the chemical and catalytic industries, the incorporation of nanoparticles into consumer goods and textiles demonstrates the direct impact of nanotechnology on everyday life, extending beyond industrial processes. Nanomaterials enable the development of stronger, lighter, and more durable fabrics and products with smart properties, offering functionality unattainable with

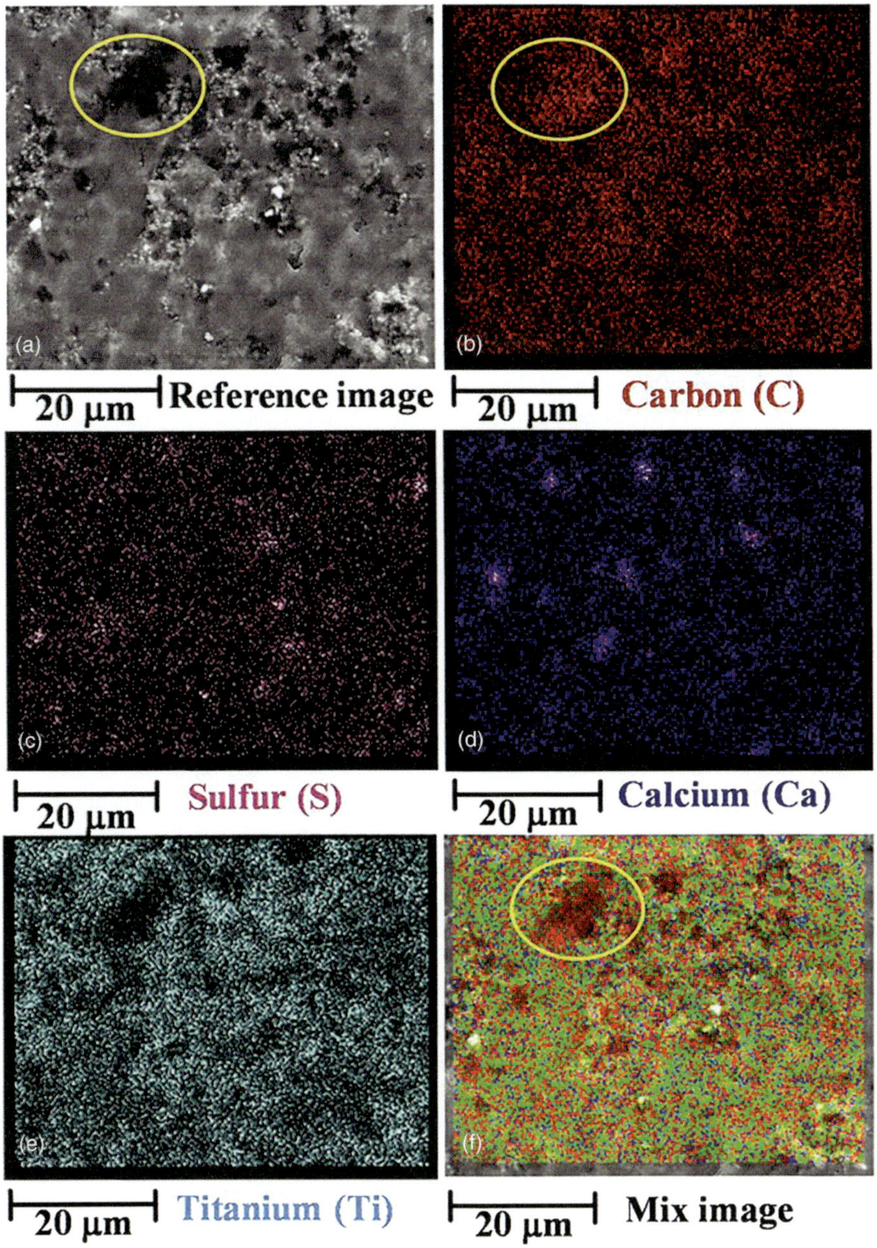

FIGURE 6.11 Fracture surface of 01GNPs sample is subjected to energy-dispersive X-ray spectroscopy (EDX) as follows: (a) control image; (b) C distribution map; (c) S distribution map; (d) Ca distribution map; (e) Ti distribution map; and (f) mixed image [8] [Open access].

conventional fibers and consumer goods. For example, the use of nanoparticles such as TiO_2, ZnO, and AgNPs in textiles has resulted in textiles with antibacterial, UV-protective, and self-cleaning properties. In industries where durability and hygiene are critical, such as healthcare, sportswear, and protective clothing, these developments are particularly important. Fibers and textiles treated with nanocoatings maintain their softness and air permeability while improving their stain, water, and flame resistance. This has enabled the production of smart textiles that are both comfortable and functional. Examples include textiles capable of controlling temperature, blocking odors, or even capturing energy using embedded nanogenerators. Polymer nanocomposites blended with carbon nanotubes or nanoclay enhance the barrier properties against gases and moisture in consumer goods, increasing the ability of packaging materials to preserve food and medicine. These developments contribute to sustainability and waste reduction by extending product life and reducing material use.

The use of nanotechnology in consumer electronics is also beneficial. Nanoelectrodes and conductive nanofibers are essential for energy-storing textiles, flexible displays, and wearable technology to maintain performance without compromising flexibility. Similarly, nanocomponents used in cosmetics and personal hygiene products, such as ZnO and titanium oxide nanoparticles in sunscreens, provide effective UV protection without compromising skin transparency. These applications demonstrate how nanomaterials can combine performance, aesthetics, and customer requirements.

Smart textiles for healthcare monitoring are a new area in this field. Nanosensors embedded in textiles can track physiological indicators, such as temperature, heart rate, and hydration levels, in real time. This combination of electronics, consumer product design, and nanotechnology enables personalized and interactive consumer experiences. Overall, the use of nanoparticles in consumer goods and textiles demonstrates the practical advantages of nanotechnology, combining advanced materials research with consumer-oriented innovation. These innovations demonstrate how nanoparticles can improve product performance while advancing the overarching goals of sustainability, efficiency, and user-centric design. Building on these developments, current studies have focused on integrating carbon-based nanomaterials into future electrical and consumer goods. With its exceptional mechanical strength, electrical conductivity, and large specific surface area, graphene—a single sheet of sp^2-hybridized carbon atoms arranged in a two-dimensional honeycomb lattice—has become one of the most promising nanomaterials. Researchers are exploring ways to combine graphene and nanocellulose to produce high-performance and flexible nanoelectronic composites. Nanocellulose has received significant attention as a sustainable material for consumer and biomedical applications due to its renewable nature, biodegradability, and biocompatibility. To overcome this challenge, graphene is increasingly being added to nanocellulose composites to improve their electrical performance while maintaining their mechanical and environmental benefits [23]. These two materials work in concert to produce lightweight, flexible, and sustainable electrical components. These components could be used in sensors, smart fabrics, and portable electronics. The fundamental processes underlying their synergistic effects—such as charge transport channels, interfacial bonding, and structural stability—remain unclear despite ongoing research [24]. In the rapidly evolving field of flexible electronics, further research into graphene-nanocellulose composites offers a potential path that bridges sustainability and advanced capabilities.

Petroleum-derived products have long been used as flexible substrates for electrical devices [25]. However, these materials pose significant environmental problems due to their polluting nature and the difficulty of disposing of them in an environmentally acceptable manner. Furthermore, their mechanical properties are often substandard. Amorphous regions have been removed from cellulose nanocrystals (CNC) [21,22].

Figure 6.12a and b shows AFM images of CNCs and cellulose nanofibers (CNFs), highlighting their structural differences [9]. Additionally, a stage in bacterial cellulose (BC) synthesis called glucose polymerization leads to the development of β-1→4 glucan chains, which are then expelled from the cell through the pores of the bacterial cell wall. These chains assemble to form 1.5 nm-wide clusters. The BC network is further strengthened as these clumps develop into primary fibers, and then into flat ribbons, as shown in Figure 6.12c [9].

By integrating graphene with nanocellulose, researchers hope to develop next-generation flexible electronic materials that boast high mechanical strength, electrical conductivity, and environmental sustainability. Further research is needed to fully understand how this interacts with thermal stability. This highlights the urgent need for conductive materials with superior mechanical and electrical properties that are renewable, biodegradable, and environmentally friendly.

Graphene is a two-dimensional nanomaterial with a single-layer disc lattice structure, consisting of densely packed carbon atoms connected via sp^2 hybridization [26]. Due to its exceptional electrical, mechanical, and thermal properties, as well as its large specific surface area and superior biocompatibility, graphene has been extensively studied in several fields, including energy storage, biomedicine, sensors, catalysis, and optoelectronics [27]. Furthermore, due to its conductivity and flexibility, exceptional mechanical and electrical robustness make graphene ideal for flexible electronic devices such as electron shells and flexible energy storage systems [28].

Despite these advantages, graphene has its drawbacks. The significant increase in grain size when magnetized at high temperatures can affect its performance. To solve this problem, researchers have considered combining graphene with nanocellulose, which can be divided into two main categories based on processing methods. CNFs, made mechanically, are characterized by flexibility and a high aspect ratio. On the

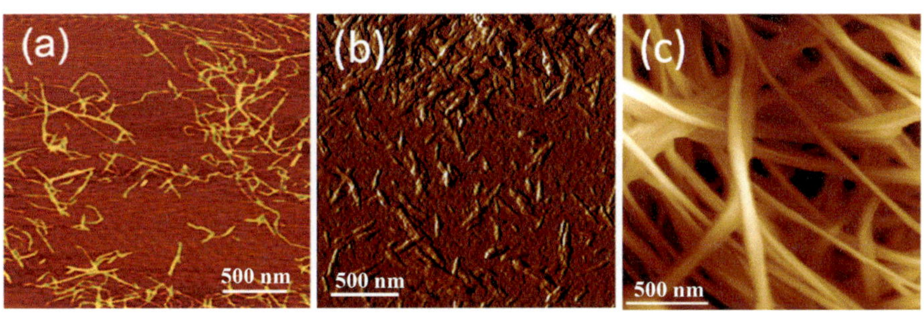

FIGURE 6.12 AFM images of (a) CNF, (b) CNC, and (c) BC [9] [Open access].

other hand, CNCs, also known as nanowhiskers, are produced by acid hydrolysis and feature a lower aspect ratio and higher crystallinity. These materials improve the functionality of the fibers in nanoelectronics applications while preserving the amorphous parts of their structure through simple chemical and mechanical procedures used to separate the carbon nanofibers.

The term "thermal conductivity" describes a material's ability to transfer heat. According to a study conducted on chitosan–CuMnO nanocomposites, thermal conductivity increases with increasing temperature (from 21 to 80 °C), while thermal resistance decreases. This was discovered during fabrication at a pressure of 312 MPa, with the maximum temperature sensitivity measured at 347.5 K [26].

Furthermore, a study using inkjet printing technology to monitor temperatures up to 200 °C using a nickel oxide matrix on a PI substrate revealed that curvature had no significant effect on performance [27]. A highly sensitive dual-parameter thermomechanical monitoring sensor was also developed.

Figure 6.13 shows the spatial distribution of nanoparticles (NPs) in 3 and 5 wt% SiO_2, MgO, and Al_2O_3 nanocomposites. The scanning electron microscopy (SEM-EDS) mapping technique facilitates the observation of dispersion. Figure 6.13a–c , respectively, display the SEM-EDS maps of SiO_2, MgO, and Al_2O_3 composites containing 3 wt% epoxy, showing a nearly uniform dispersion of nanoparticles in the epoxy nanocomposites.

6.9 MATERIALS FOR FLEXIBLE SUBSTRATES

Any flexible sensor must include substrates, as substrates greatly influence the flexibility of the sensor. Without affecting the sensing mechanism, it gives the sensing material flexibility and stability. There has been extensive research into the potential use of flexible electronics in low-cost electronic products for several interesting applications, including electronic skin, health monitoring patches, and other lighting and sensing applications. When selecting suitable metal substrates, it is important to consider factors such as cost, impurity content, surface roughness, and coefficient of thermal expansion. Exceptionally high efficiency (about 18%) is found using stainless steel and (expensive) titanium foil. For industrial steel, a reasonable efficiency of about 13% has also been achieved using suitable insulating layers to stop the spread of impurities, especially the spread of iron from the substrate [2,6]. This growth mode results in the formation of columnar V-shaped grains as shown in SEM cross-section obtained by focused ion beam (FIB) milling in Figure 6.14a [12]. In this case, the column boundaries typically intersect the entire film. Before continuous films can completely cover the substrate surface, a certain thickness is required for formation of islands, which depends on the mixture of the substrate material and the thin film, as well as on the growth conditions [12].

Because of organic field-effect transistors are a fundamental component of most of these systems and can be processed at low temperatures and with inherent flexibility, they are being extensively explored for using in wearable electronics. These flexible gadgets are demonstrated using a variety of new materials. However, before this technology can be applied in practical and credible ways, several challenges need to be addressed. In

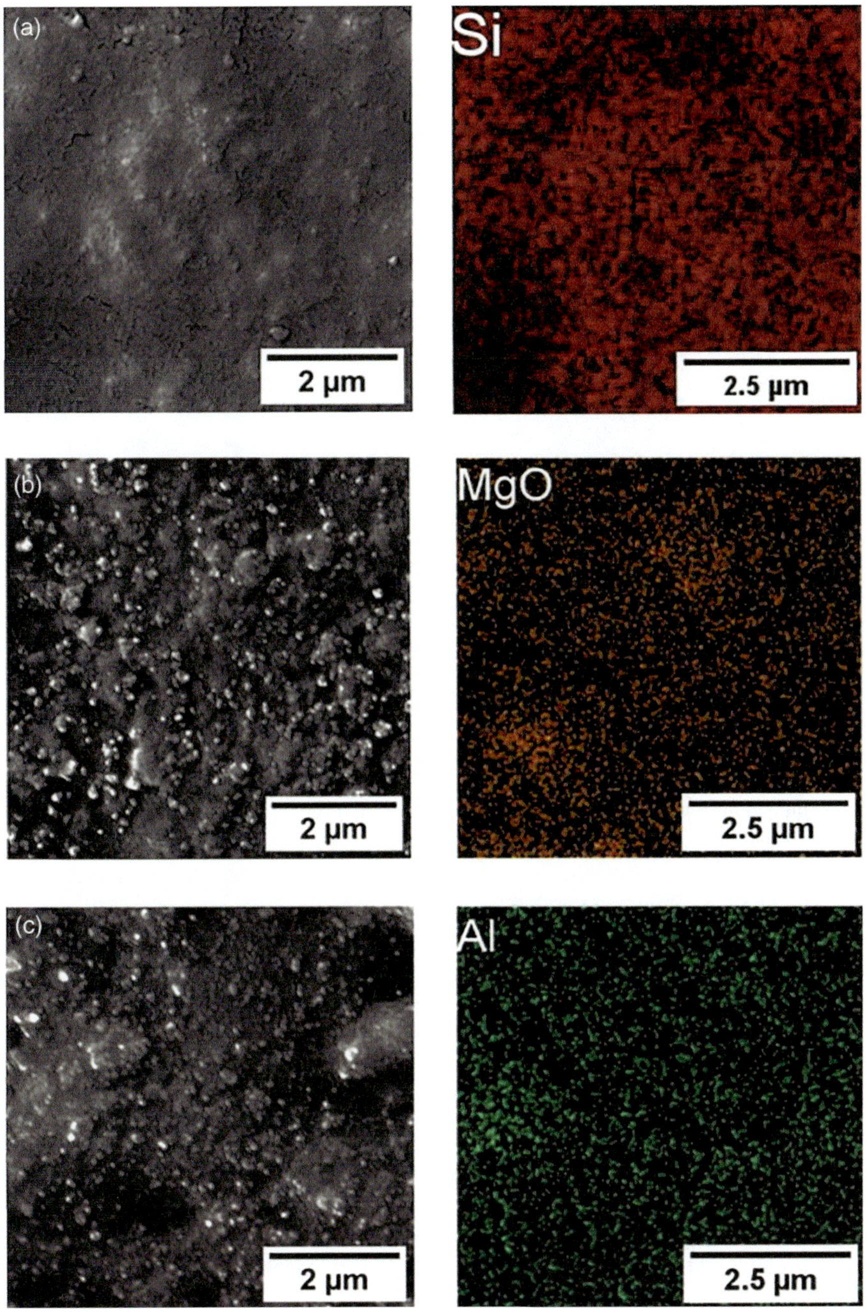

FIGURE 6.13 SEM mapping and EDS image (a) Epoxy/3 wt% SiO_2. (b) Epoxy (3 wt% MgO). (c) 3 wt% Al_2O_3 epoxy [11] [Open access].

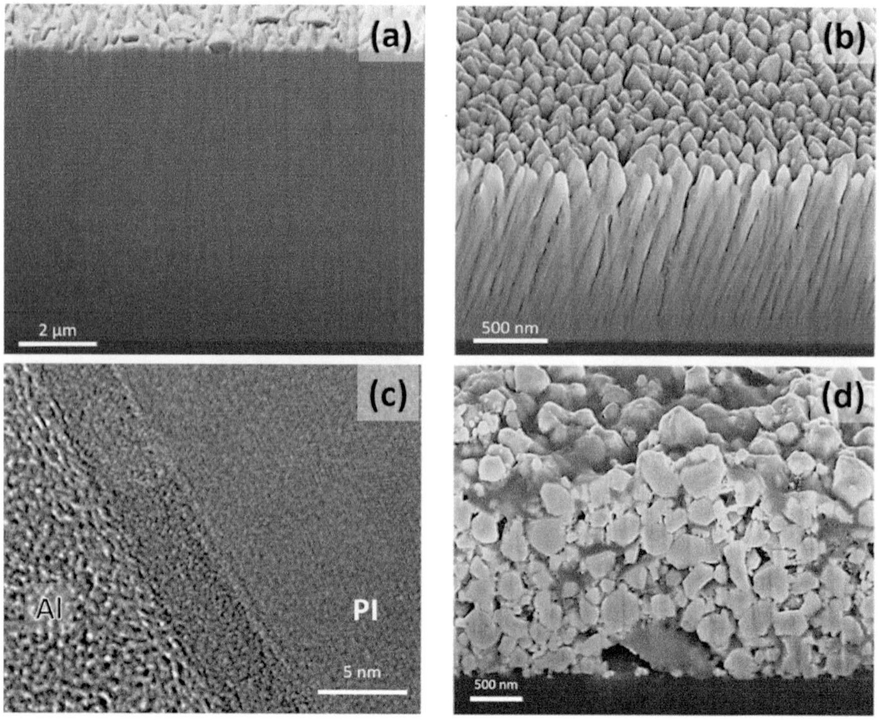

FIGURE 6.14 Cross-sections of thin films. A rotating molybdenum–aluminum–titanium target is placed in front of a molybdenum–aluminum–titanium composite film, and SEM image (a) thin vertical film growth with vertical columns on the substrate surface; (b) vertical growth of an open-air film grown from the same target and using the same deposition parameters under tilted conditions, resulting in a molybdenum–aluminum–titanium composite [19]; (c) TEM cross-section of an aluminum film thermally deposited on a polyimide (PI) substrate, clearly showing the variable region near the surface that formed during the film deposition process; (d) SEM image of copper film deposited [12] [Open access].

addition, to reduce the environmental impact of e-waste, it is necessary to develop and optimize manufacturing procedures suitable for green electronics [2,4]. Compared to rigid glass substrates, flexible substrates have many advantages, including low weight, ease of coil fabrication and new applications such as building-integrated photovoltaic systems. Copper indium gallium selenide (CIGSe) solar cells use a variety of metal substrates, including industrial steel and titanium foil. Steel and stainless steel are used.

6.9.1 Pigments and Nanocomposite Coatings

The process of creating hybrid nanomaterials involves combining layers of artificial or natural clay with copolymers or surfactants to create new nano-additives. The exceptional adhesion of these materials, which improves their mechanical and thermal properties, allows them to form a strong bond with the polymer matrix. As their molecules are smaller than the wavelength of light, these materials can be used as nano pigments, making materials colored with plant dyes completely transparent. While the coatings containing hydrogenated oils exhibited multi-level roughness with open porosity, improving the water contact angle to 128° in the steady state and 138° in the advanced state, the coatings based on unsaturated oils generally demonstrated higher water resistance. The use of this technology with plant oil-coated nanoparticles deposited on paper surfaces is shown in Figure 6.15, demonstrating the high efficiency of these materials in improving the performance of smart coatings and materials [13].

In addition to their high oxygen content, which increases their long-term stability, these compounds have exceptional qualities such as bright fluorescence, resistance to UV radiation and high temperatures. An important factor in increasing the effectiveness of nanocoatings is vegetable oil, which improves the ability to withstand water (hydrophobicity) and serves as a natural binding agent between nanoparticles, preventing cracking and contributing to the creation of uniform, heat-resistant coating layers. Modified cellulose microsheets (MFC) fiber structure with multi-scale morphology forms thin wax spots (Figure 6.16) as a result of combined wax release and migration, which determines the resulting surface hydrophobicity. As mentioned earlier, the barrier properties of MFC/NFC films are significantly reduced in the presence of moisture, but the current design allows for a combination of hydrophobicity and oxygen barrier resistance. Alternatively, modified MFC fibers can be blended as nanofiller in biopolymer matrix to create a barrier coating that is completely protective. Thermal wax release and increased surface hydrophobicity during processing (e.g., extrusion) ensure interface compatibility with the non-polymer matrix [13].

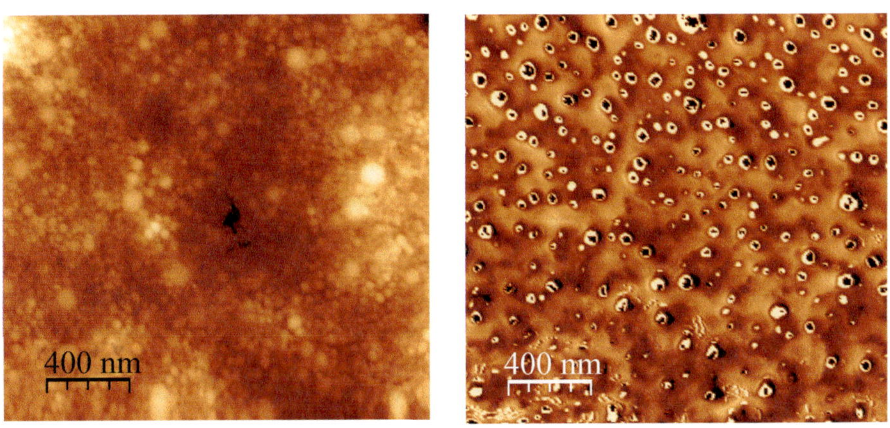

FIGURE 6.15 SMI/oil nanoparticles are applied using encapsulated vegetable oils such as (a) 50% soybean oil and (b) 70% palm oil, creating a continuous coating layer [13] [Open access].

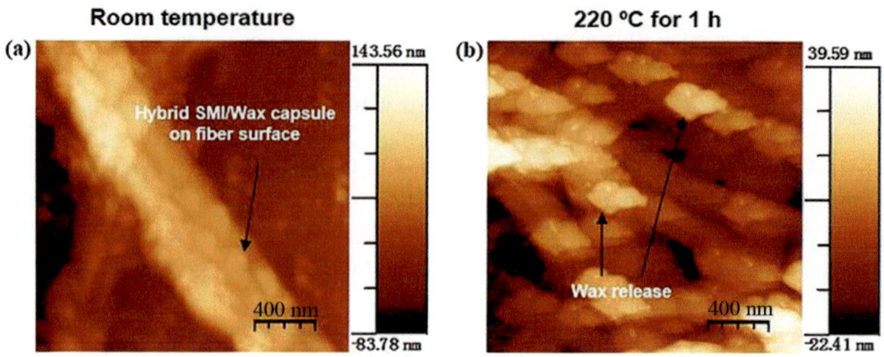

FIGURE 6.16 Morphology of MFC obtained using AFM; (a) untreated, (b) treated at temperatures up to 220 °C.

It is expected that hazardous dyes such as cadmium (Cd) and palladium (Pd) will be replaced by plant dyes or nanoparticles composed of clay and organic dyes. In bulk polymers and coatings, plant pigments are common at the nanoscale. By selecting appropriate pigments from a large variety of organic pigments, pigments of different colors can be created

6.10 FUTURE TRENDS AND CHALLENGES

The delicate balance between scientific advancement, economic viability, environmental sustainability, and public acceptance will determine the future of industrial nanomaterials. Several factors are expected to influence the trajectory of nanotechnology in the coming decades as it continues to spread across various industrial sectors. Developing sustainable manufacturing methods that utilize renewable resources, reduce energy consumption, and eliminate harmful byproducts will be a key priority. As companies shift toward environmentally friendly practices that support global sustainability goals, green nanotechnology approaches—such as the use of plant extracts, microbial processes, and biodegradable polymers—are expected to gain momentum. The development of multifunctional hybrid nanostructures is another important trend. Researchers are developing composite materials that combine excellent electrical, mechanical, thermal, and catalytic capabilities by blending metallic, ceramic, polymeric, and carbon nanoparticles on individual platforms. These hybrid systems will be essential for next-generation applications in biomedical engineering, smart coatings, flexible electronics, and renewable energy devices.

The push toward scalable production technologies is also critical. Nanomaterials can now be produced in large quantities without compromising quality or functionality, because of improved processes such as continuous flow synthesis, roll-to-roll processing, and additive manufacturing. There are still some obstacles to these promising paths. Ensuring the long-term safety and biocompatibility of nanomaterials is essential, particularly for their uses in consumer goods, food, and healthcare, where human exposure is unavoidable. Expanding toxicological research and regulatory

TABLE 6.1
Sources' Impact in AM and Industry 4.0 and 5.0

Element	h_Index	g_Index	TC	NP	PY_Start
Journal of Intelligent Manufacturing	26	55	3256	118	2014
AM	20	32	1299	32	2018
International Journal of Advanced Manufacturing Technology	17	29	909	49	2016
Journal of Manufacturing Systems	13	18	643	18	2017
IEEE Access	10	19	387	30	2019
Materials	9	17	406	17	2018
Polymers	9	12	172	19	2021
Procedia CIRP	9	16	305	33	2017
Rapid Prototyping Journal	9	15	238	20	2014
Robotics and Computer-Integrated Manufacturing	9	9	323	9	2017
Applied Sciences (Switzerland)	8	16	519	16	2019
IFAC Papers Online	8	13	211	28	2016
International Journal of Production Research	8	11	1583	11	2018
Journal of Manufacturing Technology Management	8	10	622	10	2018
Materials Today: Proceedings	8	18	354	18	2021
Procedia Manufacturing	8	11	142	13	2018
Sensors	8	20	429	28	2017
Sustainability (Switzerland)	8	14	513	14	2018
30th International Conference on Flexible Automation and Intelligent Manufacturing (FAIM, 2021)	7	11	141	22	2020
Applied Sciences—Basel	7	14	210	18	2018

Source: [29] [Open access].

frameworks is essential to address concerns about nanoparticle accumulation, degradation, and environmental discharge. The high costs of producing and integrating advanced nanomaterials represent another significant obstacle. Cost-effective synthesis and processing technologies must be developed for widespread industrial use.

Furthermore, the lack of international regulations and standardized testing procedures hinders cross-border marketing and trade. To foster widespread industry and public confidence, it will be essential to establish international standards for nanomaterial characterization, safety assessment, and quality assurance. As stakeholders must ensure that the benefits of nanotechnology are shared fairly and any potential concerns are disclosed, ethical issues are becoming increasingly important. The future of industrial nanomaterials depends on solving these issues while leveraging the tremendous potential they offer.

Nanomaterials are expected to be at the forefront of industrial innovation, with further research into scalable technologies, multifunctional hybrid systems, and sustainable synthesis. Their integration into other industries will also enhance the product [28]. Table 6.1 presents a comprehensive bibliometric analysis of several journals and conferences with a significant impact on manufacturing and technology [29]. For each source, this study includes several metrics, such as h-index, g-index, total citations (TC), number of publications (NP), and publication start year (PY_start).

PROBLEMS

1. What is the primary function of nanomaterials in the aerospace and automotive industries?
2. How do nanomaterials improve fuel cells and batteries?
3. What is the role of nanomaterials in environmental protection?
4. What is the importance of metal nanoparticles in chemical catalysis?
5. How do consumer goods and textiles benefit from nanomaterials?

REFERENCES

1. Lai, Q. T., Sun, Q. J., Tang, Z., Tang, X. G., & Zhao, X. H. (2023). Conjugated polymer-based nanocomposites for pressure sensors. *Molecules*, 28(4), 1627.
2. Goto, T., Morita, J., Maekawa, Y., Kanehashi, S., & Shimomura, T. (2020). Semiconducting properties of the hybrid film of elastic poly (styrene-b-butadiene-b-styrene) block copolymer and semiconducting poly (3-hexylthiophene) nanofibers. *Polymers*, 12(9), 2118.
3. Raimondo, M., Naddeo, C., Vertuccio, L., Lafdi, K., Sorrentino, A., & Guadagno, L. (2019). Carbon-based aeronautical epoxy nanocomposites: Effectiveness of atomic force microscopy (AFM) in investigating the dispersion of different carbonaceous nanoparticles. *Polymers*, 11(5), 832.
4. Werner, E., Güth, U., Brockhagen, B., Döpke, C., & Ehrmann, A. (2023). Examination of polymer blends by AFM phase images. *Technologies*, 11(2), 56.
5. Jeanmaire, D. L., & Van Duyne, R. P. (1977). Surface Raman spectroelectrochemistry: Part I. hetero-cyclic, aromatic, and aliphatic amines adsorbed on the anodized silver electrode. *J. Electroanal. Chem. Interfacial Electrochem.*, 84, 1–20.

6. Lai, W. F., & Wong, W. T. (2022). Design and practical considerations for active polymeric films in food packaging. Int. J. Mol. Sci., 23(11), 6295.
7. Jensen, T. R., Van Duyne, R. P., Johnson, S. A., & Maroni, V. A. (2000). Surface-enhanced infrared spectroscopy: A comparison of metal island films with discrete and nondiscrete surface plasmons. *Appl. Spectrosc.*, 54, 371–377.
8. Guadagno, L., Naddeo, C., Raimondo, M., Speranza, V., Pantani, R., Acquesta, A., & Monetta, T. (2019). UV irradiated graphene-based nanocomposites: Change in the mechanical properties by local harmoniX atomic force microscopy detection. *Materials*, 12(6), 962.
9. Oprea, M., & Panaitescu, D. M. (2020). Nanocellulose hybrids with metal oxides nanoparticles for biomedical applications. *Molecules*, 25(18), 4045.
10. Tan, N. P. B., Lee, C. H., & Li, P. (2016). Green synthesis of smart metal/polymer nanocomposite particles and their tuneable catalytic activities. *Polymers*, 8(4), 105.
11. Alsoud, A., Daradkeh, S. I., Al-Bashaish, S. R., Shaheen, A. A., Jaber, A. M., Abuamr, A. M., & Holcman, V. (2024). Electrical characterization of epoxy nanocomposite under high DC voltage. *Polymers*, 16(7), 963.
12. Cordill, M. J., Kreiml, P., & Mitterer, C. (2022). Materials engineering for flexible metallic thin film applications. *Materials*, 15(3), 926.
13. Mladenović, I. O., Vuksanović, M. M., Dimitrijević, S. P., Vasilić, R., Radojević, V. J., Vasiljević-Radović, D. G., & Nikolić, N. D. (2023). Mechanical properties of electrolytically produced copper coatings reinforced with pigment particles. *Metals*, 13(12), 1979.
14. Mladenovic, I., Lamovec, J., Vasiljevic-Radovic, D., Radojevic, V., & Nikolic, N. (2022). Determination of the absolute hardness of electrolytically produced copper coatings by application of the Chicot-Lesage composite hardness model. *J. Serb. Chem. Soc.*, 87, 899–910.
15. Marmur, A. (2012). Hydro- Hygro- Oleo- Omni-Phobic? Terminology of Wettability classification. *Soft Matter*, 8, 6867.
16. Song, S., Guo, M., Zhang, S., Zhan, K., Yan, Y., Yang, J., Zhao, B., & Xu, M. (2020). Plasma-assisted synthesis of hierarchical NiCoxPy nanosheets as robust and stable electrocatalyst for hydrogen evolution reaction in both acidic and alkaline media. *Electrochim. Acta*, 331, 135431.
17. Guo, M., Song, S., Zhang, S., Yan, Y., Zhan, K., Yang, J., & Zhao, B. (2020). Fe-doped Ni-Co phosphide nanoplates with planar defects as an efficient bifunctional electrocatalyst for overall water splitting. *ACS Sustain. Chem. Eng.*, 8, 7436–7444.
18. Sun, H., Yan, Z., Liu, F., Xu, W., Cheng, F., & Chen, J. (2020). Self-supported transition-metal-based electrocatalysts for hydrogen and oxygen evolution. *Adv. Mater.*, 32, 1806326.
19. Pu, J., Shen, Z., Zhong, C., Zhou, Q., Liu, J., Zhu, J., & Zhang, H. (2020). Electrodeposition technologies for Li-based batteries: New frontiers of energy storage. *Adv. Mater.*, 32, 1903808.
20. Liu, J., Li, P., Bi, J., Zhu, Q., & Han, B. (2022). Design and preparation of electrocatalysts by electrodeposition for CO2 reduction. *Chem. Eur. J.*, 28, e202200242.
21. Ji, L., Wang, J., Teng, X., Meyer, T. J., & Chen, Z. (2020). CoP nanoframes as bifunctional electrocatalysts for efficient overall water splitting. *ACS Catal.*, 10, 412–419.
22. Gupta, S. K., & Misra, R. D. (2023). Experimental pool boiling heat transfer analysis with Copper–Alumina micro/nanostructured surfaces developed by a novel electrochemical deposition technique. *Int. J. Thermophys*,, 44, 112.
23. Triantou, K. I., Pantelis, D. I., Guipont, V., & Jeandin, M. (2015). Microstructure and tribological behavior of copper and composite copper+alumina cold sprayed coatings for various alumina contents. *Wear*, 336–337, 96–107.

24. Requena, G. C., & Degischer, H. P. (2022). Effects of particle reinforcement on Creep behaviour of AlSi1MgCu. *Int. J. Mater. Res.*, 96, 807–813.
25. Chen, P., Ye, J., Wang, H., Ouyang, L., & Zhu, M. (2021). Recent progress of transition metal carbides/nitrides for electrocatalytic water splitting. *J. Alloys Compd.*, 883, 160833.
26. Huang, Z., Chen, B., Ren, B., Tu, D., Wang, Z., Wang, C., Zheng, Y., Li, X., Wang, D., & Ren, Z. (2023). Smart mechanoluminescent phosphors: A review of strontium-aluminate-based materials, properties, and their advanced application technologies. *Adv. Sci.*, 10, 2204925.
27. Nemani, S. K., Annavarapu, R. K., Mohammadian, B., Raiyan, A., Heil, J., Haque, M. A., Abdelaal, A., & Sojoudi, H. (2018). Surface modification of polymers: Methods and applications. *Adv. Mater. Interfaces*, 5, 1801247.
28. Jabbarzadeh, A. (2019). The origins of enhanced and retarded crystallization in nanocomposite polymers. *Nanomaterials*, 9(10), 1472.
29. Dehghan, S., Sattarpanah Karganroudi, S., Echchakoui, S., & Barka, N. (2025). The integration of additive manufacturing into industry 4.0 and industry 5.0: A bibliometric analysis (trends, opportunities, and challenges). *Machines*, 13(1), 62.
30. Jabbarzadeh, A. (2019). The origins of enhanced and retarded crystallization in nanocomposite polymers. Nanomaterials, 9(10), 1472.

7 Principles of Composite Materials

7.1 INTRODUCTION

One of the most important developments in contemporary engineering has been the creation of composite materials that combine the advantages of two or more different components to enhance their overall properties. The matrix, which may be ceramic, metallic, or polymeric, is the main component of a composite material. As a continuous phase, the matrix transmits loads, maintains the integrity of the structure, gives it its shape, and protects it from weathering. The reinforcing phase, embedded within the matrix and typically consisting of fibers, particles, or whiskers, is responsible for enhancing certain properties, including stiffness, tensile strength, abrasion resistance, and thermal resistance. By carefully selecting and customizing these phases, engineers can produce composite materials that perform better than traditional homogeneous materials. One of the fundamental ideas of composite materials is the idea of load transfer between the reinforcement and matrix. A composite material can withstand higher mechanical loads because efficient stress distribution is ensured by effective bonding at the interface. Strength, fracture toughness, and toughness are directly affected by the quality of the interfacial bonding. The rule of thumb, which predicts the effective properties of a composite material based on the properties and proportions of its components, is a crucial concept. While this method provides a useful approximation, microstructural factors, such as fiber orientation, particle dispersion, and failure distribution, often determine actual performance [1–4].

Composites fall into several different groups. Fiber-reinforced composites, such as glass fiber or carbon fiber composites, are widely used in the automotive and aerospace industries and are known for their excellent strength-to-weight ratios. Particle-reinforced composites are suitable for protective coatings and cutting tools because they increase durability and wear resistance. For large-scale engineering construction, structural composites—such as insulating panels and laminates—are designed to maximize stiffness and stability. Each type is designed to meet specific performance requirements in a range of industrial applications. Composite materials have specific advantages, such as corrosion resistance, thermal stability, and the ability to combine multifunctional properties with mechanical advantages. For example, advanced polymer composites can be engineered to have self-healing properties, thermal insulation, or electrical conductivity. Due to their adaptability, these composites are

DOI: 10.1201/9781003758556-7

essential in many industries, including biomedical devices, sports equipment, marine engineering, and civil construction.

Sustainability and lightweight design are key factors in the growing importance of composite materials. Composite materials can reduce material use in construction and increase energy efficiency in transportation systems by reducing weight without compromising strength. However, research is still ongoing on issues such as cost, recyclability, and long-term durability. Careful selection, blending, and design of matrix and reinforcement phases form the foundation of composite materials. Because composite materials offer tailored solutions to modern engineering challenges through the use of surface bonding, load transfer, and microstructural optimization, they are a key component of innovative technologies and sustainable development. Their design concepts enable the construction of strong, lightweight, and multipurpose structures that meet the growing needs of contemporary industries. Meanwhile, advances in materials science have enabled the use of nanotechnology to create more advanced inventions [5–7].

Nanotechnology, also known as molecular nanotechnology, is the industrial application of materials at the atomic, molecular, and supramolecular levels. It was initially described as the precise control of atoms and molecules to produce large-scale products. However, over time, this definition has changed. The phrase was expanded by the National Nanotechnology Initiative to include the manipulation of materials at scales between 1 and 100 nm in at least one dimension. This expansion has made nanotechnology a broad field of study aimed at discovering the special properties of matter at the quantum level, in addition to achieving specific technological goals. Consequently, improved materials design and nanotechnology are now inextricably linked. Nanotechnology offers size-dependent phenomena, including quantum confinement, surface reactivity, and tunable conductivity, while composite materials rely on microstructure optimization and interface engineering. Today, the terms "nanotechnology" and "nanotechnologies" are widely used to refer to a wide range of studies and applications where nanoscale effects dominate material performance. By facilitating the production of stronger, lighter, more efficient, and more sustainable materials, the integration of composite materials and nanotechnology is impacting the future of engineering, ultimately bridging the gap between grand engineering problems and quantum breakthroughs.

Design and manufacturing are among the most promising areas of nanotechnology, offering significant benefits. Using nanotechnology can significantly extend the lifespan of automotive parts and other mechanical components. The unique physical and chemical properties of nanomaterials improve their quality, enhancing their mechanical strength, optical quality, and magnetic properties. Furthermore, researchers are constantly developing new raw materials that impart specific properties to materials, enabling the production of more reliable and efficient products.

Another important aspect of nanotechnology is its ability to control material properties at the nanoscale. Every material on our planet is composed of individual atoms and molecules, and advances in nanotechnology enable precise control of these basic units. Specifically, atomic force microscopy (AFM) and lateral force microscopy (LFM) studies of molybdenum disulfide (MoS_2A) have revealed a smooth surface.

However, when the material is examined using conductive atomic force microscopy (C-AFM), irregular side current injection is observed. Upon interaction of the MoS_2 surface with a boron-doped diamond conductive tip at the nanoscale, some dark circular patterns with sizes between 3 and 4 nm were detected.

Due to their higher current levels compared to surrounding locations, these properties indicated a lower tip/MoS_2 contact resistance. These high-current structures are nanoscale in nature, with areal densities ranging from 10^{10} cm² to 10^{11} cm², according to other research. To better understand these high-current locations in the scanning tunneling microscope (STM) maps, additional tests are being conducted by overlaying a MoS_2 network over the circular pits depicted. In light of all this, the use of nanotechnology in materials science and engineering continues to create new opportunities for the development of more durable, efficient, and highly flexible materials [8–10].

As research in this field advances, the potential for groundbreaking discoveries in industrial applications will increase. Figure 7.1a shows the SBH distribution map, obtained from a set of 128 × 128 I–V curves on a freshly sectioned MoS_2 surface using a conductive diamond tip. More specifically, as shown in Figure 7.1c, local ΦB values were determined by fitting the low forward bias region of each I–V curve to a thermionic emission model. On this high-resolution SBH map, subsurface defect locations exhibited minimum ΦB values of around 0.3 eV, while the surrounding regions exhibited a relatively constant SBH of around 0.53 eV.

The surface density of the detected pits, which can be measured as $(8 \pm 3) \times 10^{10}$ cm², closely matches the density of the conductive structures shown in the C-AFM images. The average size of these lateral pits is 5 nm. Based on experimental studies, the researchers hypothesized that the features shown in the STM maps are caused by electronic processes rather than topographic changes. This conclusion is based on the absence of these pits in the topographic and lanthanum strontium manganite (LSM) images, suggesting that subsurface defects, rather than surface deformations, are the cause of these anomalies.

Although nanoscience and nanotechnology are relatively new, the use of microscopic materials has been known for generations. However, this has led to advanced microscopic methods that enable scientists to observe objects at the extremely precise nanoscale. The distinctive properties of nanoparticles, which offer advantages such as enhanced strength, light weight, improved control of the light spectrum, and stronger chemical reactivity compared to their larger counterparts, are driving the growing interest in nanotechnology. To optimize their properties, modern scientists and engineers continue to research a wide range of nanomaterials for various applications. The study of electrical conductivity at the nanoscale is one of the most important areas of nanotechnology research. For example, in a single-layer MoS_2 film, synthesized by chemical vapor deposition (CVD) on a SiO_2/Si substrate, the conductivity loss associated with grain boundaries (GBs) is evaluated using nanoscale current mapping using C-AFM [5]. An optical micrograph of synthetic MoS_2 domains, which are typically triangular in shape (Figure 7.2a), shows that their sizes reach 50 μm. This image illustrates how annealing of MoS_2 domains leads to the formation of GBs. As illustrated in Figure 7.2b, the experimental setup for C-AFM measurements uses a

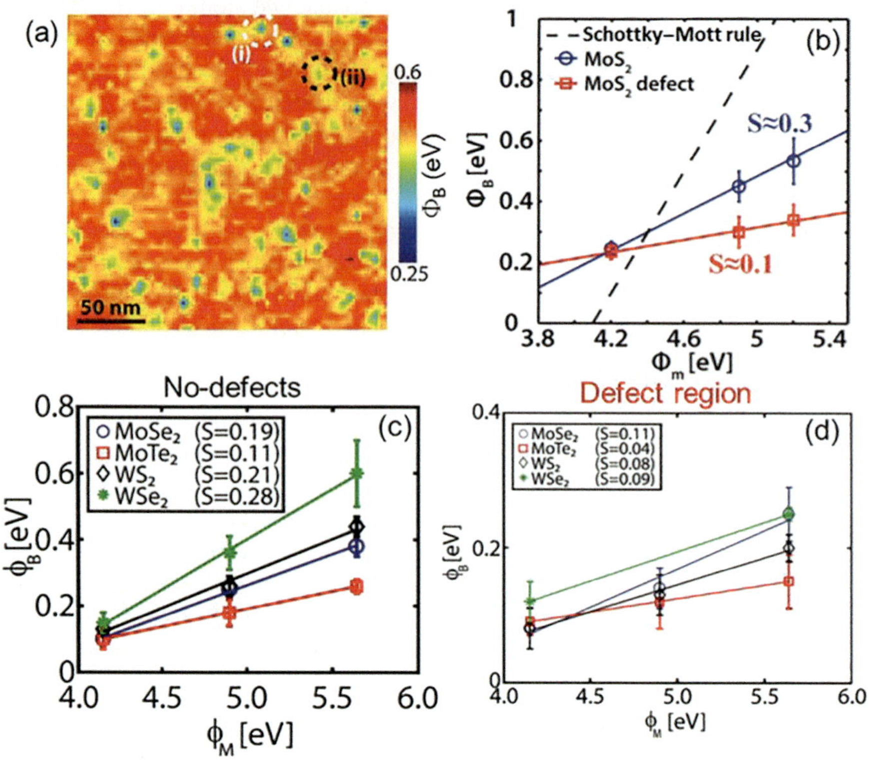

FIGURE 7.1 (a) Schottky barrier height (SBH) map on the MoS_2 surface, produced using the thermionic emission model to fit a set of local current–voltage (I–V) curves with dimensions of 128 × 128. (b) SBH versus metal tip work function (ΦM) for defect-free MoS_2 regions (blue) and on subsurface defects (red). The slope of the linear fit was used to calculate the pinning factor S. The dotted line represents the typical Schottky–Mott rule. Metal tip work function versus SBH for several TMDs ($MoSe_2$, $MoTe_2$, WS_2, WSe_2) for defect-free (c) and defective (d) regions. (Reproduced from reference [45]. With permission, panels (a) and (b) are copyright 2017 The Chemical Society. Panels (b) and (c) [1] [Open access]).

platinum–iridium alloy (PtIr) tip to create a nanoscale contact with a macroscopic electrode partially overlapping several MoS_2 domains. This contact has proven to be vital for current measurements. With an applied tip voltage bias (V_{tip}) = 1 V, the surface morphology and associated current map are shown in Figure 7.2c and d. The area of interest, indicated by the red rectangle in Figure 7.2b, contains two MoS_2 domains (denoted D1 and D2). The importance of the two-domain contact for electrical performance is highlighted by the intriguing observation of direct contact between the D1 domain and the macroscopic electrode.

This demonstrates how nanotechnology is increasingly capable of understanding and altering electrical properties at the atomic level. With advances in nanoimaging and materials manufacturing techniques, new opportunities for technological

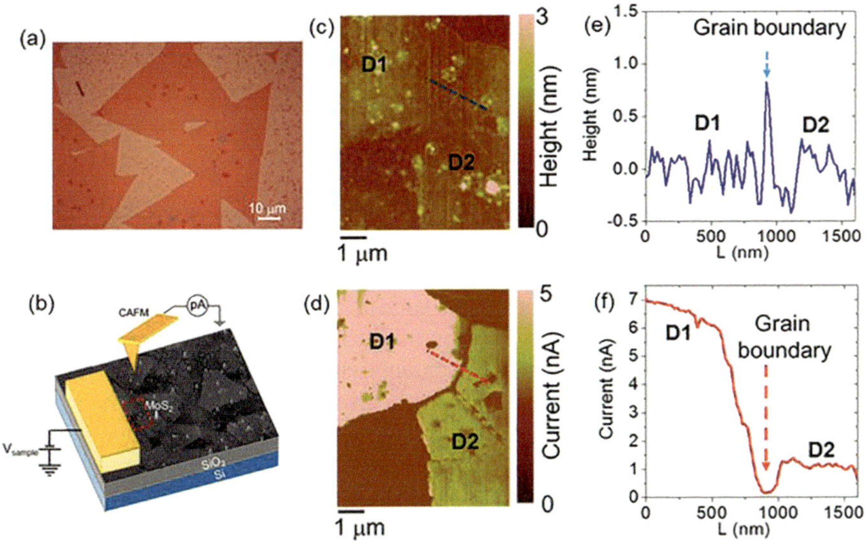

FIGURE 7.2 (a) Single-layer MoS_2 domains grown by CVD on a SiO_2/Si substrate as observed by optical microscopy. (b) Schematic illustration of the experimental setup used to detect the local current using C-AFM on MoS_2 domains. (c) C-AFM morphology measurements and (d) current map in a region containing two MoS_2 domains (D1 and D2) separated by a grain boundary. The macroscopic contact is directly connected to the D1 domain. The height scan (e) and the current line across the grain boundary (f) are shown [1] [Open access].

development will open up, increasing the potential for producing high-performance nanomaterials for a range of applications, such as electronics, energy storage, and optoelectronic devices [11,12].

The GB separates the two domains. As shown in the GB map, both appear in the image (as a sharp peak about 1 nm high). There is a marked decrease in conductivity from domain D1 to domain D2, which the GB divides, even though the current change is nearly constant between the two domains. This is also illustrated by the two scan lines in Figure 7.2e and f, taken from the image and the current map, respectively. In many applications, producing nanomaterials with suitable properties and high performance is essential. Generally, there are two approaches to producing nanomaterials: top-down and bottom-up. In the top-down approach, structures and tools are fabricated from bulk materials without precise atomic control. A major advantage of this approach is its ability to be mass-produced in an industrial setting [3]. However, its major drawbacks are its high cost and energy consumption, long processing times, and surface structure distortion [4]. Methods for this top-down approach include mechanical milling, cutting, chemical etching, laser ablation, electroblasting, and lithography. In the top-down approach, a modifiable nanostructure is constructed by assembling atomic-scale components.

7.2 CLASSIFICATION OF COMPOSITES

Depending on the internal structure, reinforcement type, and matrix type, composite materials can be classified in several ways. The matrix material and reinforcement type are the most common classification systems, and each directly impacts the strength, performance, and applications of the final product.

- *Polymer Matrix Composites (PMCs):* Because polymers are lightweight, easy to handle, and more affordable than metals or ceramics, PMCs are the most common type. PMCs typically use thermoplastics such as nylon and polypropylene, or thermosetting resins such as epoxy, polyester, and phenolic. To improve strength, stiffness, and thermal stability, the reinforcements are often made of glass fibers, carbon fibers, aramid fibers, or nanoparticles. Due to their excellent strength-to-weight ratio and corrosion resistance, PMCs are highly useful in the automotive, sporting goods, marine, and aerospace industries. High-performance automobiles, wind turbine blades, and aircraft structures now use advanced PMCs, particularly carbon fiber-reinforced polymers.
- *Metal Matrix Composites (MMCs):* MMCs focus on applications requiring high strength and stiffness, thermal conductivity, and corrosion resistance at high temperatures. Common matrices consist of copper, magnesium, titanium, and aluminum alloys reinforced with continuous or short fibers or ceramic particles (SiC, AlO_3). MMCs are used in aircraft construction, brake rotors, engine components, and military applications that require strong, lightweight, and thermally stable materials. MMCs are suitable for automotive and aircraft propulsion system components because they are heavier than PMCs but can withstand higher loads and temperatures.
- *Ceramic Matrix Composites (CMCs):* CMCs are chosen for their high hardness, corrosion and oxidation resistance, and thermal stability. Their main drawback is their brittleness, which can be mitigated by adding fibers such as carbon, alumina, or silicon carbide to reinforce the matrix. Composite multi-surface composites (CMCs) are widely used in nuclear reactors, gas turbines, and aircraft engines, where high temperatures make traditional metals and polymers incompatible. These composites perform better under high temperatures and high stress conditions, have a longer service life, and weigh less than metals.
- *Classification by Reinforcement Type:* Composites are classified according to the reinforcement form and material. Fiber-reinforced composites (whether short or continuous fibers) offer superior stiffness and strength, making them suitable for load-bearing applications. Particle-reinforced composites, commonly used in cutting tools or automotive brake pads, use ceramic or metal particles to improve uniformity, stiffness, and wear resistance. Laminated composites offer customized anisotropic properties by being composed of multiple layers bonded in different directions. Short, high-aspect-ratio crystals are used in whisker-reinforced composites to significantly increase strength and stiffness. The classification of composites demonstrates their versatility across a wide range of sectors. Glass fiber-reinforced composites (MMC)

perform effectively in high-temperature and high-performance environments, while polycarbonate-reinforced composites (PMCs) are the industry standard for lightweight structures, and polycarbonate-reinforced composites (CMCs) are the industry standard for harsh applications such as reactors and turbines. Similarly, engineers can create composites with properties tailored to specific mechanical, thermal, or functional needs using reinforcement classification. Understanding these categories aids in material selection and ensures the continued importance of composite materials in innovating modern engineering solutions. Meanwhile, nanotechnology, through the classification of nanomaterials, adds a new and powerful level of interest to materials science. Organic and inorganic nanomaterials are two main categories, each with unique properties that enhance the functionality of composite materials and open up new opportunities for multipurpose systems.

Because of their extraordinary properties, organic nanomaterials—particularly those based on carbon—have received significant attention. Single-walled carbon nanotubes (SWCNTs), graphene, fullerenes, and buckyballs are examples of carbon nanostructures characterized by exceptional mechanical strength, high electrical conductivity, and superior thermal stability. These materials are ideal reinforcements for polymeric, metallic, and ceramic composites due to their high aspect ratios, light weight, and good interaction with matrix materials. For example, carbon nanotube (CNT)-based composites combine flexibility with exceptional mechanical performance, while graphene-enhanced composites boast enhanced durability and electrical conductivity. Conversely, inorganic nanomaterials, including silica, clay nanoparticles, and metal oxides, offer unique advantages such as high stiffness, chemical stability, and heat resistance. These properties are essential in applications including power systems, catalytic reactors, and aircraft structures, where composite materials must withstand harsh operating conditions. Furthermore, the incorporation of organic and inorganic nanoparticles into hybrid composite materials enables the design of new materials that combine lightweight performance with multifunctional performance [13–15].

Thus, linking the classification of nanomaterials and composites demonstrates how the two fields converge to offer tailored engineering solutions. By improving durability, conductivity, and sustainability, the incorporation of nanoparticles into composite matrices keeps industries evolving and ensures that advanced materials remain at the forefront of technological advancements.

One of the most significant barriers to sustainable development is ensuring efficient energy conversion and storage, which requires the development of low-cost electrochemical devices. Although the performance of these devices can be significantly improved using nanomaterials, large-scale manufacturing remains a significant barrier [15]. To address this problem, researchers have developed several synthesis techniques, including solution-based synthesis, template-based synthesis, and microwave synthesis. These techniques have the potential to reduce manufacturing costs while producing high-quality nanomaterials in large quantities.

Although nanoparticles offer numerous benefits, electrochemical devices present a significant obstacle to their use. To improve performance and efficiency, the interactions of nanomaterials with other device components must be carefully optimized, as their properties differ significantly from those of their raw material counterparts [16]. In electrochemical devices, the way nanoparticles interact with other materials is critical, as it has a direct impact on stability, charge transport, and overall performance. Without proper integration, the full potential of nanomaterials in energy conversion and storage technologies cannot be realized.

Integration with Large-scale Barriers: The production of these materials for practical applications is one of the major barriers to the widespread adoption of nanoparticles. Although much research has been conducted to improve manufacturing techniques, such as solution-based manufacturing, template-based manufacturing, and microwave manufacturing, problems of scalability, consistency, and cost-effectiveness still exist [17]. Furthermore, as the performance of electrochemical devices is highly dependent on the interactions between nanomaterials and other device components, their integration needs to be improved [18].

To better understand the properties of nanomaterials, characterization is performed using a variety of microscopic techniques. Electrostatic force microscopy (EFM) and scanning electron microscopy (SEM) are used to analyze the size and shape of silver nanoparticles. Most of the silver nanoparticles appeared spherical in the SEM images, as shown in Figure 7.3b. In addition, energy-dispersive X-ray spectroscopy (EDS) demonstrates the effectiveness of the silver nanoparticle synthesis (Figure 7.3a).

To analyze silver nanoparticles coated with conductive material PCPDTBT, a scanning electron microscope (SEM) was used. Due to its excellent photovoltaic properties, poly[2,6-(4,4-bis(2-ethylhexyl)-4H-cyclopenta[2,1-b;3,4-b′] dithiophene)-alt-4,7-(2,1,3-benzothiadiazole)] is a semiconducting organic polymer often used in organic solar cells (OPVs) and organic electronics. The topography of silver nanoparticles/PCPDTBT was obtained using a scanning Kelvin microscope (SKPM) in conjunction with SEM and EFM images, as shown in Figure 7.3c and d. As the silver nanoparticles in the SKPM and EFM images remain the same size, the results show that the resolution of AFM, one of the most widely used scanning microscopy techniques, is very similar to that of SEM. This consistency demonstrates the flexibility and reliability of the data [19].

Electrochemical devices, particularly those for energy conversion and storage, can be greatly enhanced using nanomaterials. However, to realize their full potential, issues such as efficient integration and large-scale manufacturing must be addressed. To overcome these challenges, characterization and synthesis techniques must be improved. As research advances, nanotechnology will continue to revolutionize materials science, paving the way for more efficient and sustainable electrochemical technologies.

To fully utilize nanomaterials in electrochemical systems, their integration must be optimized to achieve optimal performance. Nanoparticles can significantly enhance the performance and efficiency of these devices due to their unique properties [20]. Nanomaterials can be divided into different categories according to their dimensions. One-dimensional (1D) nanomaterials, such as CNTs, can be wires, fibers, tubes, or rods. 2D nanomaterials, often found in sheets or films, are graphene and graphene

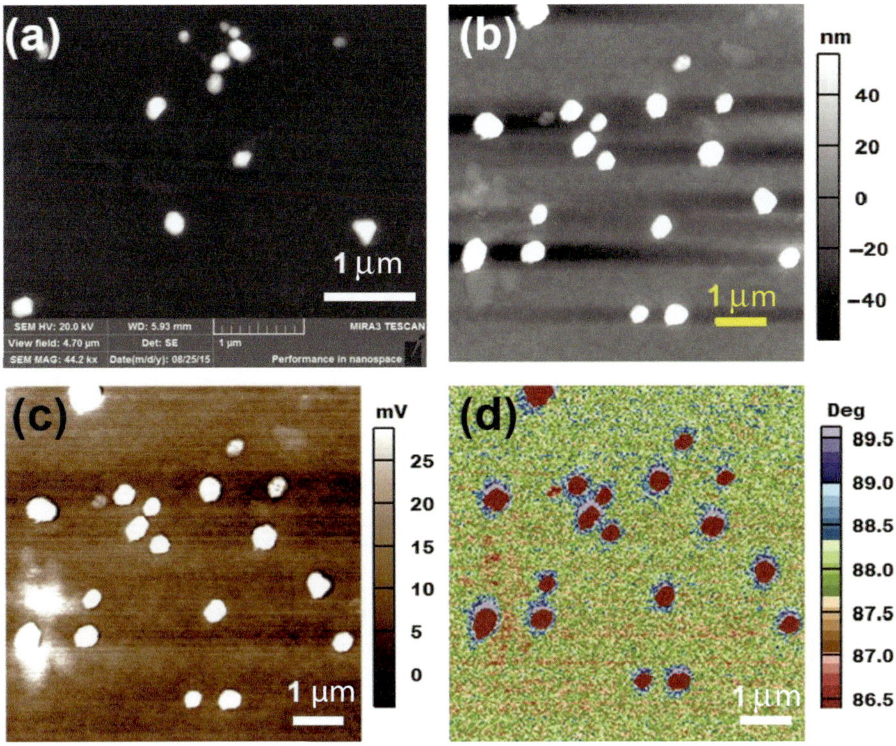

FIGURE 7.3 (a) Silver nanoparticles imaged by SEM, (b) AFM, (c) surface potential and (d) EFM image of silver nanoparticles/poly[2,1-b;3,4-b′]dithiophene)-alt-4,7-(2,1,3-benzothiadiazole)][2,6-(4,4-bis-(2-ethylhexyl)-4H-cyclopenta] (PCPDTBT) sample at 488 nm. VEFM is equivalent to 6 V. The scale bar is 1 µm [2] [Open access].

oxide (GO). Furthermore, 3D nanomaterials, such as graphite and diamond, feature completely 3D structures. These nanoparticles come in a variety of sizes and can be either crystalline or amorphous, as shown in Figure 7.4.

This section focuses on carbon nanomaterials, which have received significant attention due to their multiple applications in a range of industries. Figure 7.4a shows SEM analysis of silver nanoparticles on PCPDTBT films. The nanoscale size of the silver nanoparticles is approximately 120 nm, as shown in Figure 7.4d. The surface potential of the Ag–PCPDTBT composite is measured using SKPM. Figure 7.4b shows the surface potential of the composite in the absence of light excitation, while Figure 7.4c shows the surface potential in the presence of 488 nm radiation. The results demonstrate that the high surface potential of the films cannot be protected from the surface charge separation of the Ag–PCPDTBT composite alone, as the conductive polymer is continuously coated with silver nanoparticles. This indicates that the interface directly reflects light-induced charge separation and recombination at the reported surface potential. The brighter regions of silver nanoparticles in the dark-field SKPM images (Figure 7.4b) show that the particle surface potential is

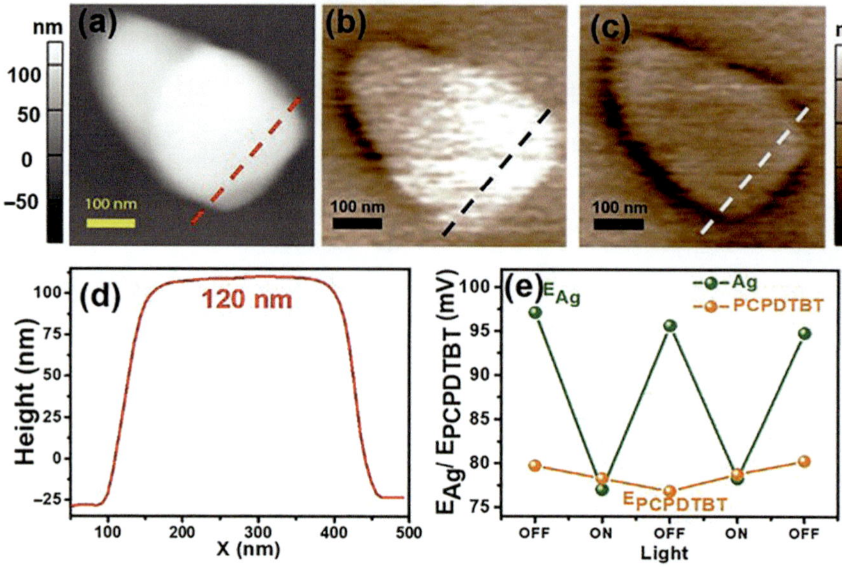

FIGURE 7.4 (a) AFM topographic of Ag NPs on PCPDTBT films; (b) Surface potential AFM images of an Ag nanoparticle on a PCPDTBT film without light excitation and (c) with light excitation at a wavelength of 488 nm; (d) Table 120 nm; (e) The light-modulated surface potential of the Ag nanoparticle on PCPDTBT films [2] [Open access].

higher than that of the surrounding medium. Specifically, the silver nanoparticle surface potential is 96 ± 3 mV, while the PCPDTBT polymer surface potential is slightly lower, 79 ± 2 mV. This demonstrates the critical role that surface potential plays in the behavior of nanomaterials in electrochemical devices and emphasizes the importance of optimizing their surface interactions to improve device performance [22].

7.3 TYPES OF NANOMATERIALS

The two main groups of nanomaterials are inorganic and organic. Carbon structures, including fullerene, graphene, SWCNTs, and buckyballs, are the most common types of organic nanomaterials. The remarkable mechanical, electrical, and thermal properties of these materials have attracted much interest. The development of affordable electrochemical devices is essential to ensure efficient energy conversion and storage, which is one of the major challenges in sustainable development. Nanomaterials can significantly improve the performance of these devices, but large-scale production still remains a major challenge [5]. The researchers have created a number of synthesis methods, such as solution-based, template-based, and microwave synthesis, to address this issue. High-quality nanomaterials can be produced in large quantities using these methods, which also have the potential to reduce manufacturing costs.

While nanoparticles have several advantages, their incorporation into another significant challenge is electrochemical devices. As the characteristics of nanomaterials are very different from those of their bulk counterparts, it is essential to carefully optimize their interactions with other parts of the device to maximize performance and efficiency [6]. The interaction of nanoparticles with other materials is critical in electrochemical devices as it directly affects stability, charge transport, and overall performance. The full potential of nanomaterials in energy conversion and storage technologies cannot be achieved without proper integration.

7.3.1 Fullerene (0D)

Fullerenes are a special type of 0D nanomaterial consisting only of carbon atoms grouped in a hollow, closed configuration. Because it resembles a soccer ball, the best-known fullerene, C_{60}, is often called a buckyball. With 20 hexagons and 12 pentagons, this molecule has a perfectly symmetrical spherical shape. Since their discovery in 1985 by Richard Smalley, Robert Curl, and Harold Kroto [19], fullerenes have attracted much interest due to their exceptional mechanical, optical, and electrical properties. Their unique structure allows them to exhibit high stability, electrical conductivity, and the ability to interact with a variety of chemical compounds, making them valuable in many scientific and industrial applications.

Despite their fascinating properties, fullerenes face certain challenges, particularly their low solubility in polar solvents, which limits their immediate use in some situations. However, fullerene structures have been modified by advances in nanotechnology, increasing their potential in areas such as energy storage, medicine, and nanotechnology. Fullerenes are currently being studied for their potential use in superconductors, drug delivery systems, and even nanomachines. Along with CNTs and graphene, fullerenes continue to be key components in the development of carbon-based nanomaterials and modern nanotechnology. The buckyball is the most famous of many interesting carbon-based nanostructures known as fullerenes. Buckyballs are fullerenes that take on a spherical shape. Richard Smalley [21] discovered the fullerene molecule, which has attracted much attention due to its special properties. These molecules are large, triangular in shape, reminiscent of spheroids, and can form C–C bonds. The fullerene appears symmetrical due to its structure, which consists of 12 pentagons and 20 hexagons. Fullerene is an interesting molecule to study due to its unique symmetry and reactivity with many substances. For example, the spherical shape of C_{60} fullerene is often used as a “wheel” in the construction of nano cars and nanovesicles due to the peculiar geometry of this structure. However, the insolubility of fullerene in polar solvents limits its wider application in various industries, which in turn limits its practical usefulness. Fullerene derivatives have received less attention in recent years as research has shifted toward CNTs and graphene-based materials [22–24]. However, fullerene research is still ongoing. Phenyl-C70-butyric acid methyl ester (PC70BM)-rich domains undergo phase separation when fullerene derivatives are dissolved in a poorer solvent, as shown in Figure 7.5. Combined with the decrease in PC70BM solubility, this effect is particularly noticeable when using a 1:3 mixture of ortho-dichlorobenzene and chloroform (oDCB and CF), which shows higher separation than a 1:1 mixture. Further analysis of the emission spectra of fullerene mixtures, as shown in Figure 7.5, reveals some intriguing results. A primary

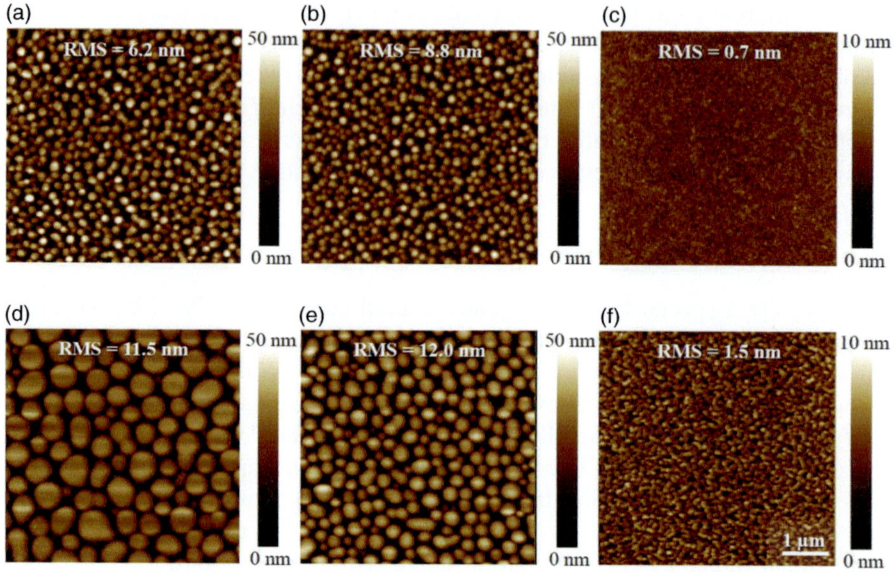

FIGURE 7.5 TQ1:PC70BM (a–c) and 1:3 TQ1:PC70BM (d–f) thin films spin-coated with chloroform (CF) (a, d), chlorobenzene (CB) (b, e), and ortho-dichlorobenzene (oDCB) (c, f). The micrographs are obtained by AFM (5 × 5 μm). The scale bar represents 1 μm. The surface roughness (RMS) has been added to each image along with the corresponding height scale [3] [Open access].

TQ1 is a hole-donating polymer (donor polymer) widely sought in organic solar cells (OPVs) and organic electronics due to its photovoltaic canceling properties emission peak at 700 nm is visible when the mixture films are excited at 380 nm. A broad emission band peaking between 520 and 550 nm also appears. Interestingly, this emission band is absent when TQ1 is in its pure film form, which shows no spectrum, or when it is dissolved in solvents such as CF, CB, or an analog-to-digital converter (ADC). This demonstrates the different interactions that fullerene derivatives exhibit under different solvent conditions [3].

By changing the spin coating rate (Figure 7.6). As can be seen from AFM images (Figure 7.6a), thicker films with slower spin coating exhibit larger PC70BM-rich domains. The obtained emission spectra (Figure 7.6b) show that the emission maximum changes from 520 to about 540 nm with increasing film thickness from 90 to 130 nm. The emission range does not change significantly with decreasing film thickness from 90 to 50 nm. This indicates that the domains must grow to a certain size for co-emission to become noticeable. The relationship between the morphology of TQ1:PC70BM blends and their in-depth emission spectrum analysis is investigated. Greater phase separation occurs when solvents with lower PC70BM solubility and higher PC70BM concentrations are used. PC70BM aggregates are expected to form in these larger domains, resulting in a fairly broad and soft emission at 380 nm excitation between 520 and 550 nm. The fullerene derivative remains well mixed with the

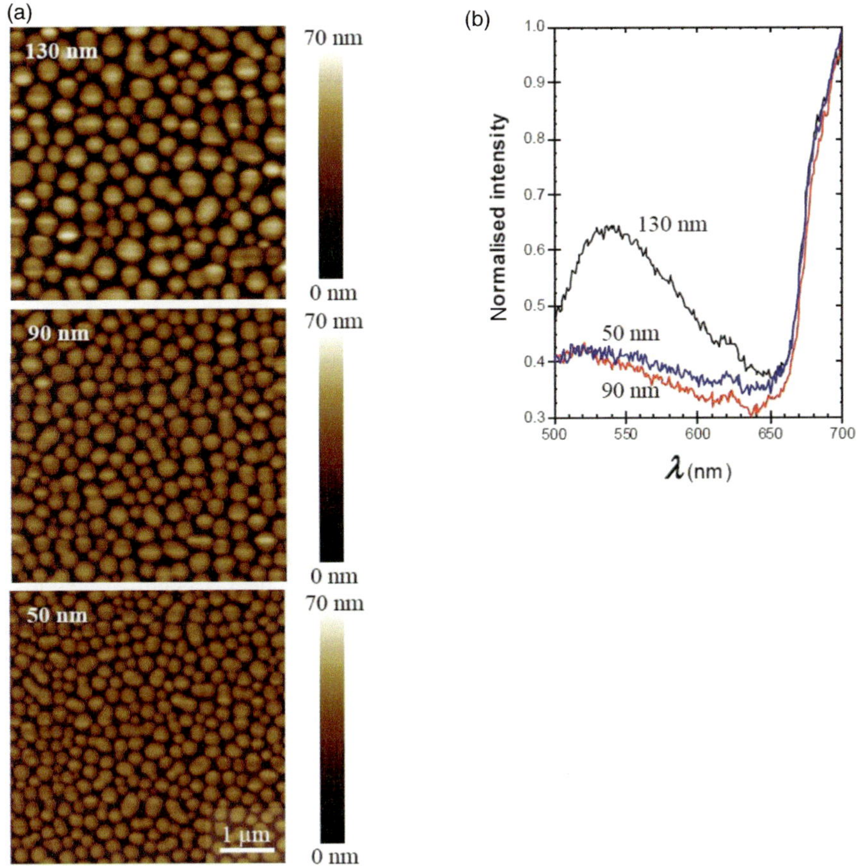

FIGURE 7.6 (a) 5 µm AFM micrographs of 1:3 spin-coated TQ1:PC70BM CB thin films with the indicated film thickness. A height scale is added to each image. (b) Fluorescence emission spectra corresponding to the films with the corresponding thickness [3] [Open access].

polymer in superior PC70BM solvents and at low PC70BM concentrations, and no emission is observed between 500 and 550 nm [3].

Depending on the size of different shapes. C70 is ovoid rather than a sphere. Fullerene contains up to 2200 carbon atoms. In 1970, Osawa [8] was the first to mention fullerene in the literature after discovering that corannulene is a substructure of C60. As a result, two important isomers are formed during pyrolysis: C60 and C70. Various synthetic parameters can be modified to change their ratio. The flame process has been used to produce fullerene on a large scale. Fullerenes of different sizes are purified by another method called high-performance liquid chromatography. It contains triply degenerate lowest unoccupied molecular orbital (LUMO) orbitals that accommodate six electrons upon electrochemical reduction due to its structure, exceptional electronic and electrochemical capabilities, and specific gravity. This characteristic makes C60 a specific electron acceptor for various purposes, especially in electronics. Due to the doubly degenerate LUMO orbitals in C70, which are capable of holding up to four electrons, the nondegenerate LUMO+1 level is almost ready to accept two more electrons. C60 is used as a building block of organic electronics.

7.3.2 Nanotubes of Carbon (1D)

Sheets of graphene are rolled to create hollow cylindrical structures that constitute a special class of 1D nanomaterials known as CNTs. These nanoparticles are among the most widely studied materials in nanotechnology due to their remarkable mechanical, electrical, and thermal capabilities.

SWCNTs, which have a single layer of graphene, and multi-walled carbon nanotubes (MWCNTs), which have multiple concentric layers of graphene, are the two main types of CNTs that were first identified in 1991 by Sumio Iijima [9]. CNTs have many applications in electronics, medicine, energy storage, and composite materials due to their high aspect ratio, low weight, and exceptional strength.

Depending on their structural configuration, the electrical conductivity of CNTs can be either metallic or semiconducting, making it one of their most amazing characteristics. This makes them very promising for conductive coatings, transistors, and nanoelectronics. In addition, their exceptional mechanical strength—much higher than that of steel—allows them to be used in flexible electronics, aircraft technology, and reinforced materials. Despite their potential, problems remain, such as the difficulty of producing them on a large scale, purifying them, and integrating them into devices.

CNTs are among the most significant nanomaterials in modern science and technology, although further study will expand the range of their applications. Materials with 1D structures exist at the 2D nanoscale. Nanotubes, nanofibers, nanowires, nanorods, and nanothreads are just a few examples. As electrons are limited to two dimensions, they cannot move freely. Figure 7.7 (pin 2 in Figure 7.7) and Figures S1 and S2 (pins 1 and 3 in Figure 7.7) in the supplementary material illustrate the results of successful stabilization. SEM images of CNT fiber probes are shown in each of these figures before and after the treatment (Figure 7.7e), S1e, and S2e show AFM images of CNT-coated silicon surface (prepared on silicon surface using CVD as described in section 4.4), which are obtained to demonstrate the stable imaging of CNT fiber probes [4]. Similar to 0D materials, these 1D nanomaterials can be metallic, ceramic, polymeric, amorphous, mono- or polycrystalline. However, 1D nanomaterials can have pure or impure chemical composition (e.g., doped semiconductors). These elements can be displayed separately or integrated into another medium; their length exceeds the width of 1D material [9–11].

7.3.3 Graphene and Its Derivatives (2D)

As air permeates the entire world and is essential for all living things, it is one of the most important components of life. Compared to other types of pollution, air pollution has a more immediate and serious impact on human health, beause breathing is a constant and necessary process [12]. By releasing hazardous chemicals into the atmosphere, such as carbon dioxide (CO_2), nitrogen oxides (NO, NO_2), hydrogen sulfide (H_2S), sulfur dioxide (SO_2), and particulate matter (PM), the increasing pace of

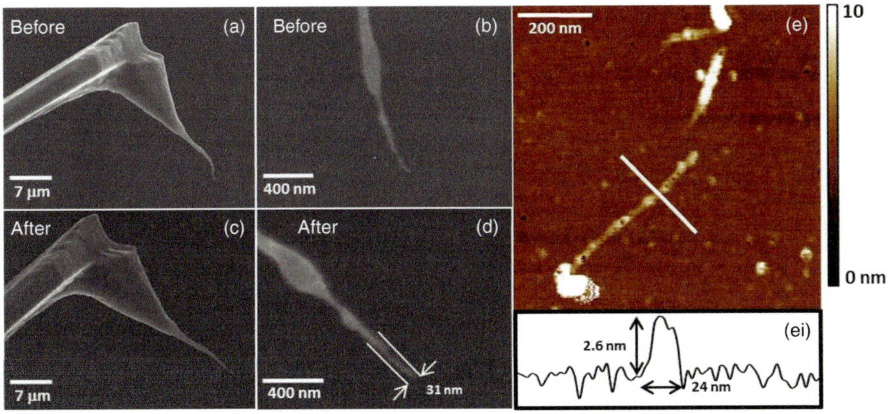

FIGURE 7.7 SEM images of Tip 2 before (a, b) and after (c, d) processing. After processing, (e) cross-section corresponding to the white line. Atomic number microscope image of carbon nanotube on carbon nanotube-coated silicon surface using Tip 2. The carbon nanotube fibers are significantly straightened as can be seen from the comparison of SEM images before and after [4] [Open access].

industrialization has significantly increased air pollution. In addition to reducing air quality, these pollutants are extremely harmful to human health. Numerous diseases, such as conjunctivitis, respiratory infections, allergies, tumors, dry eyes, vascular coagulation, and cardiovascular diseases, have been associated with air pollution [3]. While CO_2 is often considered a less hazardous gas, high concentrations of over 7.2% can cause respiratory discomfort, including rapid breathing. In addition, carbon monoxide (CO) is particularly harmful because it reduces oxygen delivery and causes toxicity by competing with oxygen (O_2) for binding to hemoglobin (Hb) in the blood.

In the initial morphological study, the molecular nucleation of submonolayer films (t = 1 nm) on a graphene substrate was studied to learn more about how air pollution affects materials and their interactions with the environment. The growth patterns and structure of organic thin films can be affected by industrial pollutants. The growth of organic thin films (Figure 7.8a and b) typically occurs along substrate defects such as graphene wrinkles, which are revealed by morphological examination as shown in Figure 7.8c. Different organic materials had different nucleation densities; PDIF-CN_2 shows more sparse nucleation sites than PDI8-CN_2, indicating a stronger molecular effect. Moreover, the average height of characteristic line profiles of PDIF-CN_2 grains is 3.3–3.6 nm (Figure 7.8e), which is consistent with previously documented organic molecular architectures [5]. We can learn more about how air pollution affects industrial materials, atmospheric chemistry, and human health by investigating it at the environmental and molecular levels. Creating effective air quality monitoring methods and advanced nanomaterials to detect and filter pollutants remains a vital scientific challenge as pollution levels continue to rise.

The main cause of climate change is CO_2, which is from flue gases, and must be captured and stored to reduce its concentration in the atmosphere. For this reason,

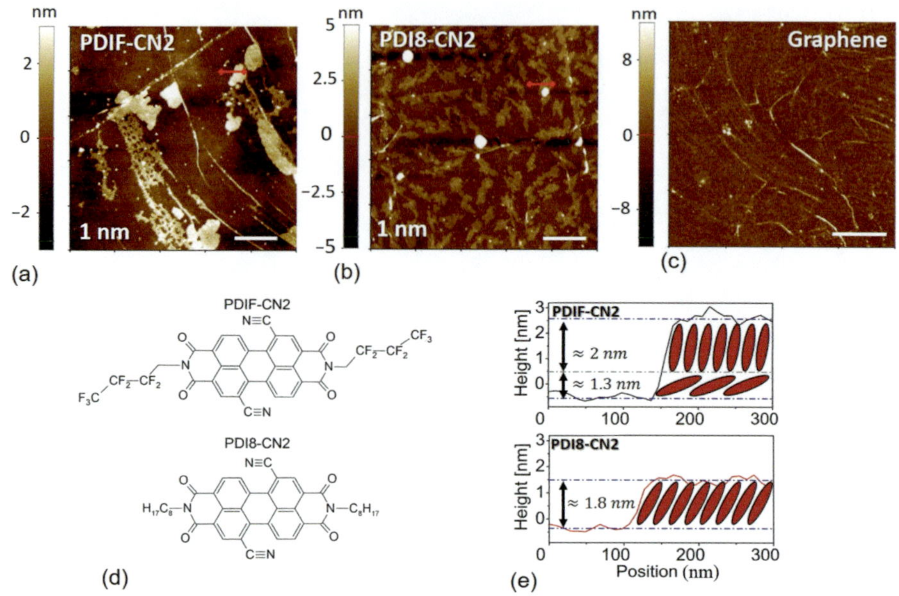

FIGURE 7.8 (a, b) 5 nm × 5 nm NC-AFM topography of PDIF-CN_2 and PDI8-CN_2 submonolayers grown on graphene surfaces with a nominal thickness of 1 nm. The white lines of the panels serve as the 1 nm reference. (c) The reference surface is pristine CVD graphene (10 μm × 10 μm, where the white bar represents the 2.5 μm reference), highlighting the discontinuous occurrence of wrinkles. (d) Molecular structures of perylene diimide derivatives used in this study. (e) Line profiles of the two materials corresponding to the red lines in panels (a, b). In PDIF-CN_2 example, the nearly downward-facing molecular wettability layer (≈1.3 nm) can be easily inferred [5] [Open access].

many techniques are being explored, including cryogenics, adsorption, absorption, and membranes [4,5]. Adsorption is one such tactic that has received a lot of attention. The key factor here is how easy it is to regenerate the adsorbent. CO_2 adsorbents should have certain qualities such as a higher degree of CO_2 adsorption capacity and selectivity, fast adsorption/desorption kinetics, moderate enthalpic heat, and good stability [6].

Many materials are reported to have the following properties: Activated carbon, CNTs, carbon spheres, metal–organic frameworks, graphene oxide, graphite surfaces, graphene-based porous adsorbents, zeolite-based molecular sieves, etc., are excellent options for CO_2 adsorption [7]. PM is a general term used to describe air pollutants that vary in composition and size as a result of a variety of anthropogenic activities [19]. PM consists of solid or liquid particles suspended in the air. PM2.5 and PM10 are two broad categories into which PM pollutants can be divided based on aerodynamic diameter. PM2.5 refers to pollutants with a diameter less than or equal to 2.5 μm, and PM10 refers to pollutants with a diameter less than or equal to 10 μm [10].

Given that PM2.5 damages the cardiovascular system immediately after exposure, it is more dangerous than PM10. In addition to its large surface area, PM2.5 contains a large number of pathogens, which increases the incidence of disease [2–4]. Their penetrating ability is great, and they are capable of transporting various harmful compounds that damage organs and cause cell death. In addition, it impairs renal function and causes renal failure [5]. In the presence of PM, the rate of oxygen transfer by hemoglobin is reduced [9].

7.3.4 Graphite (3D)

The 3D nanomaterial called graphite is made up of stacked layers of graphene, each made up of carbon atoms arranged in a hexagonal lattice. Its unique mechanical, electrical, and thermal properties make it one of the most popular and stable carbon-based materials.

In its natural mineral form, graphite is widely used in a variety of industries such as manufacturing, electronics, energy storage, and lubrication. It is a critical component in the production of pencils, batteries, and electrodes due to its layered structure, which facilitates easy interlayer sliding and makes it a good lubricant. The creation of energy storage devices such as lithium–ion batteries (LIBs) and supercapacitors is one of the most important applications of graphite. Due to its exceptional chemical stability and electrical conductivity, graphite is a preferred material for battery anodes, which are needed to increase storage capacity and energy efficiency. Its thermal stability also makes it a critical component in high-temperature industrial applications. Graphite research is still evolving despite its widespread use, especially in nanotechnology.

In the fields of sensors, energy storage, biomedicine, electronics, solar cells, optics, etc., they have found a wide range of applications [16]. By changing materials at the nanoscale, science can improve the performance and efficiency of energy storage systems. Figure 7.9a depicts an overview of vertically stacked graphene (VG) development mechanism using rf-PECVD on soda-lime glass. Initially, radio frequency (RF) discharge separates carbon source CH_4 into highly reactive carbon species (CHx, C_2Hy, C, and H atoms).

Developments in graphite exfoliation techniques have produced graphene, a single sheet of graphite with exceptional mechanical strength, flexibility, and electrical conductivity. Consequently, research is still being conducted to improve graphite's properties and expand its use in advanced technologies, keeping it at the forefront of materials science. The manipulation and control of matter at the nanoscale (size range, 1–100 nm) are the focus of the scientific and technological field known as nanotechnology [11]. Reactive species are first adsorbed onto the glass surface at a temperature of 600 °C, slightly below the softening point of soda-lime glass. Irregular cracks, impurities, and dangling bonds serve as crystallization sites and cause the formation of a horizontal carbon buffer layer on the substrate surface [6].

Improving battery performance is one of the breakthroughs made possible by nanotechnology. Strong ion transport, high electron conductivity, fast ion diffusion, improved electrochemical capabilities, large surface area, and high current resistance are just some of the advantages that nanomaterials offer over conventional battery materials [6]. These characteristics improve safety, cycle life, and efficiency by increasing energy density, charge and discharge rates, and operational capabilities [7].

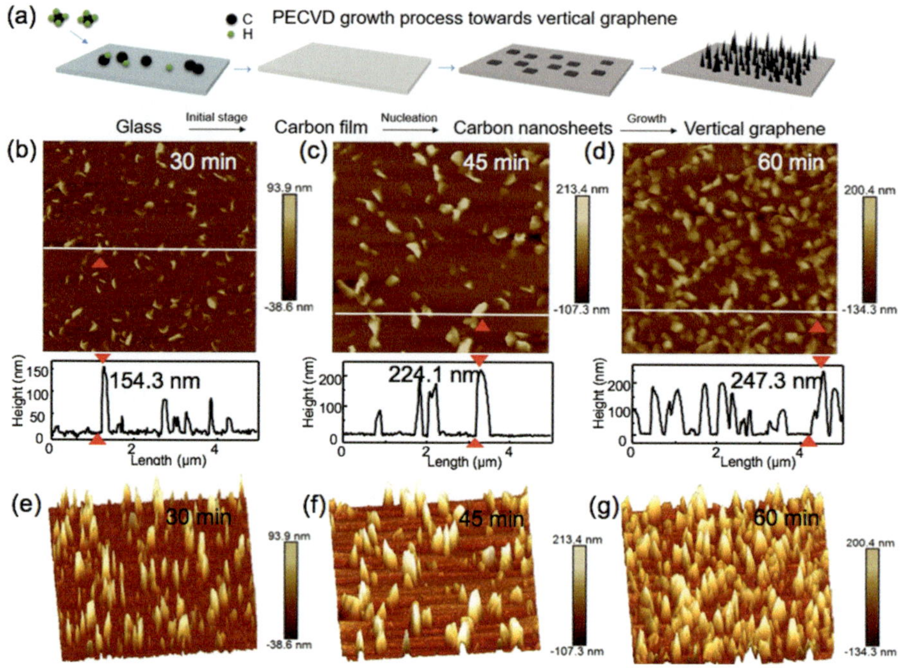

FIGURE 7.9 Growth of VG using plasma-enhanced chemical vapor deposition (PECVD). (a) Schematic representation of VG graphene growth using rf-PECVD. (b–d) AFM height images of 5 μm × 5 μm plant graphene films directly formed on soda-lime glass are acquired at three different growth times (30, 45, and 60 min), all under the same growth conditions (10 standard cubic centimeters per minute (sccm) of CH_4, 600 °C, 300 W). Portions of images in insets are shown in direction of dashed white line. (e–g) corresponding 3D AFM height images [6] [Open access].

Renewable energy research is becoming increasingly important as the world's energy needs increase. The widespread use of fossil fuels for energy production, including coal, oil, and natural gas, has worsened climate change and global warming. The development of sustainable alternatives that use nanotechnology to improve energy efficiency has increased as a result of this growing concern.

Thermoelectric (TE) heaters are constructed using samples that are collected to evaluate their TE properties and further explore these developments. Copper tapes are attached to both sides of the glass as electrodes. Analysis of TE performance of PbS monolayer shows that it outperformed bulk PbS by a large margin, with a maximum power factor of 6.1×10^{10} W/mK^2 and zT value, 1.0. This improvement is attributed to the monolayer's strong electrical conductivity and large Seebeck coefficient, which resulted from its narrow density of states (DOS) near the band gap edges and the band gap itself.

Density functional theory (DFT) and semi-classical Boltzmann transport theory were used to theoretically study the thermoelectric performance of two-dimensional monolayers of PbS [16]. According to their findings, the electronic structure is

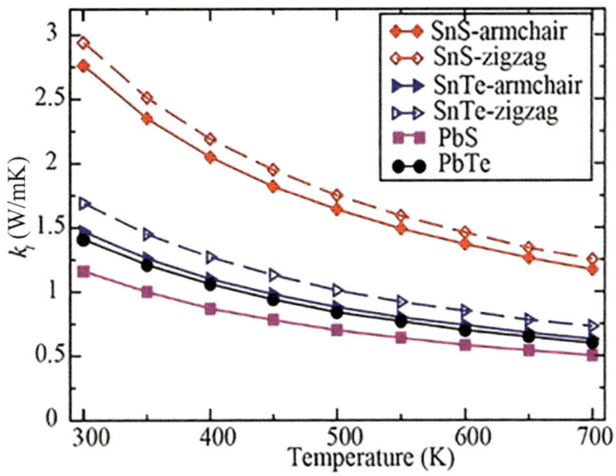

FIGURE 7.10 The lattice thermal conductivity (κl) of SnS, SnTe, PbS, and PbTe MLs is dependent on temperature [6] [Open access].

heavily influenced by spin–orbit coupling (SOC), resulting in higher *zT* values and lower thermal conductivity. It is found that one of the main reasons for the higher *zT* values of these materials is their low thermal conductivity values, as shown in Figure 7.10 [6].

To solve climate change and other environmental problems, advances in renewable energy sources such as solar, wind, hydropower, bioenergy, geothermal energy, and others are inevitable [8]. The development of renewable energy technologies represents a viable replacement for fossil fuels, serving as an endless source of energy. However, not having enough energy storage devices leads to energy wastage, so it is important to save. Energy storage technologies with high power density and energy density will inevitably be developed to provide energy for future generations [10].

Redox reactions (reduction–oxidation) are used in electrochemical energy storage devices to convert chemical energy into electrical energy. Apart from this mechanism, the main electrochemical energy storage devices that are the focus of much research in sustainable energy storage technologies are batteries, supercapacitors, and fuel cells. Supercapacitors are one of them, which have emerged as an important energy storage technology with a wide range of potential uses in many sectors. Supercapacitors are energy storage devices that have significant capacity for both rapid discharge and energy storage. They differ from regular capacitors in that they have much higher capacitance values, which allows them to store more energy. They have better power density compared to batteries [7–10].

7.4 MATRIX AND REINFORCEMENT PHASES

Composite materials consist of two or more separate phases, working together to create properties greater than the sum of their parts. The matrix phase and the reinforcement phase are the two main components of any composite material, working together to determine the mechanical, thermal, and chemical behavior of the final structure. Designing composite materials with special performance needs in the biomedical, automotive, aerospace, and construction industries requires an understanding of how these phases interact. In composites, the matrix is the continuous phase, ensuring stress transfer and maintaining the structural integrity of the composite by enclosing, supporting, and binding the reinforcement. Depending on the application, the matrix material can be ceramic, metallic, or polymeric. Polymer matrices are the most common due to their light weight, ease of fabrication, and corrosion resistance. Examples include polyimide, polyester, and epoxy resins. These matrices perform effectively in applications requiring a high strength-to-weight ratio, such as aircraft structures. When high-temperature stability is required, metal matrices, such as magnesium, titanium, and aluminum, are widely used in the aerospace and defense industries due to their high strength, good thermal conductivity, and corrosion resilience. Due to their exceptional hardness and heat tolerance, ceramic matrices such as silicon carbide (SiC) and alumina (AlO_3) are suitable for high-temperature applications such as turbine blades and engine components. The main purposes of the matrix are to provide shape and strength to reinforcements, transfer stresses between them, and protect them from environmental degradation. This also affects the composite material's ductility, impact resistance, and thermal expansion. The dispersed phase, known as reinforcement, gives the composite material its strength, stiffness, and other beneficial properties.

Depending on the design objectives, it can take various forms, such as fibers, particles, filaments, or flakes. Fibrous reinforcements, such as glass, carbon, and Kevlar fibers, are the best reinforcements for increasing tensile strength and stiffness. While short or randomly oriented fibers contribute to isotropic properties, continuous fibers provide directional strength. In metal or polymer matrices for load-bearing and wear-resistant components, particulate reinforcements, such as silicon carbide, alumina, and titanium carbide particles, enhance stiffness, wear resistance, and dimensional stability. The use of nanoreinforcements, such as graphene, CNTs, and nanoclay, is growing in the production of nanocomposites. Even at very low concentrations, their nanoscale size allows for significant improvements in mechanical, electrical, and thermal properties due to their very high surface area-to-volume ratio.

Strong interfacial contact ensures efficient stress transfer, preventing fiber separation. Adhesion of the reinforcement to the matrix is also improved through surface treatments and bonding agents. In short, the reinforcement adds strength, stiffness, and functional performance, while the matrix provides cohesion, form, and environmental protection. The synergy between these two phases determines the overall behavior of composite materials. By optimizing their selection and interfacial interactions, enhanced composite materials with specific properties can be produced for modern engineering applications. At the heart of many of the advantages of nanotechnology, which has greatly expanded the tools available to materials scientists, lies the ability to modify material structure on extremely small scales to achieve specific properties.

Materials that are stronger, lighter, more flexible, and more reactive than their conventional counterparts can be designed and engineered through nanoscale control.

Furthermore, nanotechnology can produce materials with highly selective or "sieve-like" properties and significantly improved electrical and thermal conductivity. These developments enable the production of multipurpose composite materials that perform exceptionally well in diverse industries, including electronics, healthcare, energy, and aerospace. Many commercial products currently available on the market and used daily, from advanced batteries, sensors, and medical equipment to water-resistant textiles, self-cleaning surfaces, and high-strength coatings, rely on materials and processes engineered at the nanoscale. Modern engineering continues to be revolutionized by the integration of nanotechnology and composite materials science, opening new avenues for innovation, sustainability, and efficiency in next-generation materials:

- Nano-based additives or surface treatments can help materials used in personal protective equipment resist bacterial growth, stains, and wrinkles, or facilitate the deflection of ballistic energy.
- Transparent nano-based coatings can be used to make windows, computer screens, cameras, eyeglasses, and other surfaces antireflective, self-cleaning, water- and residue-repellent, scratch-resistant, antimicrobial, antifog, resistant to ultraviolet (UV) or infrared light, and electrically conductive.
- Because of nanomaterials, it is now possible to create washable, long-lasting "smart fabrics" equipped with flexible nanoelectronics and sensors that can track health, capture solar energy, and regulate movement.
- If cars, trucks, airplanes, ships, and spacecraft were lighter, the fuel savings could be significant. By adding nano-additives to polymer composite materials, products such as baseball bats, tennis rackets, bicycles, motorcycle helmets, auto parts, luggage, and power tool bodies have become lighter, stiffer, more flexible, and more durable. CNT sheets are now being produced for use in future aircraft. For example, they are ideal for electromagnetic shielding and heat control due to their conductivity and light weight.
- Excellent NASA micrograph of a silicate–polymer nanocomposite.
- High-resolution image shows a silicate–polymer nanocomposite. Because of its superior thermal, mechanical, and barrier properties, this material could be used in food and beverage containers, automotive and aircraft fuel tanks, and aircraft components.
- Enzymatic nanoengineering aims to facilitate the production of ethanol as a fuel from cellulose derived from sources such as wood chips, corn stalks, and unfertilized perennial grasses. Cellulose nanoparticles have shown promising potential in a wide range of industrial fields, including electronics, construction, packaging, food, energy, healthcare, automotive, and defense. Among their many benefits, cellulose nanoparticles are expected to have a remarkable strength-to-weight ratio and be less expensive than many other nanomaterials.
- Nanotechnology-based automotive products include smart thin-film solar panels, low-cost, high-performance sensors and electronics, low-rolling-resistance tires, high-capacity batteries, TE materials for temperature control, and fuel additives for longer and cleaner emissions.

- Nano-ceramic coatings are much more resilient than conventional wear-resistant coatings when used on machine parts. The use of nano-modified lubricants and motor oils can significantly increase the wear of moving parts in all fields, from power tools to industrial equipment.
- Stabilizing, reducing, and coating agents. These synthetic tri-metallic nanoparticles are characterized using a variety of methods, including energy dispersive spectroscopy (EDS), transmission electron microscopy (TEM), X-ray diffraction (XRD), UV–vis spectroscopy, and SEM (Figure 7.11) [18].
- Nanomaterials are utilized in the manufacture of advanced household appliances, including air purifiers, filters, degreasers, and environmental sensors. Other examples include specialized paints and sealants, such as house paints, that are self-cleaning and do not scratch or stain, as well as anti-bacterial detergents.
- Based on several factors, including structure, shape, size, and distinctive physical and chemical properties, nanomaterials are classified into several groups, including organic, inorganic, and carbon nanomaterials (Figure 7.12) [18].

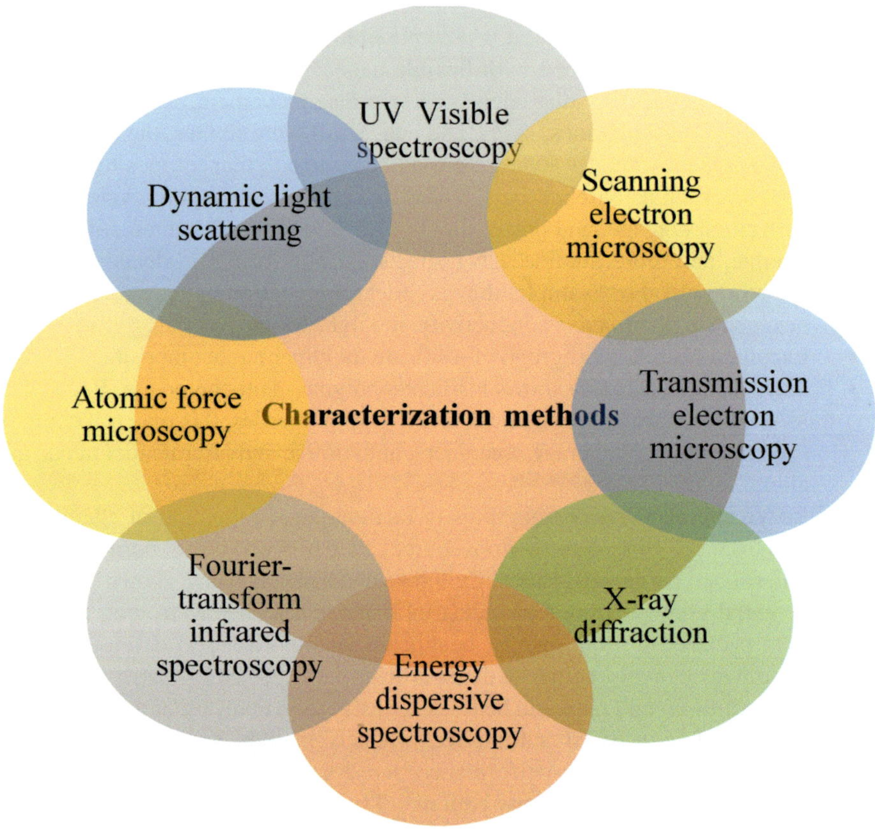

FIGURE 7.11 Various techniques for characterizing nanoparticles [18] [Open access].

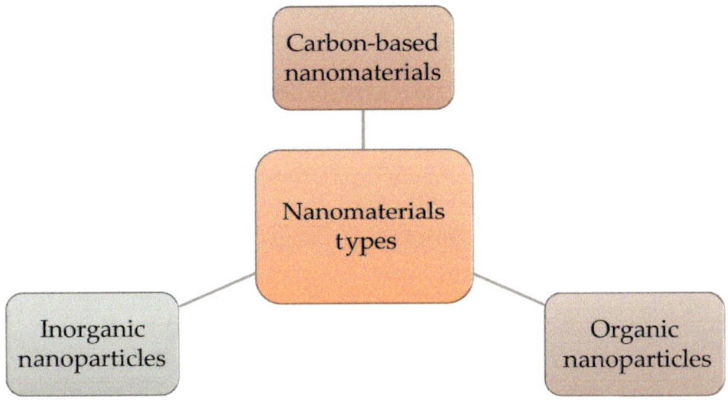

FIGURE 7.12 Various kinds of nanomaterials [18] [Open access].

7.4.1 Cosmetics and Individual Items

In addition to supporting healthy skin, makeup should offer a stunning rainbow of shades to suit all skin tones. The use of high-quality plant ingredients and finely ground minerals represents the latest achievement in classic cosmetics. These ingredients must be completely natural and 100% pure to be suitable for the most delicate skin. However, most dyes used in mass-produced cosmetics are synthetic. In fact, the Food and Drug Administration has outlawed all-natural color additives in the United States unless they are explicitly approved (like annatto seeds) to prevent allergic reactions and irritation. Color additives intended for cosmetic use, both inorganic (powders) and organic (dyes and pigments), are safe for human consumption and do not contain heavy metals such as cadmium, lead, and hexavalent chromium.

Inorganic pigments include ultramarine, iron oxide, manganese, chromium oxide, titanium dioxide, and ferric ferrocyanide. Cosmetic pigments, both organic and inorganic, do not dissolve in either water or oil. However, they can be dispersed microscopically in a liquid medium, giving the appearance of dissolution. Because soluble dyes can be harmful to your health, especially when used in excessive concentrations, very few of them are used in cosmetic products. For example, as different metal salts (sodium, calcium, barium, and aluminum ions) impart different shades, lakes are pigments produced by precipitating a soluble dye with these salts.

Bismuth oxychloride, guanine, or titanium dioxide can be used with color additives. When mixed with titanium dioxide-coated mica, it produces incredibly beautiful pearls in two colors. A soft substance used as frost in lipsticks is bismuth oxychloride. The main use of guanine obtained from herring scales is in nail paints. Apart from their technological applications, titanium dioxide and zinc oxide are the most common mineral powders used in cosmetic products sold in the market. These powders may have soothing and anti-inflammatory properties [23]. These nanomaterials do not create an aesthetic translucent composition but also effectively block UV rays. In consumer products, silver nanoparticles are used as antibacterial and preservative agents. Cosmetic products in the United States cannot make antibacterial claims because these claims involve physiological processes; as a result, they are used in medicines. Moreover, nanotechnology has had a major impact on

TABLE 7.1
Current Research on Cosmetics That Use Nanoparticles

Nanomaterial	Description	pH	Penetration
Quantum dot-655/COOH	Shape: Ellipsoid diameter: 6–12 nm Hydrodynamic diameter: 18 nm Skin: Pig	9	Negative
Quantum dot-655/PEG	Skin: Human stratum corneum	8.3	Negative
TiO_2 nanoparticles	Particle size: 200 nm Skin: Pig	—	Negative
TiO_2 nanoparticles in oil/water	Particle size: 10 × 50 nm Skin: Pig	—	Negative
ZnO nanoparticles	Skin: Human epidermal Size: 15–30 nm Duration: 24 h Skin: Human	—	0.03%— Negative

renewable energy sources, namely energy storage and conversion. Table 7.1 lists the applications of nanomaterials in various fields, and some of these applications are briefly described in the next part [25].

7.4.2 Transportation and Automobiles

Due to the special material, physical, and chemical properties of products, nanotechnology is used, and these properties greatly benefit the global materials industry. The manipulation of matter between 0.1 and 100 nm in size is known as nanotechnology [22]. The distinctive physical and chemical characteristics of nanotechnologies could lead to significant improvements in (photo) catalysis, optical sensitivity, and mechanical strength, enabling their use in energy storage, sensing, and other applications. Thus, through controlled manipulation of shape and size measured in nanometers, nanotechnology helps in the development of tools, instruments, and structures [23]. In fact, due to the properties of materials, particle size is very important and inevitably affects the nanoscale. This is because enhanced material properties are achieved at the nanoscale, where the electrostatic forces and quantum phenomena cancel out the effects of gravity. Many disciplines, including chemistry, biology, and engineering, have undergone revolutionary changes as a result of the nanotechnology revolution [24]. Due to its higher efficiency, nanotechnology is increasingly being applied in various sectors, including biomedicine, robotics, electronics, automobiles, and civil engineering (including transportation) [25]. The following is a quick description of nanomaterials associated with cars.

7.4.2.1 Body Parts with Low Weight

Reducing your vehicle's weight offers amazing benefits that benefit your passengers as well as your business's bottom line. It is critical to adapt to new manufacturing regulations to make your vehicles lighter and overall, more efficient. Therefore, how can you effectively shorten this method? Some useful tips you should know about. This is the information you will need to produce premium cars by learning about car weight reduction.

Realize the importance of car weight loss first of all. Why should losing weight in your car be a top priority for you? Lighter cars use less fuel and produce fewer emissions. Your customers will be grateful for the reduced burden on their pockets, as well as the fact that your vehicles will comply with the latest environmental regulations. By taking into account the weight reduction of the vehicle during construction, you can create a more economical and environmentally friendly vehicle. In addition, these modifications improve vehicle safety, which benefits all drivers [23].

In addition, polymer nanocomposites such as clay nanocomposites reinforced with polyamide, polycarbonates, and polypropylene are widely used. These nanoparticles will become less flame-retardant and more heat-resistant when mixed with polymers [24]. Using nanocomposite and hybrid solutions, a car that normally weighs 900 kg of steel can weigh as little as 300 kg. On average, weight savings when using nanocomposite plastics are 25% compared to densely filled plastics and 80% compared to steel. Vanadium oxide (VW) uses long fiber-reinforced thermoplastic (LFRT), which provides superior design flexibility, high strength-to-weight ratio, impact resistance, and less corrosion compared to metals [24].

7.4.2.2 Paint-Coating

Manipulating matter at the nanoscale is the focus of nanotechnology. At this level, materials undergo significant physical and chemical changes due to a high surface-to-volume ratio. With its applications in science and technology to create new materials at the nanoscale useful for various fields such as economics and the environment, nanotechnology has become increasingly popular in recent years. Recent studies [23] have identified potential uses of TiO_2 nanoparticles in some industries, including sensor technologies, photocatalysis, solar energy conversion, pigments, and biological activities such as antifungal, antibacterial, and anticancer properties. Despite their widespread use, TiO_2 nanoparticles have many other characteristics, including high chemical stability, high redox potential, and availability and environmental friendliness [24]. In addition, the process has used to create nanoparticles that have a significant impact on these characteristics, including the specific surface area, crystal structure, particle shape, and crystallite size. There are three different polymorphic crystalline forms of TiO_2 nanoparticles: rutile (tetragonal structure), brookite (orthorhombic structure), and anatase (tetragonal structure). Anatase and brookite can be heated to high temperatures to produce rutile, which is a stable phase [23]. Depending on the structure and particle size, each crystalline phase has unique physical characteristics that make it suitable for different applications [24]. For example, band gap, 3.2 eV, means that exposure to UV light can activate the anatase phase, so it is often used for photocatalysis [25].

7.4.2.3 Scratch–Resistance

Nowadays, the automotive coating system—a multilayer polymer system—plays an important role in bringing the appearance of cars in line with the evolving aesthetics of people. In addition, the system protects key components from radiation and environmental degradation. Each layer in a multilayer system has a specific function and is used in a sequence of drawing commands. A previous study [21] has provided detailed descriptions of functional features of each layer. Overall, the resistance of coating systems to damage directly affects the long-term appeal of a vehicle and the durability of its beautiful body. Over the past few decades, maintaining the integrity and appearance of a vehicle has become a top priority for original equipment manufacturers (OEMs), especially for premium vehicles.

However, used vehicles are easily scratched during daily use, and resulting damage to the finish can seriously degrade the appearance and lifespan of the vehicle. Customers now consider scratch resistance one of the most important requirements for automotive coating systems. However, research on automotive coatings, which typically have more complex structures, is rather sparse.

Divided scratches into two extreme groups depending on their diameter. Just a few microns in width and depth constitute one category of scratch that is considered scratch damage [22]. Usually, the cause of deposits is wear and tear on the automatic car wash brushes when washing the car. However, due to its microscale, a single mark is difficult to detect with the naked eye except under special viewing conditions. Moreover, if dirt damage accumulates in large quantities, the gloss of the vehicle's paint surface will decrease, even if it cannot penetrate the coating [19].

Over 20 years ago, two fundamental methods were widely used to determine the abrasion resistance of coating surfaces: the meter test [20] and advanced sensors and measurement technologies at AMTEC and Kistlercar wash test (ISO/FDIS 20566) [21]. In addition, several studies [22] have been published to date on the subject of nano scratch testing, which has been used to quantitatively measure the scratch resistance of automotive coatings and give some idea of their mechanical properties [23]

7.4.2.4 Tires

Commercially produced nanoparticles, known as carbon black (CB), are produced by partial combustion and vapor-phase pyrolysis of hydrocarbons under carefully controlled conditions, resulting in a variety of target grades. It consists of pure elemental carbon (EC) in the form of a fine black powder. CB is mainly used as a black pigment in inks, a reinforcing agent in vehicle tires, and as a way to improve the electrical conductivity of plastics. Eight-hundred thousand tons of CB are added to plastic every year, and the demand for plastic is growing every year worldwide.

As automobile tires typically contain CB as a hardening agent, these tires may emit PM during braking, and tire wear is an example of a non-emission mechanism. Most rubber particles that form non-exhaust PM resulting from tire wear on road surfaces include CB, zinc oxide, and various heavy metals. It has been shown that inhalation of non-exhaust PM from worn tires and exhaust PM from road-going vehicles can lead to respiratory diseases caused by oxidative stress and inflammatory pathways. The non-exhaust PM emission factor associated with tire wear on gasoline-powered vehicles is twice the exhaust particulate emission factor of 6.1 mg/km. Although PM risk management and toxicity testing have been extensively studied, nothing is known about the health effects of CB-containing

PM that is generated during tire wear. The main particles that make up CB are typically nanosized and can aggregate to sizes ranging from a few microns to 100 nm. Due to their special characteristics and small size, these nanoparticles may have negative effects on the environment and human health. Therefore, an adequate risk assessment should be carried out regarding the clean and free particle-related wear of CB tires. Many studies [20] have been conducted in laboratory settings or on CB production workers to investigate the possible health effects of CB. According to others [21], CB may be directly genotoxic and harmful to reproduction. However, some toxicological data mistakenly identify CB with black carbon (BC). Characteristics are comparable to CB in terms of component content, aggregate shape, and black color. Given their similar names, it is not surprising that CB and BC are often misused [22].

7.4.2.5 Other Uses in Automobiles

Due to their unique chemical and physical properties, nanomaterials have attracted much attention. The result of nanotechnology and nanoscience is nanomaterials. While nanomaterials are a vast field in themselves, they have the potential to completely change the way materials and products are produced. The purpose is to report on the latest advances in the use of nanotechnology in the automotive industry. The results have demonstrated that using more nanomaterials in automobiles improves vehicle performance and makes transportation more environmentally friendly. Nanomaterials, including graphene, CNTs, aluminum–silicon carbide composites, and aluminum–CNT composites, are often used in automobiles. This chapter looked at how these resources are used for future applications as well as in many automotive industries [9]. Fluid viscosity increases as a result of the orientation of magnetic nanoparticles in magnetorheological fluids, which are dispersed in a solvent and respond to small changes in pressure and magnetic fields.

7.4.3 Agriculture

The use of agrochemicals has increased agricultural production, but at the expense of the health of farmers and the general population who consume chemically produced food, as well as the soil and water systems with their associated flora and fauna. Applications of nanoparticles in the automobile sector are listed in Table 7.2 [25].

The risk assessment of this technology occurs much later than its use. Therefore, this chapter will cover hazards, uncertainties, and nanoecotoxicology. The use of nanotechnology in agriculture raises ethical issues. We will talk about how the public views and understands nanotechnology at the moment [23]. Thus, the development of nano agrochemicals such as nano pesticides, nano fertilizers, and nanosensors to improve solubility, bioavailability, and targeted delivery, as well as their controlled release, will have many potential benefits, such as precise dosage of fertilizers, enhanced control of disease vectors and pests, reducing chemical pollution, and ultimately increasing agricultural productivity. Regarding the toxicity and safety of these systems for human health and the environment, many unresolved problems remain. Graphene oxide (GO), titanium dioxide (TiO_2), silver (Ag), and zinc (Zn)

TABLE 7.2
Applications of Nanomaterials in the Automotive Industry

Nanomaterials	Merits	Demerits	Applications
Long fiber-reinforced thermoplastic	Extensive impact resistance, high stability	Production is challenging	Body parts
Silicon dioxide (SiO_2)	Lower paint toxicity, supports self-healing	Uniform dispersion is challenging	Paint coating
Copper dioxide (CuO_2), molybdenum disulfide (MoS_2), titanium dioxide (TiO_2)	Coefficient of friction can be lowered up to 50%	Lifespan is shorter	Engines
layered double hydroxides (LDH)	Absorbs heat, enhances char growth	Stability concerns	Paint coating
Carbon nanotubes (CNTs)	High mechanical stability and lightweight	Expensive	Body parts
Soot, carbon black, silica	Extended durability, higher fuel efficiency	Uniform dispersion is challenging	Vehicle tires
Copper oxide (nano-fluids)	Upgrades heat transfer coefficient	Agglomeration issues	Engines
Carbon nanotubes (CNTs), copper (Cu), zinc oxide (ZnO), gold (Au)	Zero toxicity, good efficiency	Relatively expensive	Vehicle interior

Source: [25] [Open access].

[18]. Therefore, the approaches of this promising method include protection against viruses and diseases, as well as increasing the effectiveness of pesticides, fungicides, and herbicides, which improves plant development and maintains discharges [19]. As nanomaterials apply to almost all aspects of the agricultural industry, including transportation, processing, production, and storage, they improve the efficiency, productivity, and safety of agriculture [20]. From a practical point of view, there are several options for using nanoparticles in agriculture. The main goal is to improve the delivery of fertilizers, which, in turn, leads to an increase in the absorption of elements by plant cells and a decrease in nutrient losses. They provide a synchronous introduction of macro- and microelements [20]. Because nanostructured fertilizers have a large surface area, a precise delivery mechanism, and a gradual, continuous release in response to environmental signals and biological needs, they improve nutrient use efficiency. Nanophosphorus iron (Fe), nanomagnesium (Mg), and nanozinc (Zn) are a few examples of typical nanofertilizers.

In addition, as CNTs have competitive thermal, chemical, electrical, and mechanical capabilities, they can be used to produce micro- and macronutrients while reducing inputs, which can yield good results in agriculture [21]. The second goal is to control insect pests. Some nanoparticles, such as zinc oxide, titanium dioxide, and gold, are effective in pest control due to slow and active release of active chemicals as an affordable and reliable alternative to industrial drugs that have adverse side effects. Nanopesticides reduce the movement of unwanted pesticides and the release of organic solvents. In addition, nanoparticles can detect, identify, and report crop diseases using rapid analytical methods and smart field systems, prompting producers to apply the necessary ingredients before symptoms appear [21]. Nanofungicides perform another function. Chemical fungicides damage plants, which are sensitive to the environment and health. Considering the size and shape of nanoparticles, nanofungicides provide a valuable answer to the abovementioned problems. The antibacterial properties of Ag nanoparticles make them widely known nanofungicides. While TiO_2 nanoparticles have antibacterial and photocatalytic activities that provide plant protection, ZnO nanoparticles also exhibit antifungal, antibacterial, and antimicrobial properties [22]. The latest application is nanosensors, which are highly sought after in the agricultural industry for their ability to monitor environmental, aquifer, and soil contamination, easily and accurately [23]. Ag nanoparticles are a well-known nanomaterial used as a sensor. Typical nanomaterials used in agriculture are listed in Tables 7.3 and 7.4 [24].

Nanotechnology is said to be the next big thing in chemistry, medicine, materials science, engineering, and other fields. It is necessary to talk about the various risks associated with using nanotechnology. Breakthrough developments can come from molecular manufacturing, a complex form of nanotechnology. As the production of nanomaterials increases worldwide, toxicologists and regulatory experts are

TABLE 7.3
A Few Instances of Nanoparticles Used in the Field of Agriculture

Nanomaterial	Function of Nanomaterial	Industrial Name	Application
Zinc (Zn) nanoparticles, copper (Cu) nanoparticles, boron (B) nanoparticles, iron (Fe) nanoparticles, bio humus	Fertilizer	Motawazen, Saula, Solocross, Green Earth NanoPlant	Nano-fertilizers
Copper-doped montmorillonite	Detection of propineb fungicide	–	Nano-sensors
Copper (Cu) nanoparticles, polymeric nanoparticles, zinc (Zn) nanoparticles	Bactericide, controlled release, and fungicide	Bio-nano technology	Nano-pesticides
poly-epsilon caprolactone	Controlled release	—	Nano-herbicides

Source: [25] [Open access].

TABLE 7.4
A Few Other Uses of Nanomaterials in Variety of Domains

Applications	Examples
Biomedical and healthcare	• Gold nanoparticles for targeted delivery of drugs directly to cancer tumors • Liposome nanoparticles for controlled and targeted drug delivery • Quantum dots used for detection and diagnosis of cancer • Graphene and carbon-based nanomaterials as wearable multifunctional sensors for monitoring skin temperature, wrist pulse, breathing, and pressure map ping
Construction materials	• CNT, GO, and SiO_2 nanomaterials used as reinforced cement composites • CNT as reinforced metal matrix composite to produce stronger, ultralight, and corrosion-resistant nanocomposite materials • ZnO nanoparticles used as coatings to reduce ultraviolet radiation exposure • Silver nanoparticles embedded in paint to provide antimicrobial properties • TiO_2 nanoparticles used as glaze coating for anti-fouling properties
Electronics and electrical applications	• Carbon-based nanomaterials used as electrodes in supercapacitors, offering large storage capacities and fast charge/discharge rates
Environmental remediation	• CNT, graphene oxide, and other nanomaterials used as adsorbents to remove heavy metals and pollutants from water bodies • Functional nanomaterials such as TiO_2, Al_2O_3, CNT, and zeolites incorporated into filtration membranes to improve mechanical and thermal stability, permeability, fouling resistance, and pollutant adsorption capabilities
Food packaging	• Nanoclay used for food packaging due to its strong mechanical and thermal properties • Nanoemulsion used as an edible coating for preserving and storing food • Gold nanoparticles used as food sensors to monitor food quality and safety
Food processing	• SiO_2 nanomaterials used as anti-caking agents to maintain flow properties in powdered products • Titanium dioxide (TiO_2) used as food color additives and flavor enhancers • SiO_2-gallic acid nanoparticles used as antioxidants • Silver (Ag) and zinc oxide (ZnO) nanoparticles used as antimicrobial agents in the food packaging industry
Agriculture	• Fertilizers coated with nanocapsules or nanoparticles for improved delivery of agri-chemicals and nutrients • Nanomaterials used as pest control agents for enhanced effectiveness • Zinc oxide (ZnO) quantum dots used for pesticide detection

(*continued*)

TABLE 7.4 (Continued)
A Few Other Uses of Nanomaterials in Variety of Domains

Applications	Examples
	• Magnetic nanoparticles used for smart agrochemical and herbicide delivery systems • MWCNT nanomaterials used to increase plant production rates • TiO_2 nanomaterials used as sensors for monitoring soil conditions, pesticides, and crop growth
Renewable energy	• Carbon nanomaterials used for direct conversion of sunlight into electricity • Graphene used as a metal-free catalyst in fuel cells • Metal–organic frameworks used for hydrogen storage
Sensors	• Metal oxide nanomaterials used for various sensor applications, such as gas sensors, chemical oxygen demand (COD) sensors, and biosensors

Source: [25] [Open access].

increasingly focusing on nontoxicity and nanodisks. Despite the novelty of using nanostructures, concerns have already been raised about the possible toxicity of nanomaterials used in various aspects of human existence. To evaluate the possible harmful and toxic effects of nanomaterials on biological ecosystems and human health, more research is needed to understand the physicochemical characteristics of nanomaterials and their effectiveness. The effect of nanoparticles on the human body and associated biosystems varies depending on their type, size, shape, substituents, and coatings. The nanoparticles are generally considered harmless in specific contexts. With its ability to manipulate matter at the atomic and molecular level, nanotechnology has enormous potential in some disciplines, including electronics, environmental protection, and medicine. However, it also comes with its risks and challenges. Some of the main challenges are given below.

Ingestion, skin absorption, or inhalation are three ways nanoparticles can enter the body. Because of their small size, they can penetrate cells and tissues, which can lead to unknown health effects. Long-term exposure can cause problems such as cancer, oxidative damage, and inflammation. Plants and animals can be harmed by the accumulation of nanoparticles in the environment and ecosystems. It is not yet clear how they affect air, water, and soil quality. They can have unexpected interactions with natural processes [22–24].

Unintended Consequences: Because of their special properties, nanoparticles can cause unintended effects or interactions. For example, nanoparticles that deliver drugs may unintentionally interfere with other biological functions. The potential of nanotechnology for ethical dilemmas associated with its use. Issues include concerns about potential misuse as a weapon, privacy issues with nano-monitoring devices, and broader implications of human enhancement. Due to rapid development and significant commercial potential, nanomaterials have today become one of the highest priority areas of research. Recent developments in engineered nanomaterials have opened new possibilities for a variety of industrial applications, from electronics and energy storage to biotechnology and healthcare. Full implementation of nanotechnology in industrial applications requires addressing and successfully regulating the toxicity and other associated hazards of nanomaterials despite their superior capabilities.

The improved properties and unique characteristics of nanomaterials compared to bulk materials have made them increasingly important in the development of new technologies. Their exceptional properties are driving innovation in a wide range of industrial applications. This chapter on carbon nanomaterials provides a detailed overview of the fundamental engineering principles involved in nanomaterials. It investigates various technical applications of nanomaterials in several important sectors such as biotechnology, food processing, biomedicine, environmental remediation, construction, electronics, renewable energy, and energy storage. The chapter briefly discusses the potential impact of nanoparticles on human health, highlighting the need for continued research to ensure efficacy and safety. In short, nanomaterials have enormous potential to advance the field of nanotechnology, providing revolutionary opportunities in various industries and driving advances in advanced technological applications [25].

PROBLEMS

1. Describe the importance of nanomaterials in engineering applications.
2. List the materials of flexible electrodes in case a triple error occurs when comparing the sizes of nanomaterials.
3. What are carbon nanotubes' characteristics and characterization methods (1D)?
4. What role do nanomaterials play in engineering applications?
5. Describe the different uses used in preparing and classifying various materials.

REFERENCES

1. Giannazzo, F., Schilirò, E., Greco, G., & Roccaforte, F. (2020). Conductive atomic force microscopy of semiconducting transition metal dichalcogenides and heterostructures. *Nanomaterials,* 10(4), 803.
2. Wu, Y., Wang, D., Liu, J., Cai, H., & Zhang, Y. (2020). Atomic force microscope study of Ag-conduct polymer hybrid films: Evidence for light-induced charge separation. *Nanomaterials,* 10(9), 1819.
3. Lindqvist, C., Moons, E., & Van Stam, J. (2018). Fullerene aggregation in thin films of polymer blends for solar cell applications. *Materials,* 11(11), 2068.
4. Slattery, A. D., Shearer, C. J., Shapter, J. G., Quinton, J. S., & Gibson, C. T. (2017). Solution based methods for the fabrication of carbon nanotube modified atomic force microscopy probes. *Nanomaterials,* 7(11), 346.

5. Chianese, F., Aversa, L., Verucchi, R., & Cassinese, A. (2022). Molecular doping of CVD-graphene surfaces by perfluoroalkyl-substituted perylene diimides derivatives. *Nanomaterials,* 12(23), 4239.
6. Cui, L., Cui, K., Ci, H., Zheng, K., Xie, H., Gao, X., ... & Liu, Z. (2019). Transparent electrothermal heaters based on vertically-oriented graphene glass hybrid materials. *Nanomaterials,* 9(4), 558.
7. Wang, Z., Hu, T., Liang, R., & Wei, M. (2020). Application of zero-dimensional nanomaterials in biosensing. *Frontiers in Chemistry,* 8, 320.
8. Pandit, S., Behera, P., Sahoo, J., & De, M. (2019). In situ synthesis of amino acid functionalized carbon dots with tunable properties and their biological applications. *ACS Applied Bio Materials,* 2, 3393–3403.
9. Yan, Y., Gong, J., Chen, J., Zeng, Z., Huang, W., Pu, K., Liu, J., & Chen, P. (2019). Recent advances on graphene quantum dots: From chemistry and physics to applications. *Advanced Materials,* 31, 1808283.
10. Khan, M. B., & Khan, Z. H. (2017). Nanodiamonds: synthesis and applications. *Nanomaterials and Their Applications* (pp. 1–26). https://link.springer.com/chapter/10.1007/978-981-10-6214-8_1.
11. Dao, T. C., & Luong, T. Q. N. (2020). Fabrication of uniform arrays of silver nanoparticles on silicon by electrodeposition in ethanol solution and their use in SERS detection of difenoconazole pesticide. *RSC Advances,* 10, 40940–40947.
12. Kang, S. M., Kim, H., Choi, J. S., & An, J. H. (2021). Bioinspired omniphobic microchamber structure. *Advanced Materials Interfaces,* 8, 2100027.
13. Kolahalam, L. A., Kasi Viswanath, I. V., Diwakar, B. S., Govindh, B., Reddy, V., & Murthy, Y. L. N. (2019). Review on nanomaterials: Synthesis and applications. *Materials Today: Proceedings,* 18, 2182–2190.
14. Baig, N., Kammakakam, I., Falath, W., & Kammakakam, I. (2021). Nanomaterials: A review of synthesis methods, properties, recent progress, and challenges. *Materials Advances,* 2, 1821–1871.
15. Vollath, D. (2013). *Nanomaterials: An Introduction to Synthesis, Properties and Applications* (2nd ed.). Wiley-VCH: Weinheim, Germany.
16. Haghanifar, S., Tomasovic, L. M., Galante, A. J., Pekker, D., & Leu, P. W. (2019). Stain-resistant, superomniphobic flexible optical plastics based on nano-enoki mushroom-like structures. *Journal of Materials Chemistry A,* 7, 15698–15706.
17. David, N., Djilali, N., & Wild, P. (2012). Fiber Bragg grating sensor for two-phase flow in microchannels. *Microfluidics and Nanofluidics,* 13, 99–106.
18. Roy, A., Kunwar, S., Bhusal, U., Alghamdi, S., Almehmadi, M., Alhuthali, H. M., ... & Azlina, M. F. N. (2023). Bio-fabrication of trimetallic nanoparticles and their applications. *Catalysts,* 13(2), 321.
19. Fang, Z., Xing, Q., Fernandez, D., Zhang, X., & Yu, G. (2020). A mini review on two-dimensional nanomaterial assembly. *Nano Research,* 13, 1179–1190.
20. Hang, D. R., Huang, C. F., & Cheng, K. A. (2009). Probing semiclassical magneto-oscillations in the low-field quantum Hall effect. *Physical Review B,* 80, 085312.
21. Paramasivam, G., Palem, V. V., Sundaram, T., Sundaram, V., Kishore, S. C., & Bellucci, S. (2021). Nanomaterials: Synthesis and applications in theranostics. *Nanomaterials,* 11, 3228.
22. Wang, K., Ma, Q., Qu, C. X., Zhou, H. T., Cao, M., & Wang, S. D. (2023). Review on 3D fabrication at nanoscale. *AUTEX Research Journal*, 23(3), 350–369.
23. Gentselev, A. N., & Baev, S. G. (2022). Production of planar elements of terahertz optics by means of deep X-ray lithography. *Optoelectronics and Instrumentation Data Processing,* 58, 198–205.

24. Lohse, M., Heinrich, M., Grützner, S., Haase, A., Ramos, I., Salado, C., Thesen, M. W., & Grützner, G. (2021). Versatile fabrication method for multiscale hierarchical structured polymer masters using a combination of photo- and nanoimprint lithography. *Micro Nano Engineering,* 10, 100079.
25. Radnik, J., Kersting, R., Hagenhoff, B., Bennet, F., Ciornii, D., Nymark, P., ... & Hodoroaba, V. D. (2021). Reliable surface analysis data of nanomaterials in support of risk assessment based on minimum information requirements. *Nanomaterials,* 11(3), 639.

8 Perovskite Compounds
Structure, Modification, and Advanced Properties

8.1 INTRODUCTION

Perovskite alloys have recently garnered considerable attention due to their exceptional versatility and potential use in a variety of engineering fields. Due to their flexible crystal structure, which allows significant chemical substitution at the A and B sites, these materials, which have the general formula ABX_3, offer tailored physical properties for specific applications. Structural studies of perovskite alloys show that ion diameters, temperature, and compositional ratios can influence the ability of the crystal lattice to adopt various symmetries, including cubic, tetragonal, orthorhombic, and rhombohedral. Techniques such as electron microscopy, neutron diffraction, and X-ray diffraction (XRD) have been widely used to study grain size distribution, phase stability, and crystal defects.

Perovskite alloys exhibit a wide range of mechanical properties, including elastic moduli and hardness values, which are largely dependent on their microstructure and composition. They can be used in actuators, sensors, and structural ceramics, as nanocavities and ultrasonic testing reveal information about their mechanical strength. In addition, similar materials and synthesis conditions can be used to fine-tune the intrinsic ductility and strength of the materials [1–4].

According to magnetic studies, perovskite alloys exhibit a variety of magnetic properties, such as ferromagnetism, antiferromagnetism, and spin glass states. These properties are due to the superexchange interactions between transition metal ions at the B-site in the perovskite structure. The Curie temperature, magnetic anisotropy, and saturation magnetization all key properties of spintronic devices and magnetic sensors can be tuned by swapping magnetic and non-magnetic (NM) ions.

The topologies of the studied perovskite compounds were optimized with their cubic structure in mind to describe the geometric structure as mentioned earlier. Figure 8.1 illustrates their fully relaxed structure. The halide ion O is located at (1/2, 1/2, u) (u = 0.249), while the component atoms Cd and Cr are located at (0, 0, 0) and (1/2, 1/2, 1/2), respectively. The studied compounds were found to have the Pm3m space group in a cubic structure [1,4]. The Birch–Murnaghan equation of state [22] and the PBE-GGA functional were used to determine the ideal lattice parameters versus volume (a.u.) in the non-magnetic (NM), ferromagnetic (FM), and antiferromagnetic phases, as shown in Figure 8.2.

DOI: 10.1201/9781003758556-8

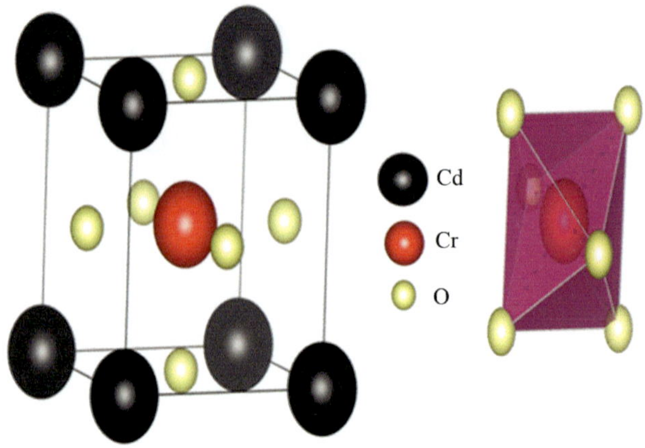

FIGURE 8.1 The crystal structure of cubic $CdCrO_3$ was calculated [1] [Open access].

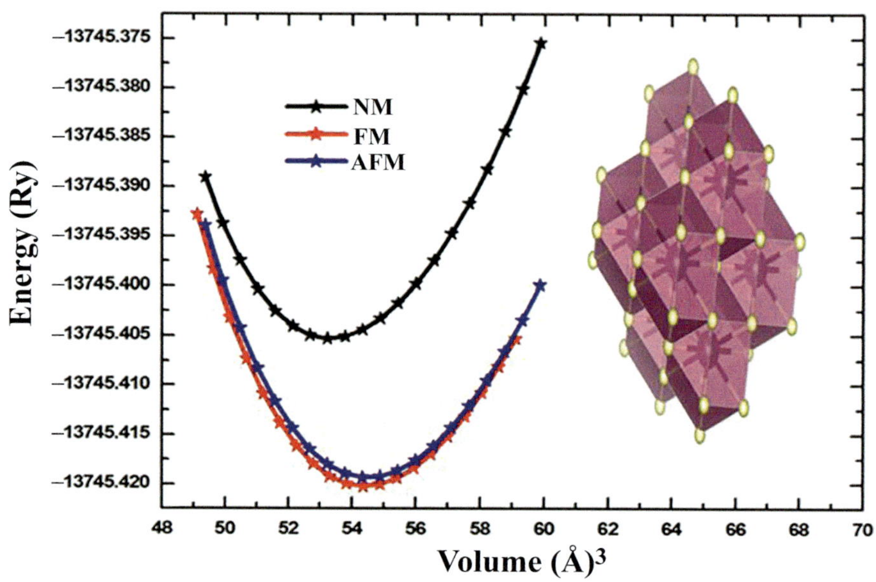

FIGURE 8.2 Total energy in the NM, FM, and AFM phases of $CdCrO_3$ with size enhancement curve [1] [Open access].

Depending on the chemical composition and electron configuration, perovskite alloys can have a wide range of electronic properties, from insulators to semiconductors and even metal conductors. A comprehensive analysis of the band structure and carrier concentration is performed using density functional theory (DFT) calculations supported by experimental methods, including ultraviolet–visible (UV–vis) spectroscopy and Hall measurements [1].

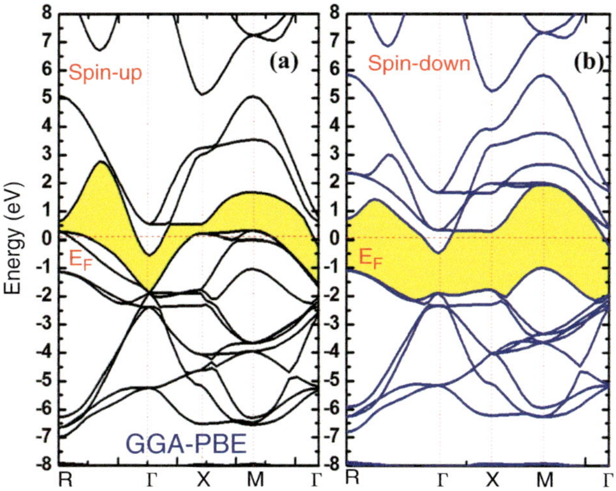

FIGURE 8.3 Using the GGA–PBE method, the spin-polarized band structures of the perovskite oxide $CdCrO_3$ are (a) spin up and (b) spin down [1] [Open access].

The development of perovskite-based thermoelectric materials, optoelectronics, and photovoltaics heavily depends on these electronic properties. Understanding the specific heat capacity, thermal conductivity, and thermal expansion coefficients of perovskite alloys is essential for thermal studies. These factors are critical in assessing the stability of perovskite-based devices under operating conditions. Thermal behavior is often assessed using techniques such as differential scanning calorimetry (DSC) and laser flash analysis (LFA). It is worth noting that some perovskite formulations exhibit low thermal conductivity, making them interesting for thermoelectric applications. At the same time, their high thermal stability makes them useful for high-temperature sensors and power electronics [5–8].

Perovskite alloys provide a unique platform for the design of multifunctional materials, allowing precise engineering of their mechanical, electrical, magnetic, thermal, and structural properties. Unleashing their full potential in advanced technologies requires ongoing interdisciplinary research combining theoretical and experimental methods [2,3]. Figure 8.3a and b shows the electronic band structures for spin-up and spin-down polarizations using GGA. Using the GGA-PBE technique, the results show that $CdCrO_3$ is a metallic compound, meaning it has high electrical conductivity and acts as an electrical conductor. Furthermore, the core band structure retains its metallic character upon the introduction of Coulomb repulsion (GGA+U) boundaries [1,4], as shown in Figure 8.4a and b; however, the core band structure exhibits a semiconducting character, indicating the semimetallic nature of $CdCrO_3$. This is an important finding because it distinguishes $CdCrO_3$ from other perovskite compounds that act as semiconductors or insulators.

Figure 8.5 shows the total and partial (projected) density of states (TDOS and PDOS, respectively) for $CdCrO_3$. The line that is broken vertically is the Fermi

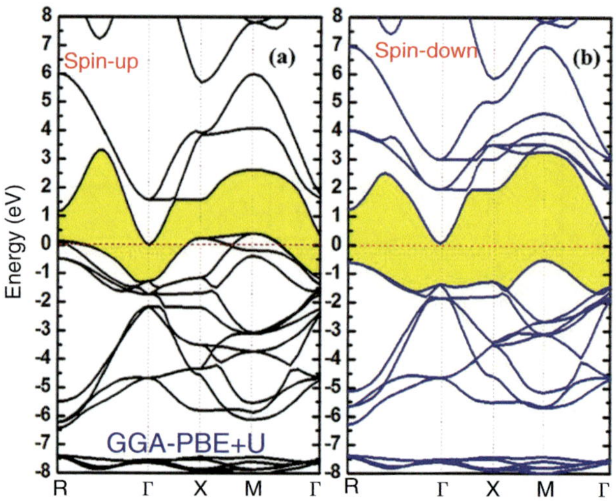

FIGURE 8.4 Using GGA–PBE+U, the spin-polarized band structures of the perovskite oxide $CdCrO_3$ are (a) spin-up and (b) spin-down [1] [Open access].

level, or EF. To better understand how each atom contributes to the total TDOS, we calculated the PDOS of Cd, Cr, and O in $CdCrO_3$. The non-zero GGA values for the total DOS at the Fermi level prove that $CdCrO_3$ must have metallic conductivity for both spin channels [1,6].

The three-dimensional (3D) architecture of the two cells is shown in Figure 8.6. The kinetic energy threshold (ecut) was optimized by varying it in 10 Ry steps from 20 to 90, and the k-points were optimized by varying them in 1 1 1 steps from 2 2 2 to 9 9 9 [2].

In various contexts, the behavior, function, performance, and processing of materials depend largely on their structural properties. Therefore, understanding and controlling the structures of materials are fundamental to the development of science, technology, and innovation in various fields [2]. The total energy of both materials is presented as a function of the volumes of standardized unit cells in Figure 8.7.

The coefficient of thermal expansion (CTE) is a key factor determining the design and performance of materials used in optoelectronic applications. A material's CTE indicates how much it expands or contracts in response to temperature changes. Figure 8.8 illustrates how the CTE of both materials changes with temperature.

This property is particularly important in optoelectronic devices, where optimal performance depends on the precise alignment of components such as lenses, waveguides, and mirrors. These components can physically move as a result of temperature changes due to thermal expansion or contraction, which can lead to misalignment [9–12].

Furthermore, the degree of misalignment can increase significantly if different materials with mismatched CTEs are used together. This can have a detrimental effect on device functionality by reducing light transmission efficiency.

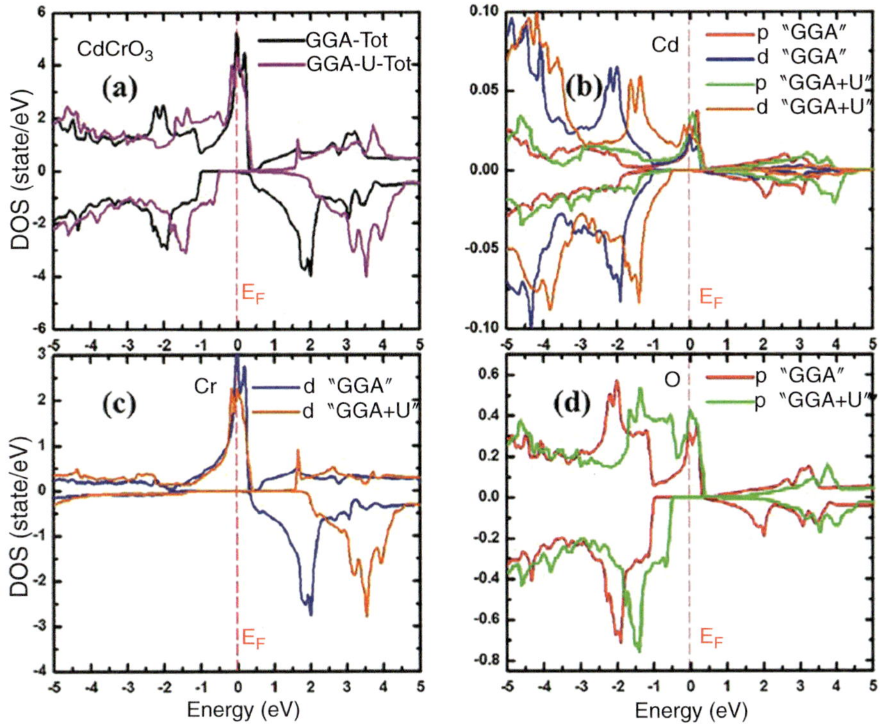

FIGURE 8.5 (a) Using GGA–PBE and GGA+U, TDOS and PDOS of the perovskite oxide $CdCrO_3$ at the equilibrium lattice constant, (b) Cd, (c) Cr, and (d) O [1] [Open access].

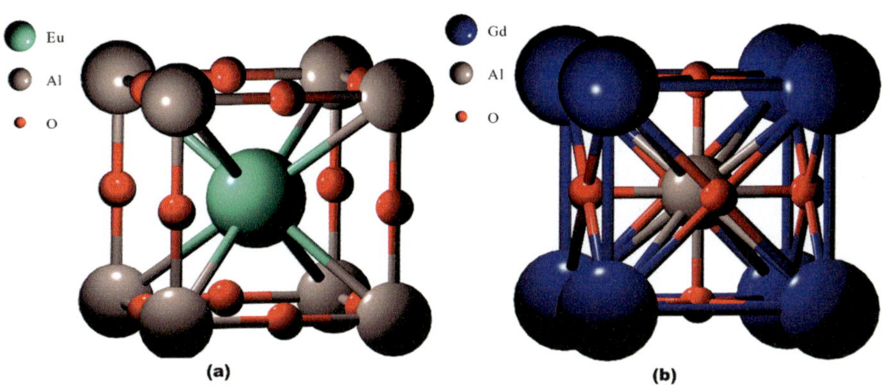

FIGURE 8.6 (a) $EuAlO_3$ and (b) $GdAlO_3$ 3D structures [2] [Open access].

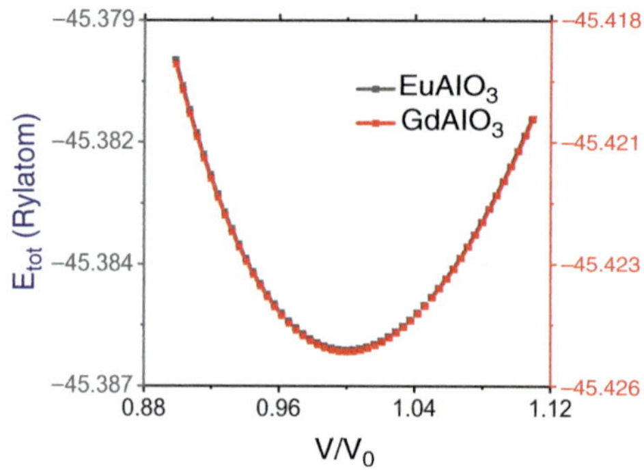

FIGURE 8.7 Total energy of $GdAlO_3$ and $EuAlO_3$ per atom versus unit cell sizes [2] [Open access].

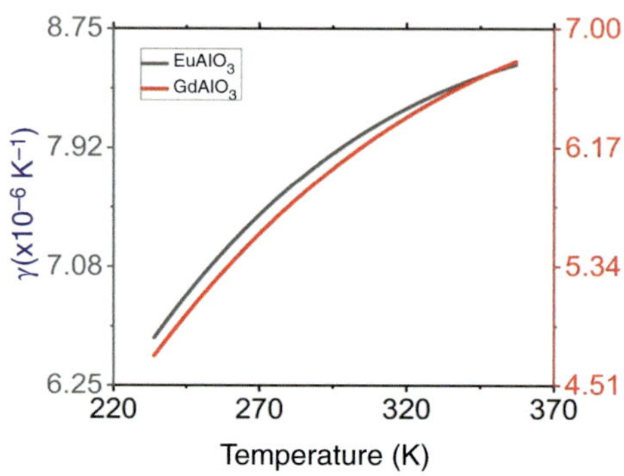

FIGURE 8.8 Thermal expansion coefficients of $GdAlO_3$ and $EuAlO_3$ depending on temperature [2] [Open access].

The mechanical stress that occurs in a material as a result of temperature changes is called thermal stress. Figure 8.9 shows that as the temperature increases, the thermal stress decreases for both materials. This is notable because it indicates improvements in the long-term reliability, manufacturing efficiency, structural integrity, and performance of the material in various applications. Thermal stress occurs when materials with different CTEs expand or contract at different rates in response to temperature changes.

From an electronic perspective, the conductive and semiconducting properties of Haussler alloys can be inferred from their band structure and density of states (DOS). Using techniques such as DFT and angle-resolved photoemission spectroscopy (ARPES) to predict and verify their electronic structure, scientists can modify these materials for applications in thermoelectric devices and magnetic tunnel junctions. The carrier concentration, Seebeck coefficient, and electrical conductivity can be modified by microdoping and heat treatment [13–16].

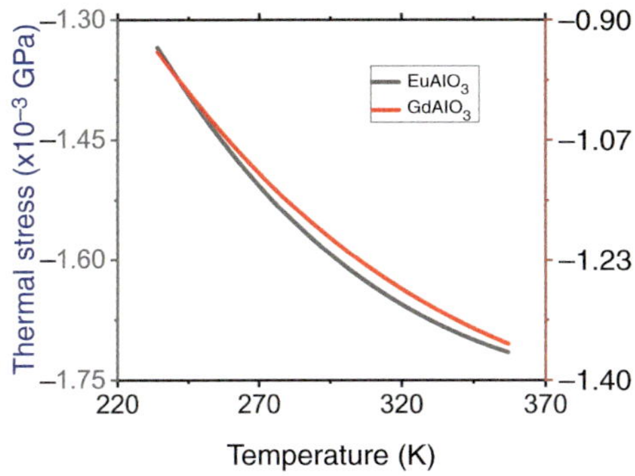

FIGURE 8.9 Thermal stress of $EuAlO_3$ and $GdAlO_3$ concerning temperature [2] [Open access].

Due to their versatile thermal behavior, which includes shape memory effects, magneto-thermal effects, and excellent thermal stability, Heusler alloys are effective options for use in solid-state cooling and thermal management systems. Thermogravimetric analysis (TGA), DSC, and thermal conductivity tests are often used to verify their thermal stability, phase changes, and durability under thermal cycling.

Based on $RInAg_2$, Galera et al. [9] investigated the electrical and magnetic properties of the entire Hessler series of rare earth elements (R = Y, La, Ce, Pr, Nd, Sm, Gd, Tb, Dy, Ho, Er, and Tm). XRD and neutron diffraction were used to create these compounds, and their crystal structures were precisely determined. In addition, Galera et al. showed that the cubic Hessler type $L2_1$ crystallizes these compounds. Based on the entire Hessler series $REInAu_2$ (RE = Y, La, Ce, Pr, and Nd), Besnos et al. [10] investigated a few rare earth elements and confirmed the phase transition from cubic to tetragonal structure, and studied these elements' structural and magnetic properties [9,10].

The specific heat, heat capacity, thermal conductivity, and electrical resistivity of two polycrystalline samples, CeInAu and $CeInAu_2$, were experimentally determined by Pleger et al. [11] using X-rays. After comparing their data with those of LalnAu and $LulnAu_2$, they discovered distortions related to FM-martensitic phase transitions and the effect of polycrystalline electric field resistance [12]. Additionally, using the rare earth material $YbInAu_2$, Alami et al. [13] calculated the effect of pressures above 8 GPa on the electrical resistivity of a full Heusler contact. A variety of full Heusler contacts, such as $YInAu_2$, $LaInAu_2$, $CeInAu_2$, $PrInAu_2$, and $NdInAu_2$, have been experimentally synthesized by Kurisu et al. [14]. The XRD technique is used to shed light on how these materials interact with mechanical forces at the atomic level. In another study, Berry et al. [15] analyzed the structural, elastic, electronic, and magnetic properties of the full-potential linear augmented plane wave (FP-LAPW) method, a highly accurate first-principles computational technique. According to their theoretical results, these compounds behave as ferromagnets and crystallize in a cubic $L2_1$ structure. To further develop theoretical knowledge, Khalfaoui et al. [16] also studied another full-potential Hössler compound, the rare-earth element-based Ag_2CeAl, using ab initio (first-principles) calculations. However, this compound belongs to the class of compounds with FM metal behavior based on analysis of its electronic and magnetic properties. The computer software WIEN2k [18], which originally used the FP-LAPW method [17], was used for the theoretical study.

Shishido et al. [19] experimentally examined the dependence of the lattice parameters and hardness of rhodium borides and carbides by varying the R atoms and detecting the boron content. They found that the mechanical strength and chemical stability of the $RRh_3B_xC_{1-x}$ phase are predominantly determined by its boron content.

Sahara et al. [20] calculated the structural and electrical properties of RRh_3B_x and $Rh_3B_xC_{1-x}$ perovskites (R = Sc and Y) using first-principles projected augmented wave (PAW) calculations. They found that with increasing boron doping, the elastic modulus consistently decreased. Kojima et al. [21] also investigated the elasticity. They investigated the effect of replacing Ce atoms with Sc, Y, and La atoms on the structural and elastic properties using ab initio simulations.

8.2 PEROVSKITES SUBSTITUTED WITH ANIONS

The general formula for perovskites is ABX_3, where A and B are cations of different sizes, and X is typically an anion, such as oxygen or a halide. The flexibility of perovskite structures provides an ideal platform for chemical substitution, allowing precise tuning of their optical, electrical, magnetic, and catalytic properties. Because it can alter the material's band structure, stability, and functional behavior without necessarily altering the cationic framework, anion substitution has recently received significant attention, among other modifications.

Anion substitution is the process of replacing the original anions (often O^- or halides, such as Cl^-, Br^-, and I^-) with additional anions, such as F^-, N^-, and S^-, or even mixed halide structures. Researchers can produce perovskite materials using this method for a variety of applications, such as photovoltaics, light-emitting diodes (LEDs), sensors, and photocatalysts. For example, partial substitution of bromide with iodide in halide perovskites used in solar cells can enhance stability and adjust the band gap, leading to increased efficiency and improved moisture resistance [22–24].

The structural stability of perovskites, which affects crystal symmetry and electrical behavior, can also be affected by anion substitution. In some cases, adding anions such as nitrogen to oxide perovskites produces oxynitrides that are more capable of absorbing visible light, making them more suitable for photocatalytic water splitting. Similarly, sulfur substitution can reduce the band gap of perovskite oxides, allowing for more efficient absorption of visible light. A variety of synthetic pathways can be used to accomplish anion substitution, including solid-state reactions, sol–gel procedures, ion exchange methods, and gas-phase processes. The intended composition, degree of substitution, and the end use of the material influence the choice of method. Despite the encouraging benefits, challenges remain in maintaining structural stability, regulating phase purity, and achieving uniform anion distribution throughout the synthesis process. In general, perovskite refers to any crystalline material similar to $CaTiO_3$ [8] with the chemical formula ABX_3, as shown in Figure 8.10 [3].

Nevertheless, anion-substituted perovskites represent a promising field of materials research, with huge untapped potential. Further research into their synthesis,

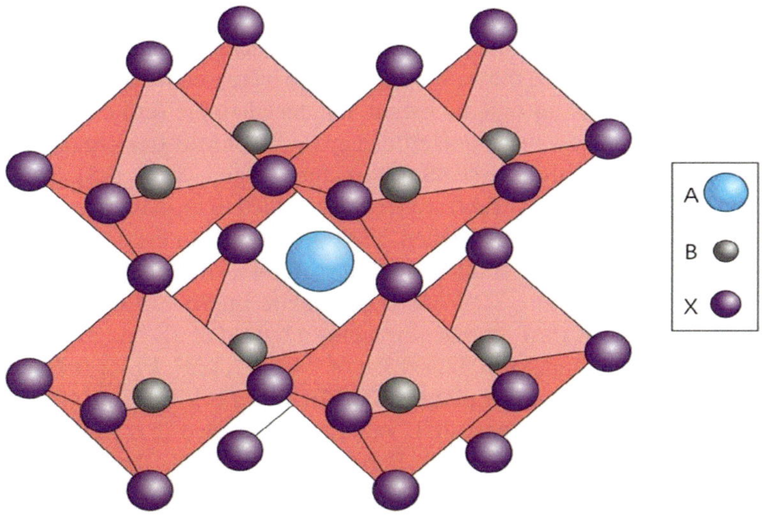

FIGURE 8.10 Typical crystal structure of perovskite [3] [Open access].

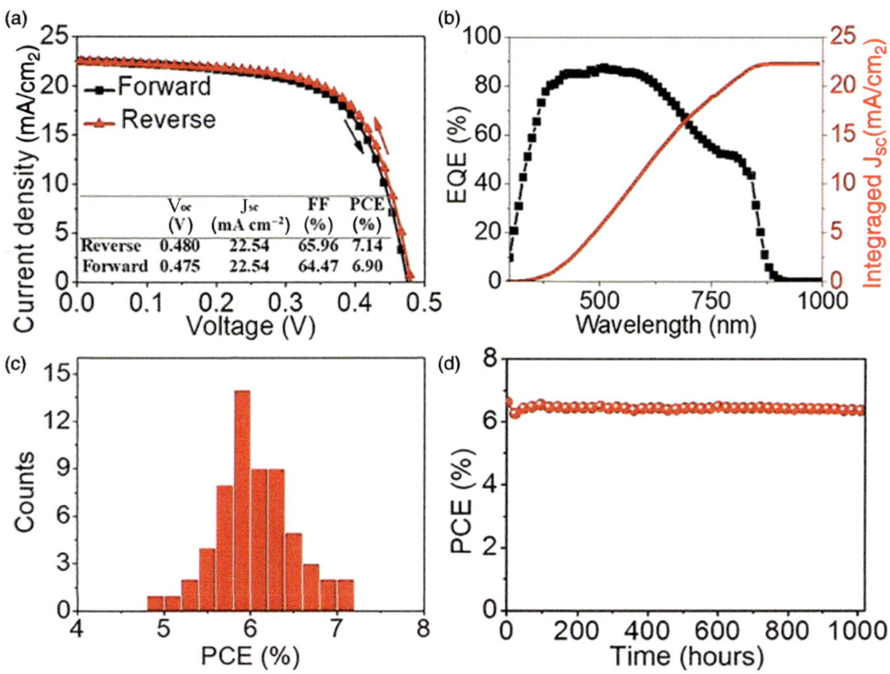

FIGURE 8.11 (a) Current density–voltage (Jsc-V) of the $FASnI_3$ perovskite absorber for the cell {en} with 10% en loading under reverse and forward voltage sweep; (b) EQE and integrated Jsc measured from solar cells with 10% en; (c) Power conversion efficiencies (PCEs) plots of 60% solar cells with 10% en; (d) Efficiency of a 10% en coated device as a function of storage time [3] [Open access].

characterization, and applications is expected to reveal new features and enhance the efficiency of electrical, environmental, and energy systems [1].

The band gap increased as the perovskite volume increased when Cs^+ was replaced by MA^+ or FA^+ [24]. As an organic cation, ethylenediamine (en) enhances air stability and raises the power conversion efficiency (PCE) of lead-free organic-inorganic perovskites to 7.14% in the {en}$FASnI_3$ structure (Figure 8.11) [25]. The organic cation 2-phenylethyl alcohol (PEAl) was recently added to tin-based perovskite solar cells. The PCE increased to 6.98% when PEAl was present in the device.

Additional recent experimental results have shown that incorporating a small amount of 2D PEA_2SnI_4 into 3D $FASnI_3$ perovskite improves the orientation and morphology of the 3D $FASnI_3$ grains, producing a hybrid of 2D and 3D perovskites. A high open-circuit voltage of 0.525 V, Jsc of 24.1 mA/cm², FF of 0.71, and PCE of 9.0% were achieved for the planar ITO/PEDOT: PSS/$FASnI_3$/C60 + BCP/Al structure [26]. This is the highest PCE value ever recorded in a lead-free organo-inorganic halide perovskite solar cell. Figure 8.12 illustrates the device performance [3].

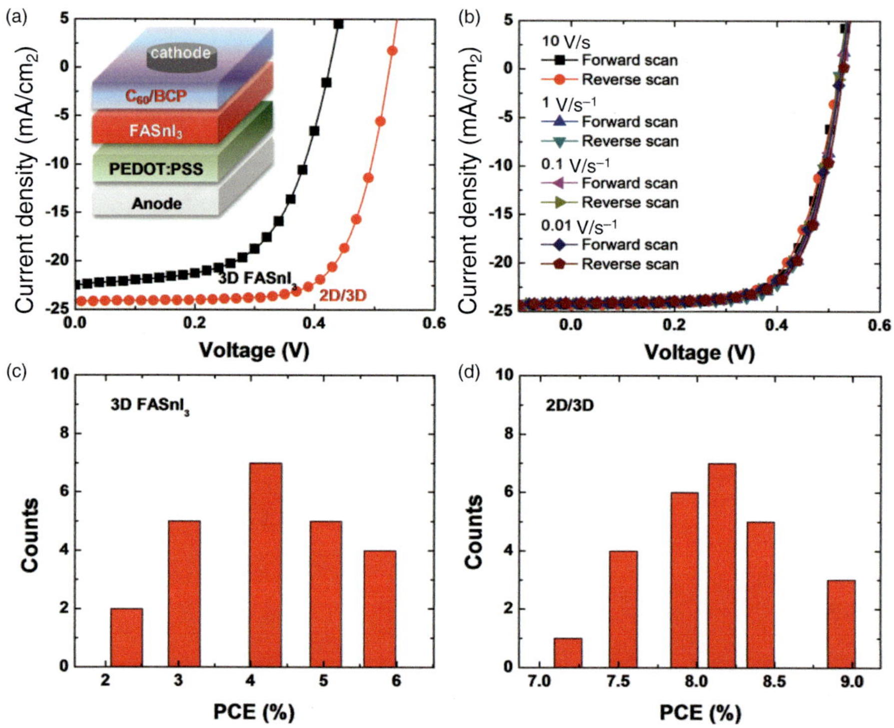

FIGURE 8.12 Device characterization and structure. Champion pure 3D and 2D (0.008 m)/3D perovskite devices (device structure is shown in the thumbnail) with J–V curves under AM 1.5 G sun conditions (a); forward and reverse scans of the J–V characteristics of a Champion-fabricated 2D/3D perovskite cell, measured at different rates (b); frequency plot of the reference cell (c); and 2D/3D perovskite devices (d) [3] [Open access].

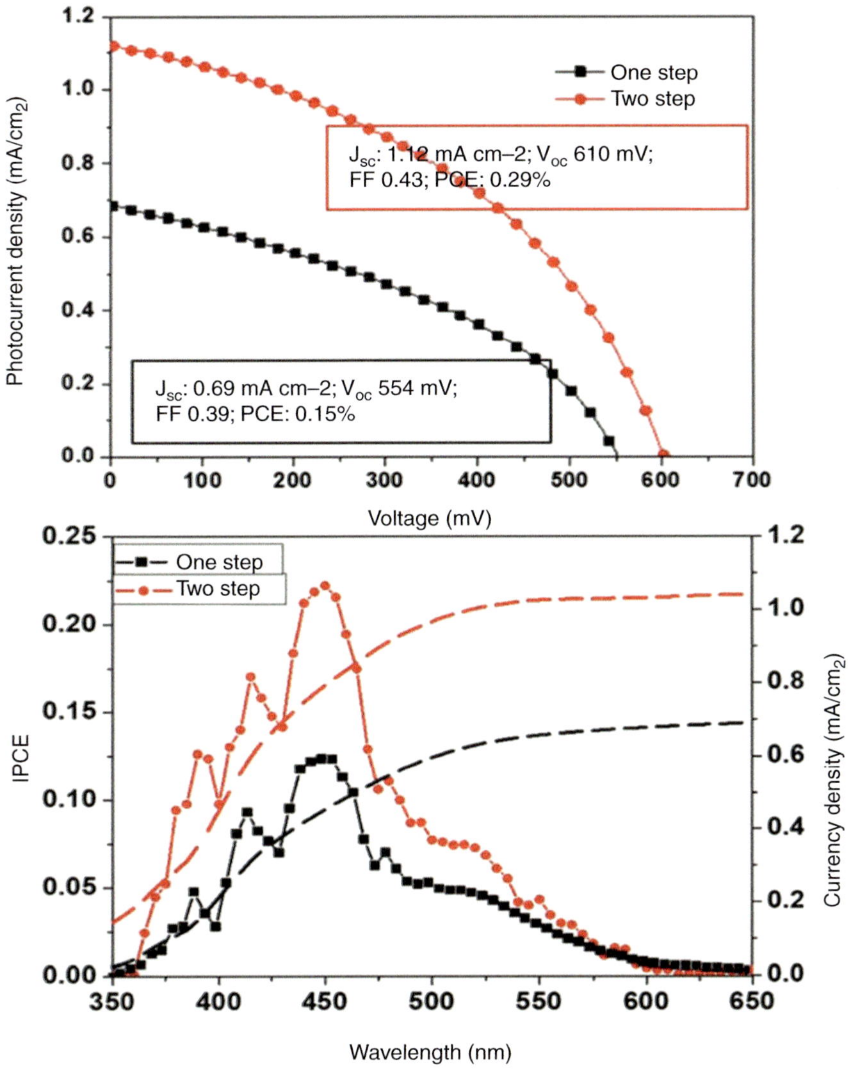

FIGURE 8.13 (a) Under simulated AM1.5 G irradiation, the J–V curves of the highest-performance devices using the one-step and two-step methods are shown, respectively; (b) corresponding IPCEs [3] [Open access].

This may be the factor preventing bismuth-based perovskites from achieving strong photovoltaic performance. Further research on highly efficient bismuth halide absorbers in perovskite solar cells showed that a two-step process, in which spin-coated BiI_3 and CH_3NH_3I were separately applied to porous titanium dioxide, increased the PCE to 0.29% compared to the previously reported PCE [6]. The PCE and incident photon-to-current efficiency (IPCE) values of the device are shown in Figure 8.13 using a two-step method developed by Galatopoulos et al. [4].

8.3 STRUCTURES OF A-SITE-DEFICIENT PEROVSKITES

In general, perovskites are highly adaptable materials due to their diverse physical and chemical properties. The anion "X," which is often oxygen or a halide, occupies the surface sites of a typical perovskite structure, while the cation "A" is a bulky ion located at the corners of the unit cell, and the cation "B" is smaller and located at the center. On the other hand, A-site-deficient perovskites are produced when the A-site is largely vacant. Due to their improved processing properties and distinctive structural characteristics, these materials have attracted significant interest [22–24].

The general formula for A-site-deficient perovskites is $A_{1-x}BX_3$, where x is the degree of deficiency. When A-site cations are removed, vacancies can cause symmetry shifts, structural distortions, and changes in the overall electronic structure. In applications such as catalysis, fuel cells, and battery materials, these vacancies often lead to more open structures, which improve ion transport and increase surface activity [22].

The tolerance factor, which is influenced by ion sizes and the extent of vacancies, is a crucial factor in the structural stability of A-site-deficient perovskites. When the tolerance factor deviates significantly from the optimal range, the structure can change from cubic perovskites to less symmetrical shapes such as rhombohedral, tetrahedral, or hexahedral. These structural changes can affect the material's conductivity, magnetism, and light absorption.

Lanthanide compounds or alkaline earth metals form the basis of a well-known class of A-site-deficient perovskites, in which the A-site cations are either completely removed or replaced by smaller ions. For example, A-site deficiency can improve oxygen mobility and performance as oxygen evolution catalysts in lanthanum nickelate ($La_{1-x}NiO_3$). Similarly, A-site vacancies in strontium titanate ($Sr_{1-x}TiO_3$) can contribute to improved electrical and electrochemical properties.

Many synthesis methods, such as sol–gel processes, soft chemical methods, and high-temperature solid-state reactions, can be used to create A-site vacancies. Because they directly influence the final structure and function of the material, controlling the number and distribution of vacancies during synthesis is critical.

A-site-deficient perovskites are an interesting class of materials in which vacancies are intentionally added to achieve structural properties. Regarding the surface topography of the nanoporous thin films, in addition to the topographic images, two important surface roughness indices were calculated: P-V (peak-to-valley), which indicates the vertical distance between the highest and lowest points on the surface, and root mean square (RMS) roughness, which provides a statistical measure of the average height changes on the surface.

The RMS values were first determined for two types of thin films: $Cu:NiO_x$ (copper-doped nickel oxide) and *β*-alanine-modified $Cu:NiO_x$, both of which were prepared on quartz substrates. With RMS values of approximately 3 nm, both films showed equivalent surface roughness, indicating smooth surfaces.

In addition, the root mean square roughness of the perovskite films remained constant at approximately 10 nm when the perovskite layer was formed on two different substrates: $ITO/Cu:NiO_x$ and $ITO/Cu:NiO_x$ treated with β-alanine. This indicates that the substrate surface roughness is not adversely affected or increased by the surface treatment with β-alanine used to modify the Cu:NiO_x layer [22–24].

These results indicate that the addition of β-alanine as a surface modification maintains the desired smoothness of the Cu:NiO_x films, which is critical for obtaining excellent perovskite layers and efficient device performance (Figure 8.14).

Photocurrent mapping (PCT) was used to create a visual representation of the degradation of untreated and β-alanine-treated Cu:NiO_x devices, as shown in Figure 8.15. Using PCT, we can obtain a visual representation of the current distribution in the device. Even in the case of modern devices, untreated Cu:NiO_x devices exhibited an uneven current distribution. Additionally, other parts of the device exhibited very low current, which may be caused by defects. The defect sites became larger and exhibited greater current loss over time. The β-alanine-treated devices exhibited a more uniform current distribution, which remained largely constant over time.

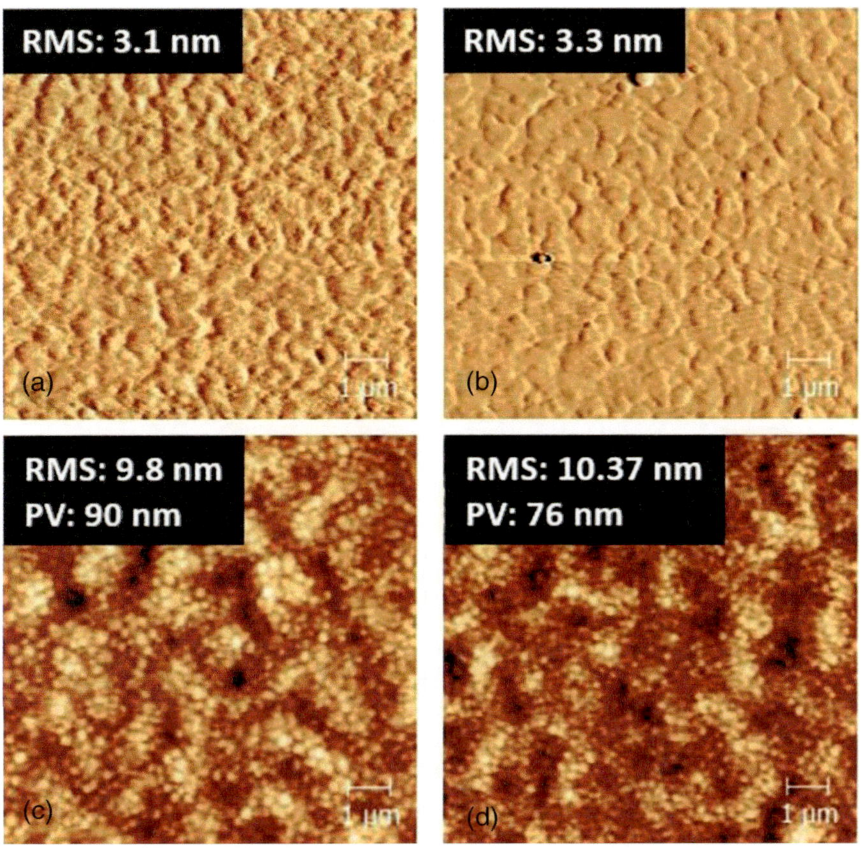

FIGURE 8.14 (a) Quartz/copper phase contrast: NiO; (b) quartz/copper phase contrast: NiO/β-alanine; (c) indium/copper oxide topography: NiO/perovskite; (d) atomic energy microscope images of indium/copper oxide topography: NiO/β-alanine/perovskite [4] [Open access].

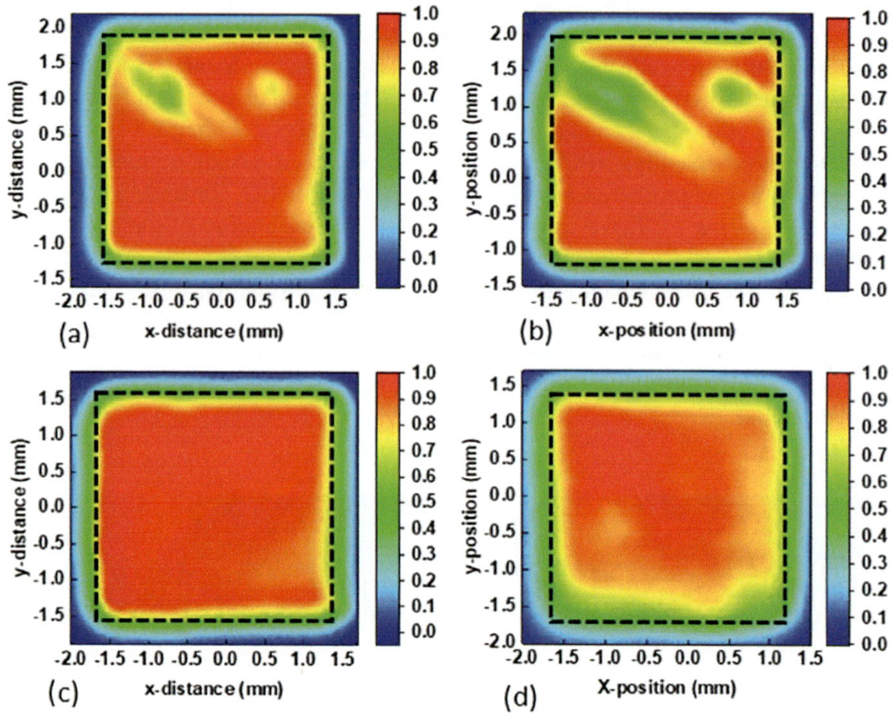

FIGURE 8.15 Photocurrent mapping of (a) fresh, (b) aged, and (c) fresh β-alanine-treated Cu:NiO_x devices. In addition, (d) aged β-alanine-treated Cu:NiO_x devices [4] [Open access].

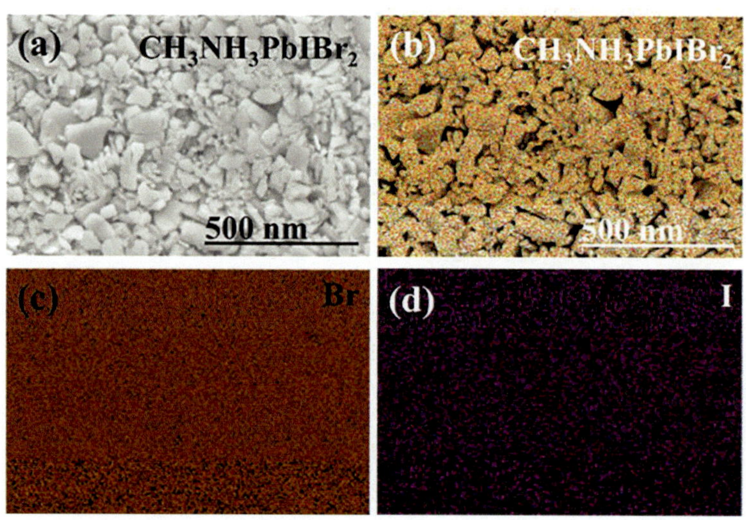

FIGURE 8.16 (a) Surface SEM analysis of the top multilayer layer of $CH_3NH_3PbIBr_2$ thin film; (b) elemental analysis of $CH_3NH_3PbIBr_2$ thin film by color scanning electron microscopy (FESEM); and dispersion of bromine (c) and iodine (d) [5] [Open access].

The scanning electron microscopy (SEM), surface analysis of the $CH_3NH_3PbIBr_2$ multilayer thin film is shown in Figure 8.16 for elemental analysis and distribution of the dihalide elements (Br, I). The SEM surface image of the $CH_3NH_3PbIBr_2$ layer, shown in Figure 8.16a, shows a loose and porous structure. According to the 2D elemental surface maps, bromine and iodine have a higher degree of homogeneity and dispersion on the surface (Figure 8.16b–d). Figure 8.17a shows the absorption spectra for the wavelength range of 300–1100 nm obtained at room temperature. The absorption spectra of the samples show two distinct patterns. In the visible region (350–700 nm), both films show a tendency toward decreased absorption, while the absorption decreases in the infrared/near infrared (IR/NIR)

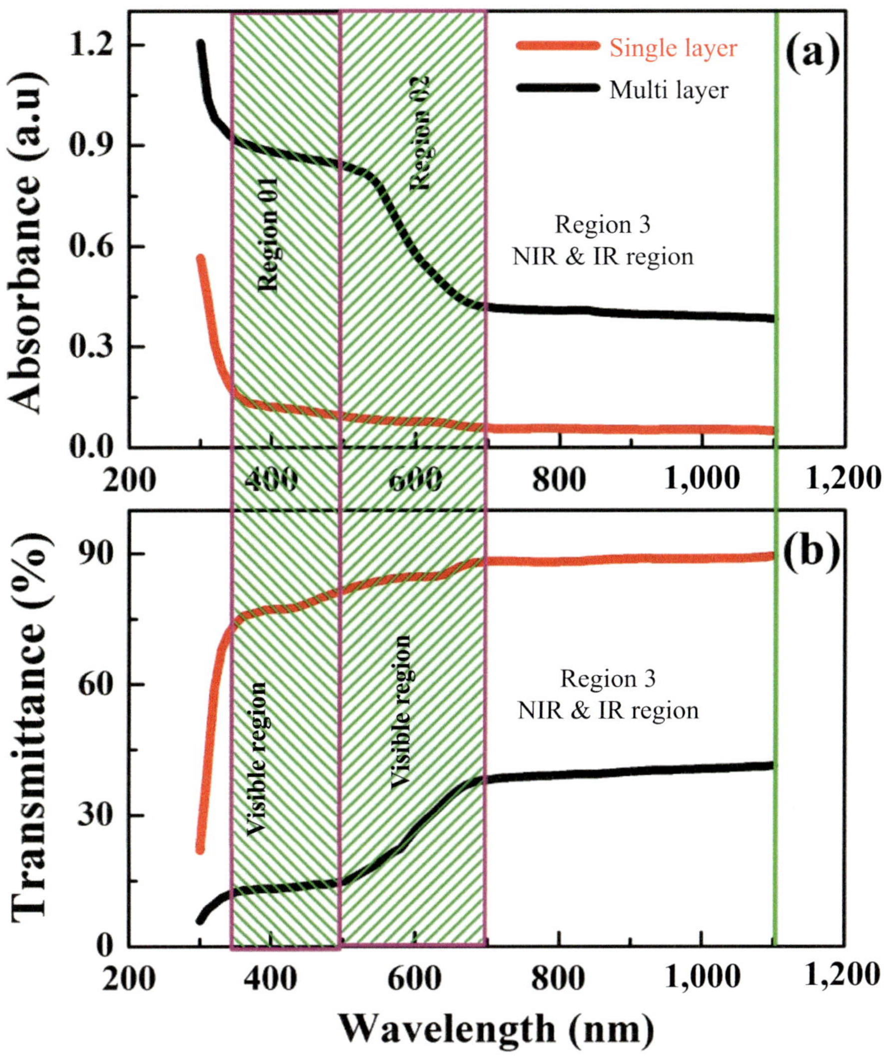

FIGURE 8.17 Spectra of single- and multilayer $CH_3NH_3PbIBr_2$ thin films showing (a) absorption and (b) transmittance [5] [Open access].

region. This decrease is due to the lower excitation energy of both samples in the band gap.

8.4 PHASES DEFICIENT IN ANION AND INCLUDING TETRAHEDRA

Anion-deficient phases are a special class of materials in solid-state chemistry where the anion concentration is lower, either intentionally or naturally, than the optimal stoichiometric value. Instead of the octahedral coordination more common in perovskites and other oxide materials, these materials often adopt complex topologies to compensate for the lack of anions. This can lead to the formation of unusual coordination environments, such as tetrahedral ones [22–24].

High-temperature synthesis conditions, redox reactions, or the presence of volatile anionic species are often the causes of this deficiency. Electron conductivity, ion mobility, and catalytic behavior are just a few examples of properties that can be significantly affected by structural voids or rearrangements resulting from partial anion addition. The crystal structure of the synthetic nickel-containing tetrahedron was determined to describe the crystal structure of the synthetic manganese tetrahedron. Recent material science research on tetrahedra may strengthen this group. The stability of the structural pattern (Figure 8.18) with each replacement of these elements is what makes the tetrahedral tennantite structure so special.

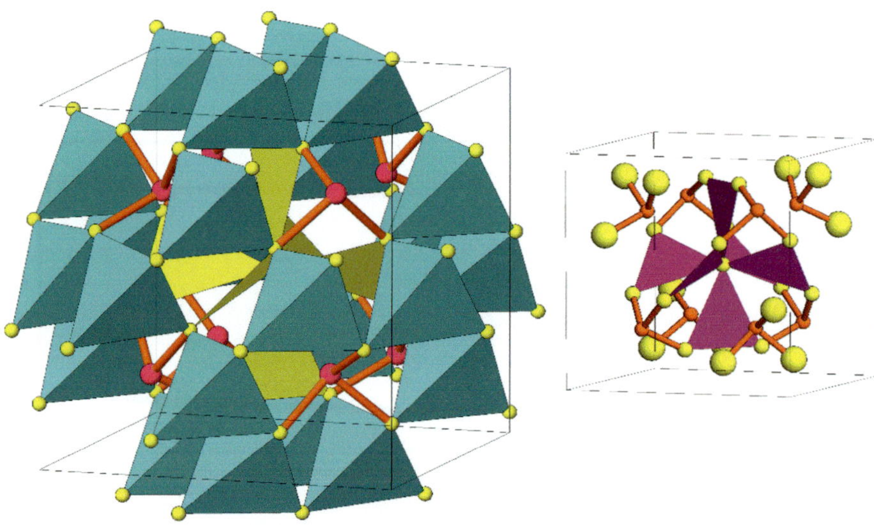

FIGURE 8.18 Crystal structure of tennantite tetrahedra. The polyhedral representation of the $Sb(As)S_3$ coordination pyramids in the form of cation–anion bonds shows the MeS_4 coordination tetrahedron (light blue) and most of the $(Cu, Ag)S_3$ coordination triangles (yellow). Note how the "vortices" made from the coordinate triangles fill the cavities of the truncated tetrahedra in the tetrahedral framework. The "vortex" of the $(Cu, Ag)S_3$ coordination triangles is shown in the inset, surrounded by the $(Sb, As)S_3$ coordination pyramids [6] [Open access].

The stability of tetrahedral units in the structure is an interesting effect of anion deficiency. In fully oxidized materials, transition metal cations usually exhibit stable electronic configurations and are essential for determining the structural, electrical, and electrochemical properties of the material.. These cations can only coordinate with four anions, creating tetrahedral geometries when available anions are scarce. When these tetrahedral structures are formed, the electron environment undergoes local distortions and changes, often resulting in new phases with unique structural and functional properties.

Brownmillerite structures are a well-known example of this type of behavior, where ordered oxygen vacancies cause alternating layers of octahedral and tetrahedral structures to form. Anion migration, which allows the remaining anions to flow more freely through the lattice, is another important feature of anion-deficient tetrahedral phases. This enhancement can enhance the material's performance in processes such as electrochemical energy conversion, gas sensing, and catalysis. Additionally, the distinct tetrahedral shape can influence the electronic band structure, sometimes resulting in a narrowing of the band gap or changes in magnetic properties [24].

In short, tetrahedral coordination in anion-deficient phases represents a rich field for materials design and research. Researchers can tailor materials for specific applications, such as energy devices or environmental technologies, by varying the anion content and understanding how it affects the coordination environment. These structures demonstrate the importance of defect chemistry in the creation of advanced functional materials, challenging traditional design principles [4,5]. The calculated electron densities in the energy range of –0.5 to 0 eV (compared to EF) in Figure 8.19 highlight this fact. As expected from the d9 electron structure and square planar environment, Cu-3d_{x2-y2} is the magnetically active orbital or the orbital that carries the magnetic moment. The predicted magnetic moment for copper in Cu_2O_{X2} is between 0.6 and 0.7 μB, which is consistent with the currently available experimental data for Cu_2OCl_2. Indeed, powder and single-crystal neutron refinements showed that the ordered magnetic moments were 0.64(5) μB and 0.66(2) μB, respectively.

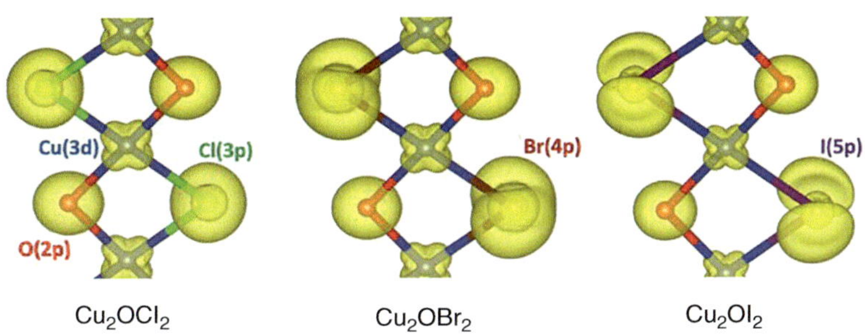

FIGURE 8.19 For Cu_2OX_2 compounds (X = Cl, Br, I), the electron density corresponding to the valence band top (from –0.5 to 0 eV vs. EF) was derived from in situ PBE0 calculations in the ground state of the collinear AFM order. The isovalue of 0.1 electron/Å^3 is indicated. Cu, O, Cl, Br, and I atoms are represented by blue, red, green, brown, and violet spheres, respectively [7] [Open access].

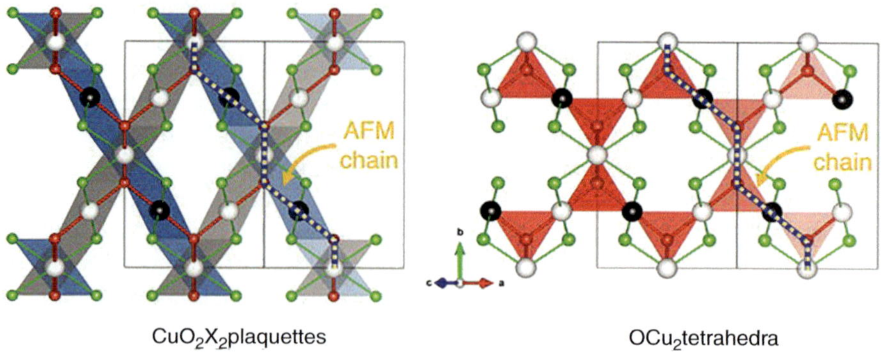

FIGURE 8.20 The basic linear magnetic structure of AFM is shown schematically in the ribbon (left) and pyrochlore (right) lattices. The up and down spins carried by the copper (Cu) sites are indicated by black and white spheres, respectively. The halogen and oxygen molecules are represented by green and red spheres, respectively. Transparent red color is used to denote the tetrahedral coordinates of OCu_4. For the up or down spin carried by the copper (Cu) center, the CuO_2X_2 plates are shown in transparent blue and gray, respectively. The yellow and blue dotted line indicates the presence of one AFM spin chain, which arises from the $J = J_1$ bond and is taken into account in the quasi-1D AFM model [7] [Open access].

In other words, the current calculations show that the ribbons are essentially FM ordered due to the first neighbor interaction (SE) J_1, and that the spins are frustrated due to the second neighbor interaction (SSE) J_4 of the atomic force microscope (AFM), whose amplitude is 1.6 times smaller than J_1. Due to J_3, there is an AFM interaction between the ribbons (J_2 is statistically insignificant). Each OCu_4 tetrahedron in the pyrochlore-like lattice is globally AFM ordered with two top and two bottom magnetic centers, emphasizing the spin-frustrated nature of the magnetic structure. Figure 8.20 illustrates the resulting more stable linear magnetic order.

8.5 PHASES DEFICIENT IN ANIONS AND INCLUDING SQUARE PYRAMIDS

Anion-free phases are an interesting area of solid-state chemistry, where compounds exhibit structural or compositional features characterized by the absence of negatively charged ions. These phases often exhibit complex crystal structures and interesting properties that make them relevant for a variety of technological applications, including magnetism, ionic conductivity, and catalysis. An interesting structural feature of many of these phases is a square pyramidal coordination environment, in which the central atom, typically a transition metal, is surrounded by five ligands: four arranged in a square plane and one axial, forming a square pyramidal geometry. This coordination is less common than the classical octahedral or tetrahedral geometry, but it is essential for stabilizing certain electronic configurations and accommodating anion vacancies. Several processes, such as high-temperature synthesis, non-stoichiometric syntheses, or redox reactions that result in the partial oxidation or reduction of cations, can produce anion-free phases. These compounds often utilize changes in the oxidation states of cations, structural rearrangements, or the introduction of defects that alter the electrical and magnetic behavior of the material to compensate for the absence of some anions.

Anion vacancies can distort the coordination environment in phases containing square pyramidal units, increasing the stability of the hierarchical arrangement. For example, asymmetric bonding conditions, where square pyramids become energetically favorable, cause a deficiency of oxygen or halide atoms in some metal oxides and halides. These patterns contribute to the diversity of physical properties observed in multilayer structures, chain compounds, and even 3D structures [23,24].

With advances in synthesis and characterization techniques, such as electron microscopy, neutron scattering, and XRD, the study of these anion-deficient square pyramidal phases continues to expand. By understanding how these structures form and behave, new materials with tailored functions can be developed, potentially impacting everything from energy storage to electronic devices. A comparative study was performed on a sample of the polystyrene–low-density polyethylene copolymer system (PS-LDPE) copolymer system to examine the tip penetration effect. With an elastic modulus of 0.1 GPa, LDPE is a softer polymer than polystyrene (PS), which has an elastic modulus of approximately 2 GPa. Using this sample, it is possible to determine whether the polymers respond differently to AFM tangential loss imaging using the first or second eigenmode frequency. An identical region of the sample excited by the first and second eigenmodes is depicted in Figure 8.21a and b, respectively. The AFM tangential loss image obtained using the first and second eigenmodes excitation frequencies is shown in Figure 8.21c and d, respectively. These figures show that as the frequency at which we excite the cantilever increases,

By changing the oscillation amplitude, it is possible to control the influence of the force interactions between the tip and the sample on the cantilever dynamics. The influence of the excitation frequency and oscillation amplitude on the measured values of the loss tangent of the PTFE sample is illustrated in a series of studies in Figure 8.22.

The dataset used in this study is shown in Figure 8.23. The sample is a thin layer of PS polymer intentionally deposited on a silicon wafer using a spin-coating process in a high relative humidity environment. This process forms discrete "islands" of polystyrene on the wafer surface. Due to the high humidity during deposition, some parts of the silicon substrate are left exposed and intentionally used as control regions for measurements. These exposed silicon regions have extremely high stiffness and low viscoelastic behavior.

Consequently, the theoretical loss tangent ($\tan \delta$) for these control regions is expected to be close to zero. The loss tangent, or $\tan \delta$, is a dimensionless measure that represents the percentage of dissipated energy.

The first eigenfrequency of the atomic force microscope (AFM) was used to obtain the height image (Figure 8.23a) and the corresponding loss shadow image (Figure 8.23c) in Figure 8.23, while the second eigenfrequency was used to obtain the same types of images (Figure 8.23b for height and Figure 8.23d for loss shadow) in Figure 8.23. The topographic data in Figure 8.23a and b do not show

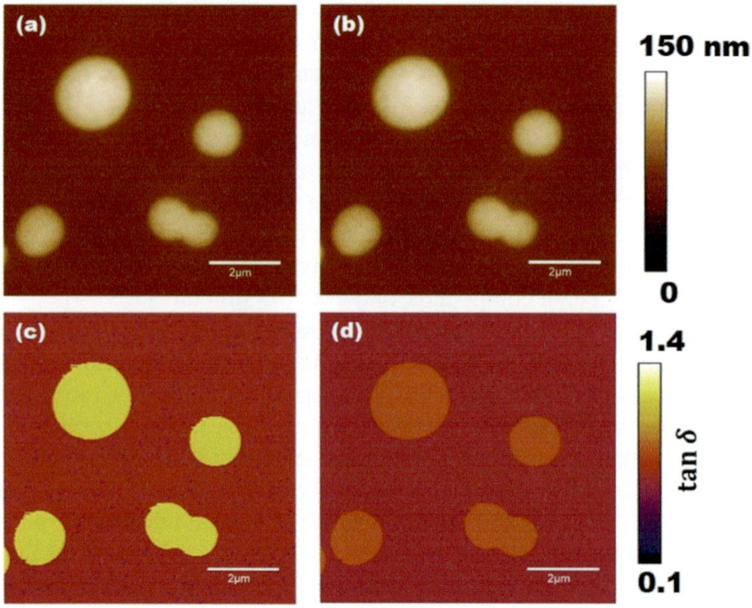

FIGURE 8.21 Atomic beam microscopy (AFM) analysis of the loss shadow of PS-LDPE: (a) first eigenmode height image, (b) second eigenmode height image, (c) first eigenmode loss shadow image, and (d) second eigenmode loss shadow image. The image dimensions are 6 μm × 6 μm. The initial free amplitude was 100 nm, while the second free amplitude was 20 nm. To ensure the visualization of the repulsive mode, the setpoint was 50% for both measurements. The frequencies of the first and second eigenmodes were 70 kHz and 458 kHz, respectively [8] [Open access].

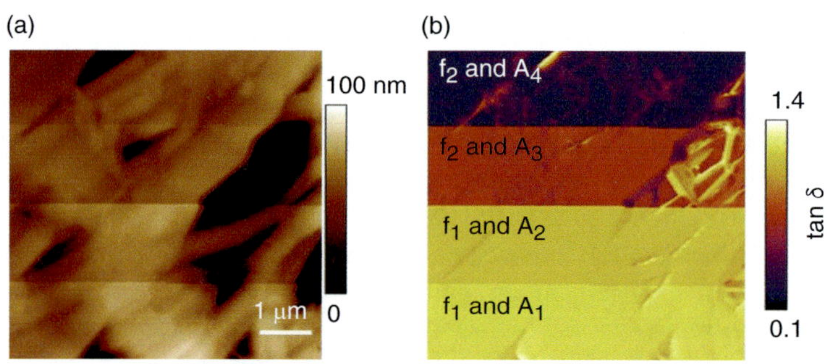

FIGURE 8.22 The topographic and tanθ images are used in an atom probe microscopy (AFM) study of the PTFE loss shadow. The excitation frequencies and free oscillation amplitudes are as follows, from bottom to top: f_1 = 72 kHz and A_1 = 50 nm, f_1 = 72 kHz and A_2 = 100 nm, f_2 = 460 kHz and A_3 = 1.5 nm, and f_2 = 460 kHz and A_4 = 3.5 nm. The scale is 1 μm [8] [Open access].

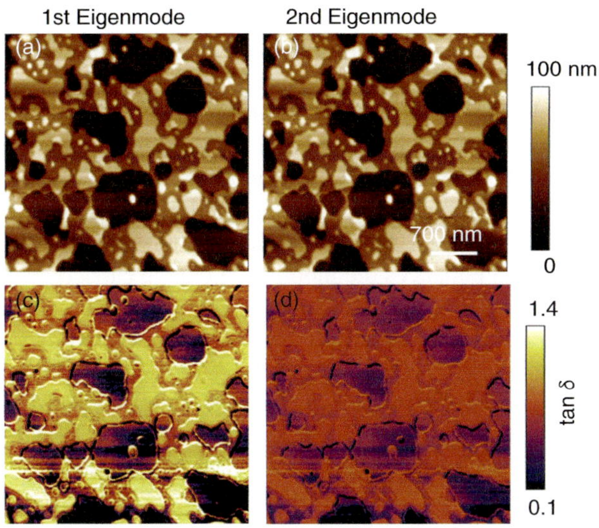

FIGURE 8.23 Atomic force microscopy (AFM) scanning of the PS loss shadow includes: (a) topography using the first eigenmode, (b) topography using the second eigenmode, (c) tanδ using the first eigenmode, and (d) tanδ using the second eigenmode. Scale bar is 700 nm. The first and second eigenmodes had free oscillation amplitudes of 100 nm and 20 nm, respectively. Both measurements were performed in the repulsion mode with a set point of 60%. The first and second excitation frequencies were 70 kHz and 460 kHz, respectively [8] [Open access].

significant differences, indicating that the surface morphology was constant across all measurements, but the loss shadow images in Figure 8.23c and d show a marked change. In particular, the average tan δ value observed in the polymer sample is significantly higher when measured using the first eigenfrequency than when measured using the second eigenfrequency.

Electronic structural investigations, including band structure and DOSs, reveal that these compounds exhibit significant hybridization between the d orbitals of Sc and the p orbitals of Z carbon near the Fermi level. This interaction is attributed to the high mobility of charge carriers, a prerequisite for improving thermoelectric efficiency, which often results in beneficial properties.

The mechanical and thermal stability of XScZ compounds are additional important properties. Formation energy calculations confirm their energetic preferences, indicating their potential for use under standard experimental conditions. Furthermore, the absence of spectral modes in the Brillouin zone demonstrates dynamic stability in phonon scattering analysis. They are also suitable for industrial energy harvesting and conversion systems due to their resistance to high temperatures.

XScZ compounds are attractive options for thermoelectric applications due to their balanced electrical conductivity and, potentially, their low lattice thermal conductivity. Additionally, when properly doped or quenched, these materials can also exhibit semi-metallic or spin-polarized states, opening the door to spintronic applications as spin filters or magnetic sensors. The thermoelectric properties of rubidium-based semi-hesselite compounds are among their most interesting features. Thermoelectric materials that are capable of converting heat into electricity and vice versa are essential for energy-efficient systems, especially in power generation and cooling technologies. Rubidium-based compounds offer an ideal balance between low thermal conductivity and electrical conductivity, two important elements for efficient thermoelectric performance. Their thermoelectric properties can be further improved by doping or alloying these compounds with other elements. In addition, it has been shown that adding rubidium to Hessler semimetals alters the electronic structure of the material in a way that enhances the mobility of charge carriers while maintaining a low DOSs at the Fermi level. This behavior leads to an increase in the Seebeck coefficient, a key measure of a material's ability to generate an electrical potential in response to a temperature gradient. High charge carrier mobility and a favorable Seebeck coefficient work together to produce efficient thermoelectric devices.

PROBLEMS

1. What is the basic structure of perovskite compounds?
2. What is the effect of doping on perovskite properties?
3. Why are hybrid perovskite solar cells used?
4. How do defects affect perovskites' performance?
5. What limits the stability of perovskite materials?

REFERENCES

1. Satapathy, S., Batouche, M., Seddik, T., Salah, M. M., & Maurya, K. K. (2023). First principle study on structural, thermoelectric, and magnetic properties of cubic $CdCrO_3$ perovskites: a comprehensive analysis. *Crystals*, 13(8), 1185.
2. Alruqi, A. B., & Ongwen, N. O. (2024). Ab initio investigation of the mechanics and thermodynamics of the Cubic $EuAlO_3$ and $GdAlO_3$ perovskites for optoelectronic applications. *Crystals*, 14(12), 1084.
3. Sani, F., Shafie, S., Lim, H. N., & Musa, A. O. (2018). Advancement on lead-free organic-inorganic halide perovskite solar cells: A review. *Materials*, 11(6), 1008.
4. Galatopoulos, F., Papadas, I. T., Ioakeimidis, A., Eleftheriou, P., & Choulis, S. A. (2020). Surface treatment of Cu:NiOx hole-transporting layer using β-alanine for hysteresis-free and thermally stable inverted perovskite solar cells. *Nanomaterials*, 10(10), 1961.
5. Jabeen, N., Zaidi, A., Hussain, A., Hassan, N. U., Ali, J., Ahmed, F., ... & Helal, M. H. (2022). Single-and multilayered perovskite thin films for photovoltaic applications. *Nanomaterials*, 12(18), 3208.
6. Makovicky, E. (2021). Twinning of Tetrahedrite—OD approach. *Minerals*, 11(2), 170.
7. Lafargue-Dit-Hauret, W., & Rocquefelte, X. (2022). Influence of the chemical pressure on the magnetic properties of the mixed anion cuprates Cu2OX2 (X = Cl, Br, I). *Computation*, 10(5), 73.
8. Eslami, B., & Caputo, D. (2021). Effect of eigenmode frequency on loss tangent atomic force microscopy measurements. *Appl. Sci.*, 11(15), 6813.
9. Galera, R. M., Pierre, J., Siaud, E., & Murani, A. P. (1984). Magnetic and electrical properties of Heusler alloys with rare earths: RInAg2. *J. Less Common Met.*, 97, 151.

10. Besnus, M. J., Kappler, J. P., Meyer, A., Sereni, J., Siaud, E., Lahiouel, R., & Pierre, J. (1985). Crystallographic and magnetic properties of REInAu2 intermetallics. *Physica B+C*, 130, 240.
11. Asma, B., Belkharroubi, F., Ibrahim, A., Lamia, B., Mohammed, A., Belkilali, W., ... & Al-Douri, Y. (2021). Structural, mechanical, magnetic, electronic, and thermal investigations of Ag2YB (Y= Nd, Sm, Gd) full-Heusler alloys. *Emergent Materials*, 4(6), 1769–1783. https://link.springer.com/article/10.1007/s42247-021-00257-8
12. Besnus, M. J., Kappler, J. P., Ravet, M. F., Meyer, A., Lahiouel, R., Pierre, J., & Sereni, J. (1986). Structural and magnetic properties of the ternary rare earth (RE) compound series REInAu2 (RE La to Lu, Y). *J. Less Common Met.*, 120, 101.
13. Alami-Yadri, K., Wilhelm, H., & Jaccard, D. (1998). Electrical resistivity of YbInAu2 and YbCuAl up to 8 GPa. *Solid State Commun.*, 108, 279.
14. Kurisu, M., Fuse, A., Nobata, T., & Nakamoto, G. (2000). Structural and magnetic properties of RInAu2 (R= La, Ce, Pr and Nd) at high pressure. *Phys. B Condens. Matter*, 281–282, 147.
15. Berri, S., Maouche, D., Zerarga, F., & Medkour, Y. (2012). Ab initio study of the structural, electronic, elastic and magnetic properties of Cu2GdIn, Ag2GdIn and Au2GdIn. *Phys. B Condens. Matter*, 407, 3328.
16. Khelfaoui, F., Ameri, M., Bensaid, D., Ameri, I., & Al-Douri, Y. (2018). Structural, elastic, thermodynamic, electronic, and magnetic investigations of full-Heusler compound Ag2CeAl: FP-LAPW method. *J. Supercond. Nov. Magn.*, 31, 3183.
17. Blaha, P., Schwarz, K., Sorantin, P., & Trickey, S. B. (1990). Full-potential, linearized augmented plane wave programs for crystalline systems. *Comput. Phys. Commun.*, 59, 399.
18. Blaha, P., Schwarz, K., Medsen, G. K. H., Kvasnicka, D., & Luitz, J. (2001). WIEN2k: An augmented plane wave local orbitals program for calculating crystal properties. Technishche University, Wien.
19. Shishido, T., Ye, J., Okada, S., Kudou, K., Iizumi, K., Oku, M., ... & Nakajima, K. (2006). Synthesis, boron-nonstoichiometry and hardness of perovskite-type rare earth rhodium borides RRh3Bx (R= La, Gd, Lu and Sc). *J. Alloys Compd.*, 408, 379–383.
20. Sahara, R., Shishido, T., Nomura, A., Kudou, K., Okada, S., et al. (2006). Mechanism of the increase in bulk modulus of perovskite Sc Rh 3 B x by vacancies. *Phys. Rev. B*, 73, 184102.
21. Kojima, H., Sahara, R., Shishido, T., Nomura, A., Kudou, N., et al. (2007). Ab initio studies of structural, elastic, and electronic properties of RRh3BX (R= Sc, Y, La, and Ce). *Appl. Phys. Lett.*, 91, 081901.
22. Behera, D., Mohammed, B., Taieb, S., Mokhtar, B., Al-Qaisi, S., & Mukherjee, S. K. (2023). First-principle investigations on optoelectronics and thermoelectric properties of Lead-Free Rb2InSbX6 (X = Cl, Br) double perovskites: For renewable energy applications. *Eur. Phys. J. Plus*, 138, 520.
23. Seddik, T., & Batouche, M. (2022). Bulk to low dimensional 2D thermoelectric materials: latest theoretical research and future view. *Progress in Nanoscale and Low-Dimensional Materials and Devices: Properties, Synthesis, Characterization, Modelling and Applications*, 571-588. https://link.springer.com/chapter/10.1007/978-3-030-93460-6_20
24. Moutassem, M., Seddik, T., Mohammed, D. E. S., Batouche, M., Khachai, H., Khenata, R., Ahmed, R., Srivastava, V., Bouhemadou, A., & Kushwaha, A. K. (2022). Metal to semiconductor transition and figure of merit enhancement of Li2CuAs compound by Na substitution. *Bull. Mater. Sci.*, 45, 110.

9 Automotive and Aerospace Applications of Self-Healing Polymer Nanocomposites

9.1 INTRODUCTION

Self-healing thin polymers have garnered significant attention within the broader field of smart polymer materials due to their ability to automatically repair damage and extend material lifespan. Numerous technological applications, including electronics, supercapacitors, batteries, membranes, and medical devices, have demonstrated the value of these materials. More recently, they have become the subject of further research in the aerospace and automotive industries, where their performance and durability are being considered [1–3].

To improve the mechanical properties of polymer matrices and provide self-healing mechanisms in these materials, nanoparticles are commonly used to produce self-healing polymer nanocomposites. Carbon nanoparticles have been incorporated into some polymers used in aircraft and aerospace vehicles, providing self-healing properties as well as structural reinforcement. This prevents crack propagation and reduces the likelihood of catastrophic failure [4]. These materials are characterized by their ability to self-heal through external mechanisms such as nanocapsules or microvascular systems, or through external stimuli such as heat or light, or internal mechanisms such as dynamic crosslinking. These methods successfully seal cracks and restore functionality by regenerating the interconnected networks within the polymer chains. Extensive morphological studies have confirmed the effectiveness of these healing processes. Self-healing nanocomposites for automotive applications offer improved performance under stress or environmental influences, longer component lifespan, and lower maintenance costs.

The development of these smart materials represents a significant step towards safer, more efficient, and more sustainable transport systems [2]. Self-healing mechanisms have been categorized as internal and external [3]. The chemistry of these self-healing materials determines their differences. External healing factors, such as capsules or vesicles, are essential for external self-healing. Therefore, the production of materials based on catalysts and healing factors is integral to external self-healing. Without the aid of external healing factors, internal self-healing can occur. Damage to the polymer network, whether internal or external, initiates the self-healing process in both approaches. Polyurethane, poly(urea-formaldehyde), poly(melamine-urea-formaldehyde), and other polymers have been used to create

DOI: 10.1201/9781003758556-9

external self-healing microcapsules [4]. In general, non-covalent chemical reactions produce self-embedding, self-healing materials that exhibit host–guest interactions, hydrogen bonding, and π–π accumulation [5]. Covalent and non-covalent reactions, such as Diels–Alder reactions, esterification, and free radical exchange reactions, can be part of exogenous self-healing. Self-healing polymers and nanocomposites outperform non-self-healing materials due to their superior structural rigidity, supramolecular attractive forces, and physical properties. Theoretical and experimental studies have been used to explore the self-healing mechanisms of polymer nanocomposites. The applications of self-healing polymers and nanocomposites are diverse, including electronics, energy, transportation, coatings, aerospace, and others [6]. As an emerging technology in mechanical engineering, the self-healing concept can be applied to composite materials, including cementitious materials.

Self-healing materials include composites, metal-matrix composites, ceramic matrix composites, and polymer matrix composites (PMCs). They can also be used in commercial coatings for corrosion protection outside designated areas [7]. Self-healing composites have a wide range of potential applications in the automotive, aerospace, battery, sealant, and textile industries. Impact loads cause significant damage in the aerospace industry [8]. Self-healing materials can extend the service life of components and repair damage caused by the external environment. Repairing dynamic damage and maintaining impact resistance are two important applications of self-healing materials [9]. A healing agent is applied to the damaged area when damage occurs. Shape-memory effects are used in shape memory alloys to seal cracks and restore machinability.

9.2 BALANCED AND INFORMATIVE STRUCTURE

Presenting scientific material clearly, coherently, and logically requires a balanced and informative structure. This structure ensures that each part of the work contributes significantly to the overall objective while maintaining smooth transitions and a depth of ongoing discussion. This structure allows readers to grasp the complex mechanisms, production techniques, and material behaviors within the context of self-healing polymer nanocomposites without distraction or loss of focus.

A comprehensive literature review highlighting important advances and current knowledge gaps typically follows a detailed introduction that defines the scope, significance, and objectives of the research. A detailed explanation of the results, mechanisms, and performance evaluation should follow the concise and comprehensive descriptions of the materials, synthesis procedures, and characterization methods in subsequent sections. An equal emphasis on theoretical foundations, experimental protocols, and practical applications is another way to maintain balance. To enhance understanding, for example, quantitative analysis and visual data should be used to complement discussions on microcapsule fabrication, nanofiller dispersion, and self-healing mechanisms. Furthermore, this chapter offers a more comprehensive and in-depth perspective by incorporating comparative analyses of internal and external healing systems [10]. This review highlights technological potential, current limitations, and research objectives, concluding with comprehensive discussions, prospects, and recommendations. In addition to enhancing the study's scientific value, its comprehensive and informative structure facilitates easy reading and makes it more coherent and impactful in the field of materials engineering. The review provides a clear and coherent overview of concepts related to self-healing polymer nanocomposites by establishing a balanced and informative framework. It also ensures readers can follow the logical sequence of ideas and understand the connections between synthesis techniques, healing mechanisms, and material performance by structuring the topic from theoretical principles to practical applications [11].

The following section, "The Self-Healing Principle," which explores the fundamental idea behind all self-healing materials—their innate ability to repair damage themselves and restore their functions without external assistance—stems from this structural balance. This concept is fundamental to the design of smart materials and forms the basis for the sophisticated systems that will be discussed in subsequent chapters.

9.3 MECHANISTIC FOCUS

Focus on Functional Nanodivisors and Their Mechanisms in Self-Healing Smart Nanoparticles. To improve the performance, efficiency, and adaptability of self-healing nanopolymers in engineering applications, understanding their underlying mechanics is crucial. The healing mechanism and the role of nanodivisors embedded in the polymer matrix are critical to the operation of these smart materials, which are designed to automatically detect and repair damage, mimicking biological healing systems [12].

9.3.1 Self-Healing Mechanisms

The self-healing process in polymers relies on two main mechanisms: internal (core) and external (external) systems. Internal mechanisms depend on the polymer's ability to reconstruct physical, chemical, or dynamic bonds within its structure. When stimulated by thermal, optical, or other catalysts, these reversible bonds—such as hydrogen bonding, Diels–Alder processes, ionic interactions, π–π stacking, or disulfide exchange—allow for multiple healing cycles. Conversely, integrated healing agents, encapsulated in microcapsules, hollow fibers, or vascular networks, are utilized by external mechanisms. These reservoirs rupture upon splitting, releasing the healing agent, which then solidifies or polymerizes to repair the damaged area. While each mechanism has its unique advantages, both aim to restore the mechanical integrity and functionality of the material. While external systems provide rapid repair of severe damage, making them ideal for structural applications, internal systems are structurally simpler and allow for multiple healing cycles [13].

9.3.2 Functional Nanostearates

Functional nanomaterials are essential for enhancing the self-healing capacity of polymer composites. Mechanical strength, thermal stability, and electrical conductivity are significantly enhanced through the use of nanomaterials such as carbon nanotubes (CNTs), graphene, nanoclay, silica nanoparticles, and MXene compounds. Acting as reinforcing agents, these nanomaterials promote stress transfer and slow crack propagation. Furthermore, their surface functionalization creates efficient pathways for thermal or electrical catalysts that accelerate the healing process by improving surface interaction with the polymer matrix. For example, local activation is possible because of the electrical conductivity provided by CNTs and graphene for Joule-heated healing. Nanoclay and silica nanoparticles improve resistance to environmental degradation by enhancing barrier properties and toughness. Healing under the influence of temperature, light, or electrical stimulation can be achieved through multiple catalytic responses, facilitated by MXene compounds [14].

9.3.3 Functional Synergy

The next generation of self-healing materials is defined by the synergy between polymer matrices and functional nanofillers. In addition to reinforcing the composite material, appropriate dispersion and interface design provide efficient energy transfer throughout the healing process.

9.3.4 Toward the Design of Smart Materials

To provide real-time sensing, adaptive interactions, and repeatable healing cycles, future developments will focus on hybrid systems that integrate multiple healing mechanisms with multifunctional nanofillers. The development of the next generation of smart materials capable of repairing, monitoring, and optimizing performance autonomously will remain heavily influenced by a mechanistic understanding of how these nanostructures interact within the polymer network [15].

9.4 APPLICATION-ORIENTED FOCUS

With the development of smart, self-healing polymer nanocomposites, new possibilities have emerged for their integration into advanced engineering systems where flexibility, adaptability, and durability are paramount. These materials are designed not only to automatically repair structural damage but also to continue functioning under harsh operational and environmental conditions. Their mechanical strength, self-healing capabilities, and multifunctional properties make them highly valuable across various industrial sectors [16].

9.4.1 Structural and Civil Engineering Applications

The use of self-healing polymer nanocomposites is increasing in structural and civil engineering applications, particularly in cementitious materials, sealants, and protective coatings. These materials have the ability to automatically restore their integrity when cracks or micro-defects appear, significantly extending the service life of infrastructure, including pipelines, bridges, and building components. Systems reinforced with CNTs and nanoclay improve corrosion resistance by reducing permeability and increasing durability. Epoxy-based composites, with precisely encapsulated curing agents, also prevent crack propagation in coatings. In civil engineering projects, these advancements lead to reduced maintenance costs and increased sustainability [17].

9.4.2 Aerospace and Automotive Engineering

Self-healing nanocomposites are highly beneficial in automotive and aerospace systems, where high-performance, lightweight materials are crucial. In addition to improving fatigue resistance and thermal stability, the addition of CNTs, graphene, and nanofibers enables self-healing after microcrack formation. These composites are used in aerospace structures to enhance safety and extend service life in aircraft panels, wings, and airframe coatings. They offer improved durability and impact resistance and are used in automotive applications for tires, fuel systems, and exterior coatings.

9.4.3 Power Systems and Electronics

Self-healing polymer nanocomposites enable the production of flexible, conductive, and adaptable materials for flexible electronics and power devices. Electrically conductive nanocomposites, such as graphene, auxins, and CNTs, facilitate self-healing through ion transport or Joule heating. In stretchable sensors, batteries, supercapacitors, and wearable devices, these materials are incorporated to maintain electrical continuity after deformation or damage through self-repair [18].

9.4.4 Environmental and Biomedical Applications

Self-healing nanocomposites capable of repairing minor damage or responding to physiological stimuli are being developed in biomedical engineering for use in drug delivery systems, tissue structures, and implants. Their biocompatibility and adaptive response make them ideal for long-term medical use. Environmentally, these materials promote green solutions such as self-healing coatings to protect the environment and marine ecosystems.

9.4.5 Industry Outlook and Performance Considerations

The performance of self-healing polymer nanocomposites is determined by healing efficiency, recovery rate, and environmental resistance. Multifunctional designs that combine real-time sensing, adaptive responses, and recyclability will be beneficial for future engineering systems. To make these materials commercially viable, the industry perspective focuses heavily on scalable manufacturing technologies and cost-effective processing. The applied development of self-healing polymer nanocomposites is revolutionizing engineering processes by enabling smarter, safer, and more sustainable systems. By bridging the gap between laboratory innovation and practical application, these materials pave the way for the next generation of intelligent structures with longer service life and self-healing capabilities.

9.5 ADVANCED VARIANT

The growing demand for high-performance materials that combine self-healing, multifunctionality, and environmental sustainability is reflected in the ongoing development of smart, self-healing polymer nanocomposites. Utilizing advanced nanostructures, hybrid systems, and intelligent design techniques, recent advancements have focused on improving healing efficiency, response, and material integration [10].

1. *Development of Smart Self-Healing Mechanisms:* From basic microcapsule-based systems, self-healing material research has evolved into complex, multi-responsive composite nanostructures. Modern systems incorporate self-dynamic chemistry, multi-stimulus response, and reversible crosslinking networks that enable repeated healing cycles, whereas previous designs primarily focused on single use and external processes. Because of these advancements, self-healing materials are now active and intelligent materials capable of sensing, reacting to, and adapting to mechanical and environmental stimuli, replacing their previous status as passive repair systems.
2. *Integrating Advanced Nanomaterials:* Recent studies have focused on designing nanomaterials specifically engineered to enhance healing activation, dispersion, and interfacial bonding. Carbon nanomaterials (CNTs, graphene, and carbon black) improve thermal and electrical conductivity, enabling electrochemical self-healing and Joule heating. Silicate and composite nanoclays enhance dimensional stability, gas barrier properties, and mechanical strength. MXene nanoparticles and metal oxides enhance multi-catalyst sensitivity, reacting with electrical, magnetic, optical, or thermal stimuli. To facilitate efficient energy transfer and stress dissipation during healing, functionalized nanoparticles improve compatibility with the polymer matrix. Dynamic polymer networks and nanomaterials work together to produce improved fatigue resistance, faster healing rates, and enhanced crack-closing efficiency—all essential for practical applications in biomedical, electronic, and aerospace systems [19].
3. *Hybrid Healing Mechanisms and Multi-Stimuli:* Multi-stimuli responsive systems are becoming increasingly important in modern self-healing nanocomposites. These systems can initiate healing through thermal, optical, electrical, or mechanical stimuli. For example, conductive fillers enable electrical activation, and thermally reversible Diels–Alder processes provide controlled healing at moderate temperatures. To achieve rapid and reproducible healing, hybrid systems combining intrinsic and extrinsic mechanisms ensure structural recovery even in severe or repeated damage [18].
4. *Sustainable and Thoughtful Material Design:* Materials development is being revolutionized by combining predictive modeling, machine learning, and artificial intelligence. These technologies save time and money in experiments

by enabling the simulation and optimization of healing behavior, filler dispersion, and performance prediction. Furthermore, the focus on recyclable and environmentally friendly polymers aligns with self-healing nanocomposite technology [10].

9.6 FUTURE PROSPECTS

The next generation of self-healing polymer nanocomposites will benefit from self-sensing, adaptive, and multifunctional capabilities, enabling them to self-heal and detect damage instantly. Sustainable chemistry, data-driven design, and nanoengineering will enable the development of materials that learn and evolve in use, in addition to their self-healing capabilities. The latest advancements in smart, self-healing polymer nanocomposites are driving materials research toward high-performance, sustainable, and intelligent systems. These achievements pave the way for self-organizing structures that can extend service life, reduce maintenance, and transform the future of engineering materials [12].

9.7 SELF-HEALING PRINCIPLE

The concept of self-healing describes the ability of a material, whether intrinsic or extrinsic, to repair structural or functional damage and return to its original properties without external human intervention. The design of next-generation smart materials is based on this idea, inspired by biological processes found in nature, such as bone or skin regeneration in humans. This concept aims to improve safety and reliability in engineering applications, reduce maintenance costs, and extend component lifespan. When damage occurs, such as microcracks or peeling, a repair process is triggered in self-healing materials, allowing the material to regain its mechanical integrity. The two main processes are well-established: intrinsic and extrinsic healing. Extrinsically cured chemicals residing in vascular networks or microcapsules form the basis of extrinsic self-healing systems. These capsules rupture as the fracture propagates, allowing the healing agent to enter the damaged area and polymerize or solidify to restore its integrity. Intrinsically, systems rely on dynamic physical bonds or reversible chemical reactions occurring within the polymer matrix, such as supramolecular assembly, Diels–Alder processes, hydrogen bonding, or ionic interactions. Under the influence of appropriate mechanical, chemical, or thermal stimuli, these reverse networks of the material allow it to repair itself repeatedly [10]. The efficiency of the self-healing process is significantly influenced by factors such as the polymer's molecular design, the distribution and compatibility of the nanofillers, and the environmental conditions during healing. The incorporation of nanomaterials such as CNTs, graphene, silica nanoparticles, or nanoclay has been shown to greatly enhance healing efficiency by improving stress transfer, providing localized heat generation, and facilitating molecular movement on the damaged surface. In high-performance industries such as electronics, automotive, aerospace, and civil infrastructure, where even small microscopic defects can lead to catastrophic failures or costly maintenance, the self-healing principle has critical implications. Engineers can create safer, more durable, and more environmentally friendly systems by integrating self-healing capabilities into structural components. The self-healing principle is expected to be crucial in the future development of smart and adaptive materials, revolutionizing the design, monitoring, and maintenance of modern engineering systems as research progresses [11].

9.7.1 Self-Healing Polymer

A class of smart materials known as self-healing polymers mimics the natural healing process of biological systems, repairing internal and external damage without external assistance. These materials are designed to activate healing processes that restore their structural integrity and function in response to mechanical stress, cracks, or fractures. The development of self-healing polymers aims to improve the sustainability, safety, and durability of a variety of products, particularly in industries requiring low maintenance and long service life. The two main categories are external and internal self-healing processes. Internal self-healing requires reversible chemical bonds or dynamic reactions within the polymer matrix, allowing the material to recover from damage. Conversely, external systems utilize built-in healing agents such as microvascular networks or microcapsules that release their contents upon damage. Electronics, coatings, medical devices, aircraft structures, and automotive components are just a few examples of the many emerging applications of self-healing polymers. Self-healing coatings in the automotive industry offer scratch and abrasion resistance, while structural components benefit from longer service life and increased reliability. These polymers are used in composite structures in the aerospace industry, where safety and weight reduction are critical factors. Their ability to bridge microcracks reduces maintenance requirements and helps prevent structural failures. Increasing the efficiency of self-healing reactions, facilitating multi-cycle recovery, and improving the compatibility of healing agents with the polymer matrix are key objectives of recent research. Nanotechnology also facilitates this progress, providing nanoparticles such as graphene or CNTs, which can enhance the polymer's mechanical properties and recovery efficiency [12]. Considering all of this, self-healing polymers represent a significant advance in materials science and promise to solve upcoming engineering challenges across various sectors [13]. Figure 9.1 illustrates the key processes of polymer self-healing mechanisms.

Physical (non-covalent) and chemical (covalent) interactions are the two main molecular mechanisms that control self-healing in polymers. These mechanisms are essential for a material's ability to heal itself from damage such as cracks and fractures. Weak, reversible forces such as hydrogen bonds, van der Waals interactions, and π–π stacking are common in physical interactions. When damage occurs, these interactions enable polymer chains to reorganize and reconnect, promoting reversible repair without irreversible chemical change. Although each of these forces is not very powerful on its own, when combined, they can produce a very efficient healing system. On the other hand, strong, long-lasting chemical bonds are formed and regenerated through chemical or covalent interactions. These interactions, which lead to more robust and long-lasting repair, are often triggered by external stimuli such as

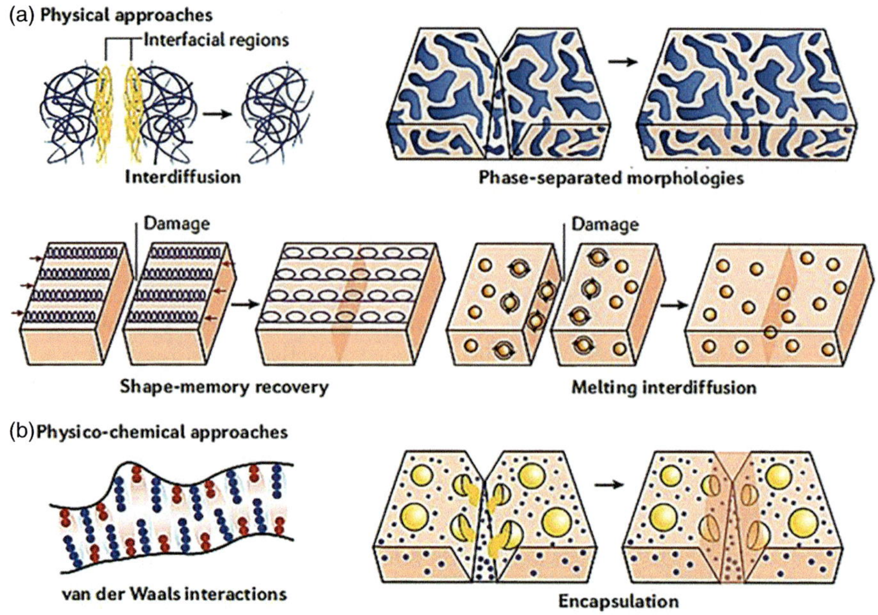

FIGURE 9.1 Self-healing mechanisms: (a) mutual diffusion of polymer chains, phase-separated shapes, shape memory, and insertion of active nanoparticles into the polymer matrix are examples of physical processes that help explain self-healing; (b) self-healing can be realized by combining physical and chemical processes [3] [Open access].

heat, light, or pH changes. Advanced self-healing polymer systems utilize dynamic covalent bonds, such as disulfide exchange or Diels–Alder processes, to robustly restore mechanical properties [14]. The development of self-healing epoxy-based adhesives with embedded microcapsules containing healing agents is an example of a practical application of this concept. The healing agent is released into the damaged area when the microcapsules rupture due to fracture. As a result, the glue hardens and polymerizes, effectively sealing the crack. This self-healing adhesive has been used in a pilot configuration. These systems demonstrate the exciting potential of combining chemical and physical healing processes in polymer matrices to produce advanced materials capable of withstanding structural damage without losing their functionality. These concepts are now increasingly being used in high-performance fields such as structural engineering, automotive, and aerospace [15]. The steel surface is coated with the adhesive (Figure 9.2).

9.7.2 Nanocomposites of Self-Healing

Self-healing polymer nanocomposites are a new family of materials that combine the outstanding mechanical, thermal, and electrical properties of nanoparticles with the self-repairing potential of smart polymers. Researchers have significantly improved the performance of self-healing systems in diverse applications such as electronics,

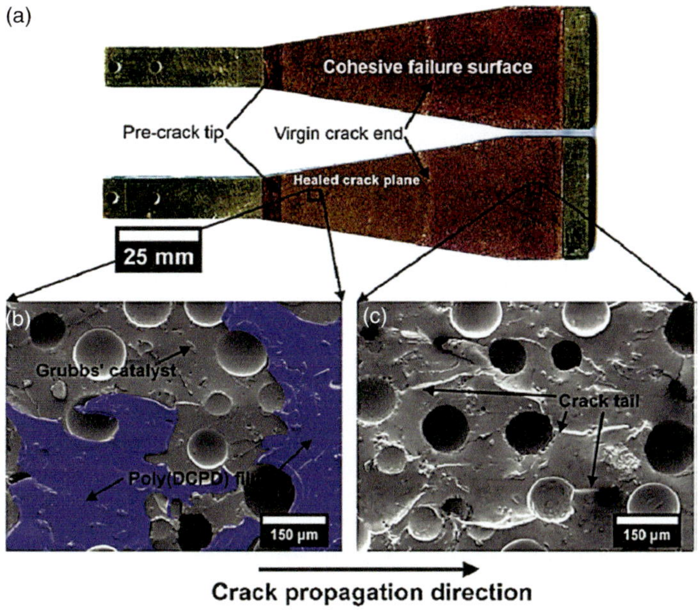

FIGURE 9.2 After the self-healing specimen has healed, the fracture plane appears nearly constant: (a) optical images of the two mated fracture surfaces reveal a cohesive fracture; (b) scanning electron microscope (SEM) image of healed fracture surface shows blue polymeric film formed during self-healing; and (c) SEM image of fracture surface at end of specimen created during imaging (unhealed) reveals crack tails indicating a strengthening mechanism [4] [Open access].

automotive, biomedicine, and aerospace, by incorporating nanofillers, such as metal oxide nanoparticles, graphene, CNTs, or nanoclays into polymer matrices. There are several reasons for incorporating nanoparticles into self-healing polymers. First, they act as mechanical reinforcements, enhancing the strength, stiffness, and toughness of the host polymer. Second, by interacting with external stimuli like heat, light, or electricity, nanofillers can act as thermal catalysts or conductive channels that promote healing. For example, carbon nanomaterials such as graphene and CNTs are electrical and thermal superconductors, capable of concentrating heat on damaged areas to accelerate self-healing processes. Self-healing of nanocomposites can occur through two main mechanisms: extrinsic and intrinsic. Reversible interactions within the polymer matrix such as dynamic covalent bonds, ionic contacts, or hydrogen bonds are essential for intrinsic healing. These interactions are made easier by the presence of nanoparticles, which encourage the re-linking and reorganization of polymer chains. Extrinsic approaches, on the other hand, involve the incorporation of vascular networks, nanocapsules, or microcapsules that release therapeutic chemicals in response to fracture formation. In these systems, the presence of nanoparticles can help regulate healing agent release and polymerization of material. The multifunctionality of nanocomposites is a notable advantage of self-healing materials. In addition to their repair potential, these composites improve corrosion resistance, flame resistance, barrier properties, and even detection capabilities. For example, self-healing nanomaterials are being researched for use in aircraft structures and structural components in the aerospace industry, where early fracture repair can prevent catastrophic failures. These materials can be used in lightweight structural components, coatings, and automotive interiors to reduce weight while increasing durability and lowering maintenance costs [16]. Improving nanoparticle dispersion, enhancing interfacial bonding, and tailoring healing kinetics for faster and more efficient performance are key goals of recent advances in materials science. Furthermore, there is growing interest in developing multi-cycle self-healing nanocomposites, which enable the material to heal itself repeatedly without losing its effectiveness. Self-healing nanocomposites can enable the development of next-generation smart materials that not only recover from damage but also enhance their functional properties. Building safer, more durable, and longer-lasting materials for high-performance applications requires continuous improvement [17].

9.7.3 Aerospace Self-Healing Nanocomposites

The aerospace industry is constantly striving to develop advanced materials that deliver superior performance, reliability, and safety in harsh environments. Despite their durability and lightweight nature, conventional materials often experience microcracks, corrosion, and deterioration over time due to mechanical stress, temperature changes, and environmental exposure. Researchers have turned to self-healing nanocomposites, a new class of smart materials designed to spontaneously repair damage and increase service life to overcome these challenges. In aerospace applications, self-healing nanocomposites are typically made from a polymer matrix supplemented with nanoscale fillers such as graphene, CNTs, or nanoclay. These nanoparticles are essential for facilitating or enhancing the healing process, as well as improving the mechanical, thermal, and electrical properties of the host polymer. The ability of these materials to repair damage at the microscopic level reduces the need for frequent inspections or repairs, improves safety, and prevents crack propagation. In aerospace nanocomposites, self-healing processes can be extrinsic, relying on healing agents contained in micro- or nanocapsules that are released upon damage, or endogenous, using reversible chemical bonds or dynamic interactions within the polymer matrix. Some systems are designed to react to external stimuli such as light and heat [18]. Self-healing nanocomposites are currently being researched for use in aircraft structures, wings, fuel tanks, insulation panels, and protective coatings in response to the growing demand for lightweight, fuel-efficient, and long-lasting aircraft and spacecraft components. Their use could lead to longer maintenance intervals, lower lifecycle costs, and safer flights. Self-healing nanocomposites in aerospace are expected to be fundamental in creating the next generation of aircraft and spacecraft as materials science advances. The self-healing mechanism of the epoxy coating is shown in Figure 9.3. The healing agent for damage repair is released when the nanocapsules burst during the repair process. An innovative and efficient self-healing mechanism at the nanoscale has been created by incorporating

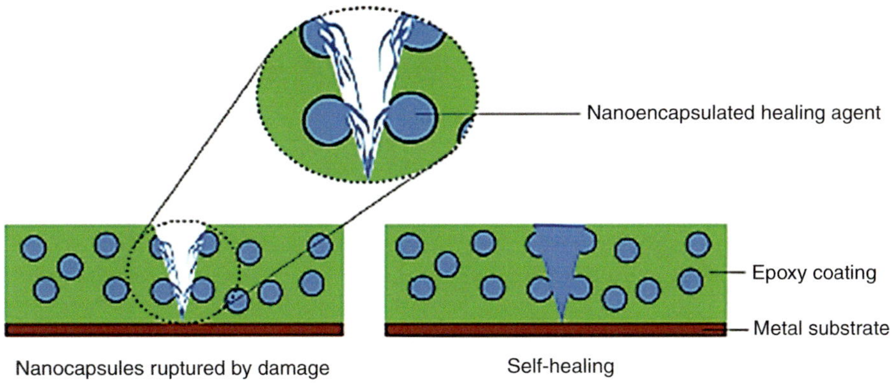

FIGURE 9.3 Self-healing method in polymer nanocomposite based on nanocapsules [5] [Open access].

β-cyclodextrin/azobenzene-functionalized graphene oxide (β-CD/AZO-GO) into epoxy nanocomposites. This technology enables self-healing of damage by exploiting clever interactions between the polymer matrix and functional molecules. The base nanofiller in this nanocomposite is graphene oxide, while dynamic host–guest molecular interactions are mediated by azobenzene and β-cyclodextrin. These interactions are ideal for self-healing because they are reversible and sensitive to changes in the environment. The β-CD/AZO-GO nanofillers improve the mobility of polymer chains in damaged areas such as surface cracks or microcracks. Without the need for external assistance, the healing process is facilitated by the increased mobility of polymer chains, allowing them to diffuse across the broken interface and reattach. As β-cyclodextrin and azobenzene can form and break bonds reversibly in response to stimuli such as temperature or light, their dynamic interactions are very effective in driving this reorganization [10]. These epoxy nanocomposites exhibit excellent mechanical properties such as increased tensile strength, hardness, and elasticity in addition to their self-healing ability. While the functional groups improve compatibility and dispersion in the epoxy system, graphene oxide helps mechanically strengthen the polymer matrix. In the field of self-healing polymers, this intelligent material system is a major breakthrough, especially for high-performance applications in electronics, automotive, and aerospace, where durability and structural integrity are critical. A promising path to materials that are strong and damage-resistant is provided by combining responsive molecular design with nanoscale reinforcement, which improves stability, safety, and reliability. The self-healing property of epoxy nanocomposites containing β-cyclodextrin/azobenzene functional graphene oxide is shown in Figure 9.4. To promote self-healing, the nanofillers activated the mobility of polymer chains in the damaged areas.

The self-healing properties of layered epoxy nanocomposites augmented with various conductive materials, with special emphasis on copper foam sheets and CNT nanosheets, are demonstrated in Figure 9.5. The results clearly show how well the CNT reinforced system recovers. In terms of fracture energy, the epoxy/CNT nanocomposite showed an average recovery efficiency of 107.7%, indicating that

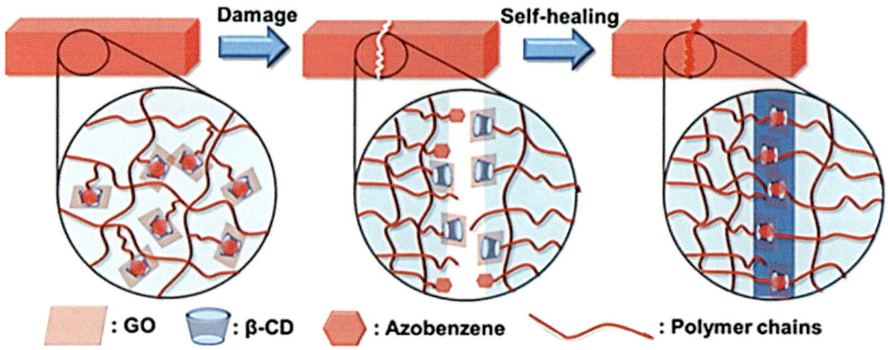

FIGURE 9.4 Schematic representation of self-healing process of epoxy nanocomposite films [6] [Open access].

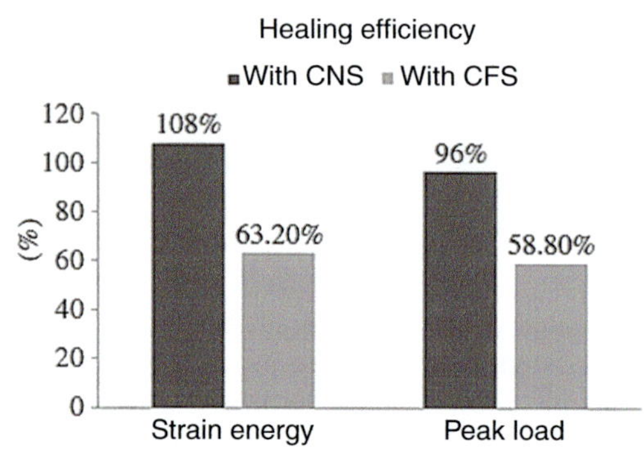

FIGURE 9.5 Effectiveness of nanocomposites, including CNS and CFS nanomaterials for healing. CFS stands for copper foam; CNS stands for carbon nanotube nanosheet [6] [Open access].

the material not only recovered its original strength after injury but also surpassed it in certain situations. It has demonstrated outstanding recovery of its mechanical strength with a recovery efficiency of 96.22% at peak load [19].

Interestingly, the maximum recovery efficiency recorded for the CNT nanosheet-based sample in terms of fracture energy is 141%. This suggests that the healing process, possibly aided by localized thermal conductivity and increased polymer chain mobility provided by the CNTs, resulted in an even stronger post-repair microstructure. On the other hand, the epoxy nanocomposite with the copper foam layer demonstrated more modest self-healing properties. For this structure, the average healing efficiency is 58.8% for peak load and 63.2% for fracture energy. Despite these lower average values, one sample showed a notable maximum healing efficiency of 128% for fracture energy, indicating the potential of copper foam as an alternative material to initiate the healing process, albeit less reliably than CNTs.

The efficiency and reliability of self-healing in epoxy-based composites are determined by the nanostructured reinforcement material, as highlighted by these data. The strong electrical and thermal conductivity of CNTs likely facilitated a more uniform heat distribution throughout the healing process, accelerating crosslinking processes and polymer chain mobility in damaged areas. This makes CNT nanosheets promising for high-performance applications such as aerospace or structural components, where fast and reliable self-healing is critical.

A comparison of the self-healing performance of many nanocomposite samples with different types and concentrations of nanofiller is shown in Figure 9.6. The sample containing 0.5 wt% thymine-functionalized carbon nanotubes (T-CNTs) has shown the greatest improvement in self-healing performance of all the formulations evaluated. The special reversible hydrogen bonding interactions made possible by the thymine groups attached to the surface of CNTs are mainly responsible for this improved behavior [12]. DNA nucleotide thymine facilitates dynamic interactions in the polymer matrix due to its strong and reversible hydrogen bonding ability. Damage such as microcracks or delamination causes the polymer chains around the functionalized CNTs to become more mobile, allowing the thymine groups to form hydrogen bonds at the fracture surfaces.

The molecular zipper function of this reversible bond facilitates rapid and efficient molecular repair. T-CNTs improve the interface compatibility between CNTs and epoxy matrix in addition to promoting self-healing. Higher mechanical integrity and uniform stress distribution after healing are further supported by improved dispersion and interaction of nanofillers with polymer chains. This finding demonstrates the promise of bioinspired functionalization methods in creating high-performance self-healing nanocomposites. It is possible to create materials that not only self-heal from damage but also retain superior structural performance over long periods of use by taking advantage of the reversible nature of hydrogen bonding, especially through biomolecular moieties such as thymine. For demanding applications where reliability, strength, and low maintenance are important, like materials hold great promise in automotive, aerospace, and smart coatings. As shown in Figure 9.7, the storage modulus of the epoxy nanocomposites with modified nanofillers is investigated at different temperatures.

One of the important measures of the stiffness of a material in its viscoelastic state is the storage modulus, which indicates how much energy is stored during deformation. It provides important information about the mechanical performance and structural integrity of polymer-based nanocomposites at different temperatures [13,14].

All nanocomposite samples showed a gradual decrease in storage modulus with increasing temperature from –50 to 120 °C. This decrease is consistent with the normal behavior of polymeric materials, which exhibit high chain mobility and a commensurate loss of stiffness as a result of the rupture of molecular bonds by thermal energy.

However, the storage modulus values in the studied temperature range changed significantly and favorably with the addition of chemically modified nanofillers. The development of reversible crosslinking networks between the epoxy matrix and the nanofillers is mainly responsible for these improvements. Stronger molecular bonds

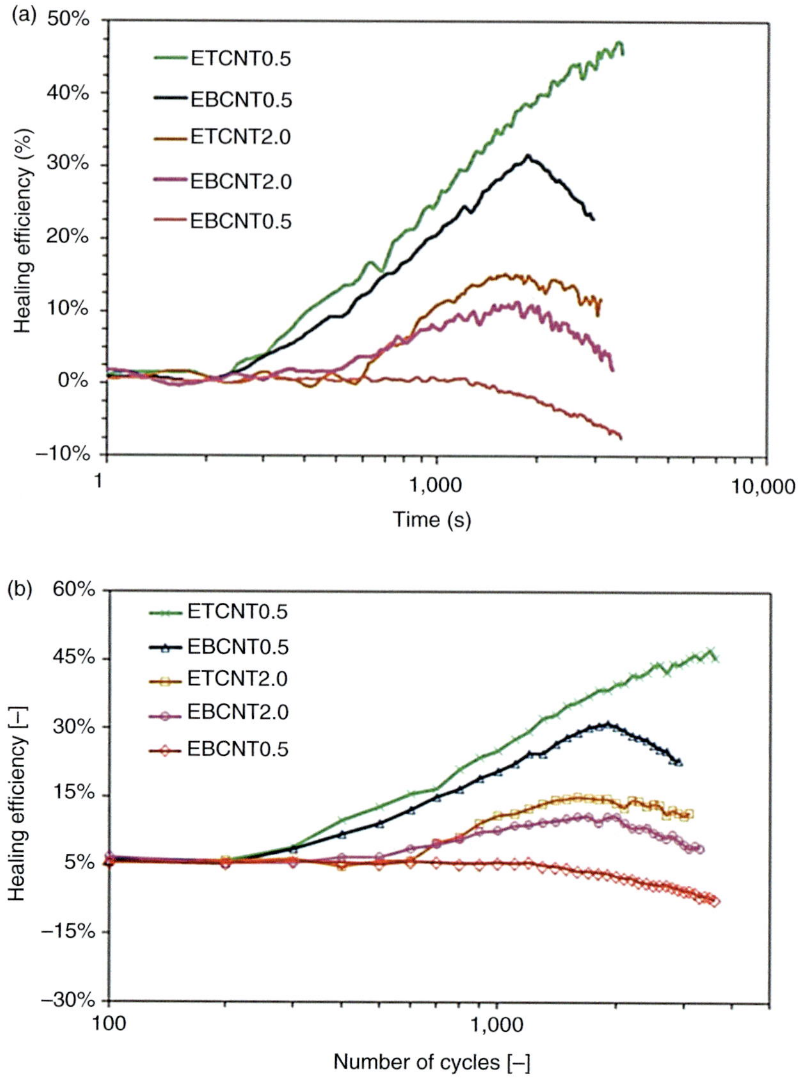

FIGURE 9.6 Self-healing efficiency as a function of (a) time and (b) number of cycles for nanocomposite samples containing unfunctionalized CNTs (ECNT 0.5 wt%), functionalized CNTs with thymine (ETCNT 0.5 wt%), ETCNT 2.0 wt%, and functionalized CNTs with barbiturate (EBCNT 0.5 wt%). Epoxy/carbon nanotube with thymine-based ligand (ETCNT) and epoxy/carbon nanotube with barbituric acid-based ligand (EBCNT) [7] [Open access].

are possible by modifying the surface of the nanofillers, which improves their compatibility at the interface with the matrix. These interactions help to maintain a higher degree of rigidity by strengthening the polymer structure. Using reversible crosslinks, which can be based on dynamic covalent bonds, hydrogen bonds, or π–π interactions, polymer chains that undergo thermal mixing can be instantly reconnected. The

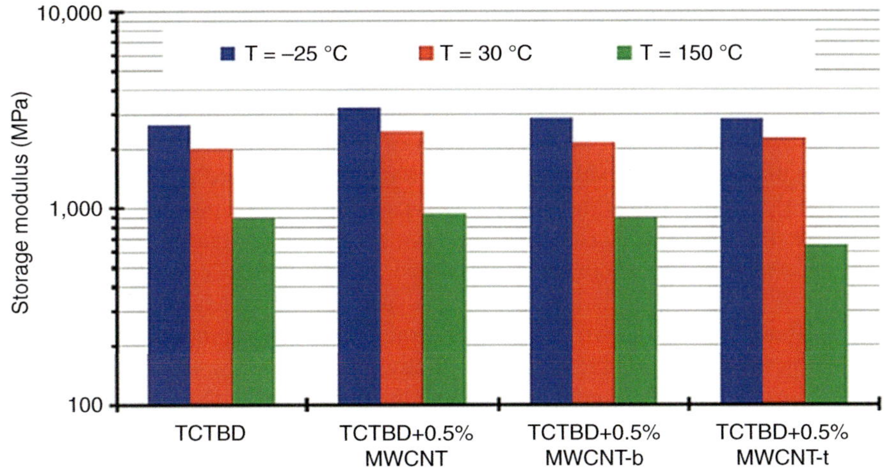

FIGURE 9.7 Elastic modulus of nanocomposites at three different temperatures. MWCNT means multi-walled carbon nanotube; TCTBD means tetraglycidylmethylenedianiline modified with carboxyl-terminated butadiene acrylonitrile, 1,4-butanediol diglycidyl ether, and 4,4′-diaminodiphenyl sulfone [8] [Open access].

structure recovery and resistance to thermal softening are provided by this dynamic linking mechanism, which functions as an integrated reinforcement system. The high mechanical strength and high thermal stability of self-healing nanocomposites make them promising for use in advanced technology fields, such as electronics, automotive, and aerospace. Surface modification of nanofillers is essential to improve their thermal and mechanical properties.

9.8 THE IMPORTANCE OF SELF-HEALING NANOCOMPOSITES IN THE SPACE INDUSTRY

Self-healing nanocomposites have become essential materials for the aerospace industry due to their ability to repair microscopic damage caused by the harsh environment of space. High temperatures, radiation, and mechanical stress are present in space, and these conditions can cause materials to deteriorate and fail. By extending the life of spacecraft components, self-healing technologies reduce maintenance requirements and improve mission safety. CNTs and graphene are examples of nanofillers that enhance the mechanical strength and repair efficiency of these smart materials. The need for durable self-healing materials is growing as space missions become longer and more complex, making self-healing nanocomposites a critical invention [15,16].

9.8.1 Aerospace Engines

Space engines are the primary source of propulsion for aircraft and spacecraft, required for flights both within and outside the Earth's atmosphere. These engines employ advanced engineering technologies to provide high thrust-to-weight ratios, fuel efficiency, reliability, and outstanding performance in harsh operating environments. Space engines can be broadly divided into two groups: rocket engines, which are essential for space travel, and air-breathing engines, which are most commonly used in aircraft. Air-breathing engines such as compressor jets, turbojets, and turbofans require oxygen from the atmosphere to burn fuel. Due to their high fuel efficiency and low noise levels, turbofans are most widely used in commercial aviation. On the other hand, turbojets are often used for military purposes due to their small size and high speed. However, rocket engines can operate in the vacuum of space because they carry both fuel and oxidizer. These engines operate according to Newton's third law, which states that if high-speed gases are expelled in one direction, the spacecraft will move in the other direction. These engines can run on solid fuels, known for their reliability and simplicity, or liquid fuels, which provide control and reusability. Hybrid engines combine the benefits of both types, which is the result of recent developments [17,18]. Innovations in materials and design have been the main focus of aerospace engine development in recent years. Engine performance, durability, and thermal stability have been improved through the use of self-healing nanocomposites, lightweight composites, and ceramic matrix composites. Engine component manufacturing has also undergone a radical change because of additive manufacturing (three-dimensional (3D) printing), which allows for complex geometries and shorter production times. The development of hypersonic engines, reusable launch systems, and environmentally friendly propulsion systems will determine the future direction of aircraft engines. For deep space missions, concepts such as ion thrusters and plasma engines are being explored, while hypersonic ramjets and combined cycle engines aim to enable next-generation aircraft to reach speeds above Mach 5. Figure 9.8 illustrates the properties of both the raw material and the engine material. Increased porosity and widespread tensile cracking are observed. In short,

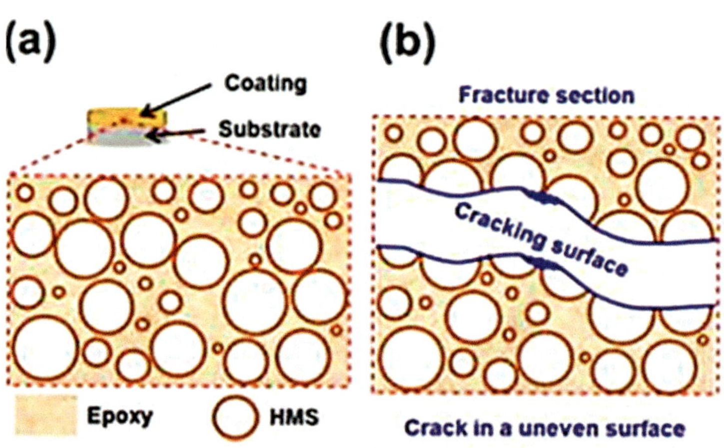

FIGURE 9.8 Schematic representation of the original material (a) and tensile fracture surface (b) after engine application. HMS is an abbreviation for "hollow microspheres of phenolic resin" [9] [Open access].

the aerospace engines are essential for both space travel and aviation. With ongoing research and technological development, these engines are becoming more efficient, environmentally friendly, and capable of achieving the noble goals of future space missions [12,13]. Atomic force microscopy (AFM) is used to closely analyze the morphological properties of the TiO_2-reinforced nanocomposite films and neat polyurethane (PU0). As shown in Figure 9.9, optical microscopy (OM) is used to complement AFM analysis to obtain micrographs of each film, which provided additional information on the surface structure and distribution of nanoparticles [10,12,13,19].

The OM photographs confirmed that there are no obvious microscopic signs of agglomeration or clustering and that the TiO_2 nanoparticles are uniformly distributed throughout the nanocomposite films. This suggests that the nanoparticles are successfully incorporated into the polymer matrix during processing, which is necessary to maintain uniform material properties throughout the composite. The PU0 film

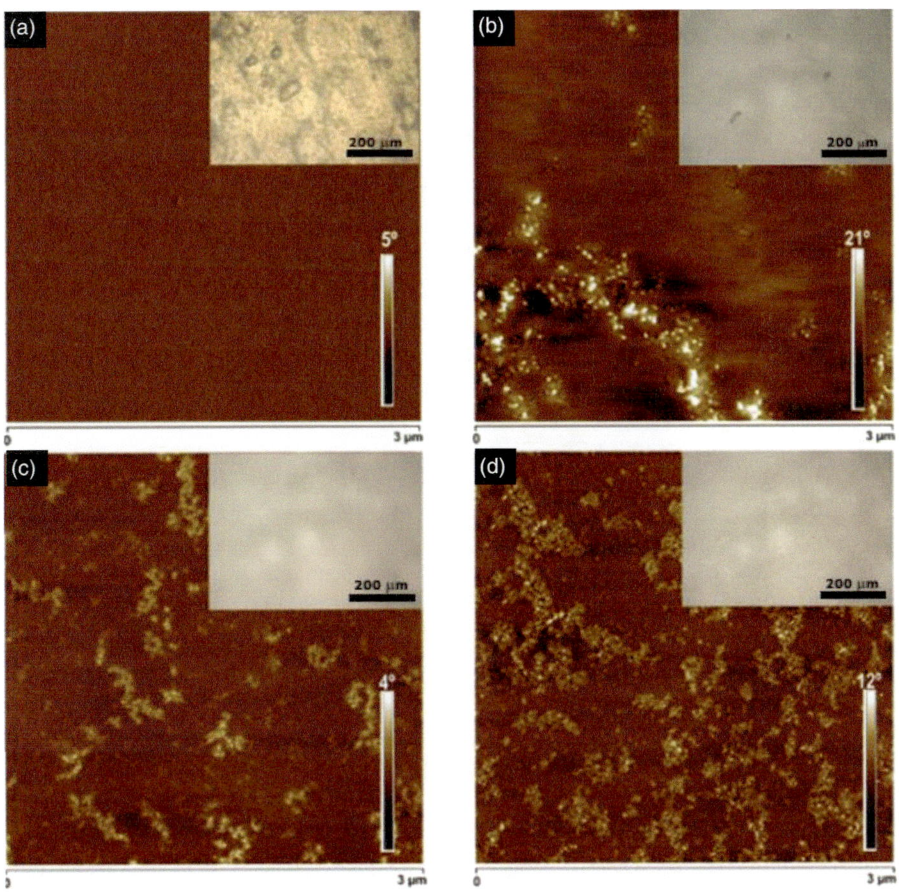

FIGURE 9.9 Atomic force microscopy (AFM) phase photographs of (a) PU0, (b) $10TiO_2$-PU0, (c) $20TiO_2$-PU0, and (d) $40TiO_2$-PU0 (3 μm′ 3 μm). Optical microscopy (OM) micrographs are shown in inset of each AFM image [10] [Open access].

exhibited a homogeneous morphology without significant phase separation between the hard segments (4,4′-methylene diphenyl diisocyanate with sodium dodecyl benzene sulfonate, MDI-SDBS) and soft segments (polyethylene oxide, PEO) as predicted in other studies [14]. The homogeneous structure depicted in Figure 9.9a and b indicates that PU0 functions more like a homopolymer and does not contain pronounced microphase-separated domains that are typically observed in segmented polyurethanes [15]. TiO_2 nanoparticles in the nanocomposite films are not localized within any particular matrix phase as a result of this morphological homogeneity. Instead, they are uniformly distributed throughout the continuous polymer phase, as seen in Figure 9.9b–d. Due to their interaction with the surrounding polymer chains, the nanoparticles tended to form spherical nanoclusters rather than larger aggregates. Electric force microscopy (EFM) results did not show any distinguishable charged domains on any of the studied material surfaces as depicted in Figure 9.10. The absence of localized electrical contrast demonstrates the suitability and effectiveness of the measurement technique used. The accuracy and reliability of EFM results are confirmed by effectively isolating the electric field response from any topographic

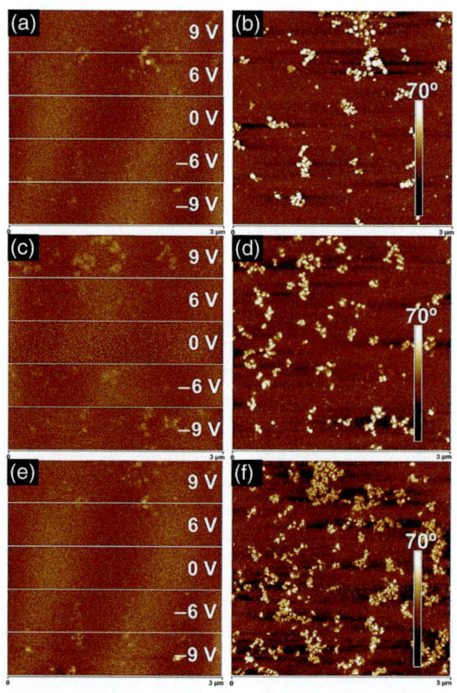

FIGURE 9.10 Electrostatic force microscopy (EFM) phase images (3 μm × 3 μm) of (a) $10TiO_2$–PU_0, (c) $20TiO_2$–PU_0, and (e) $40TiO_2$–PU_0, along with similar Atom force microscopy (AFM) phase images (3 μm × 3 μm). EFM maps surface electrostatic interactions, while AFM reveals surface morphology and phase contrast. TiO_2 represents the ceramic nanoparticle filler, and the numbers (10, 20, and 40) indicate its proportion, while PU_0 refers to the original polyurethane matrix [10] [Open access].

artifacts by applying a 0 V bias. This step is necessary to prove that any electrical events observed in later experiments would be real and not the result of surface defects or experimental artifacts. The importance of rigorous methodology in nanoscale electrical characterization is emphasized by careful control of measurement conditions.

Overall, the morphological study demonstrates the importance of matrix homogeneity to achieve uniform nanoparticle dispersion, which is essential for the reliable performance of modern nanocomposite materials, and confirms the success of incorporating TiO_2 nanoparticles into the PU matrix. A bias voltage of 0 V is intentionally applied during the study to ensure that the surface topography of the studied materials did not interfere with the EFM observations. This strategy is used to exclude the possibility of electrostatic interactions or synthetic signals that could arise due to surface morphological changes rather than real electrical properties. The measurement settings ensured that the measured signals would only represent the intrinsic electrical properties of the sample, canceling out any external voltage effect [20,21].

9.8.2 Space Structures

Spacecraft, satellites, space stations, and research vehicles rely on space infrastructure. These structures are designed to perform vital tasks such as housing instruments, supporting payloads, and maintaining the mechanical stability of space equipment in challenging extraterrestrial environments. Space structures must operate in conditions that include microgravity, vacuum, extreme temperatures, cosmic radiation, and mechanical vibrations during launch and operation, unlike terrestrial structures [21]. To balance the need for high strength and rigidity with the requirement for minimal mass, aerospace engineers create space structures that are both lightweight and incredibly durable. Aluminum alloys, titanium, carbon fiber-reinforced polymers (CFRP), and complex composites such as self-healing nanocomposites, which provide the ability to self-heal microcracks caused by thermal cycling or micrometeor impacts, are often used materials. Space structures vary in size and purpose, from robotic arms and deployable solar panels to satellite frames and antenna supports.

The ability to withstand launch loads such as vibrations, acoustic energy, and G-forces is a critical design feature. Once in orbit, the structure must accommodate thermal expansion while maintaining dimensional stability. Deployable and inflatable devices that are small at launch and grow in space are an emerging trend in space structure construction. These technologies make it possible to deploy larger surfaces such as antennas or habitat modules into orbit, further reducing launch costs. In addition, the use of complex polymers or in situ 3D-printed materials in space is being explored to directly construct or repair structural components on spacecraft or space stations [12]. The incorporation of smart materials, including piezoelectric composites and shape memory alloys, into space structures to provide adaptive behavior is another important area of innovation. For example, these materials can respond to electrical or thermal inputs by changing their stiffness or shape, allowing for more efficient vibration control and damping [13].

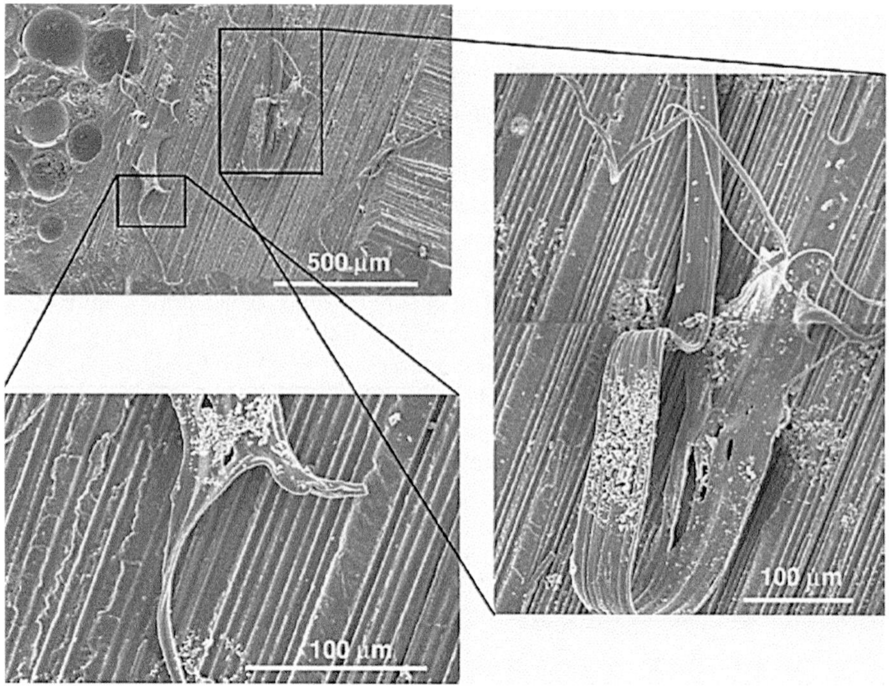

FIGURE 9.11 SEM images of the fracture surfaces of self-healing materials, showing the self-healing agent bridging the matrix strands and promoting the healing process within the delaminated matrix[11] [Open access].

The topic of space structures is an interdisciplinary one that combines materials science, mechanical engineering, and aerospace design to create resilient and flexible systems. Robust, multipurpose and self-sustaining space buildings are becoming increasingly important as human activities extend beyond Earth and space missions become more complex. Self-healing materials based on bisphenol A diglycidyl ether epoxy resin [14]. Figure 9.11 shows scanning electron microscope (SEM) images of fracture surfaces of self-healing materials. It can be seen how the self-healing agent bridges the matrix strands for the self-healing process in the delaminated matrix. Figure 9.12 shows the load curves for both bare and healed specimens. At 80 °C, self-healing is observed after 48 h. Compared to the plain epoxy specimen, crack propagation is less noticeable in the self-healing specimen. For the bare epoxy material, noticeable crack growth is observed in the midplane.

This self-healing device utilizes a specially designed 3D microvascular network embedded within the structure to deliver the healing agent into the internal cracks and fractures of the material. Because this vascular network functions similarly to biological blood systems, the healing ingredient can be automatically and precisely delivered to the areas that need to be repaired. When a crack or fracture occurs, the network allows the healing agent to be released and quickly transported

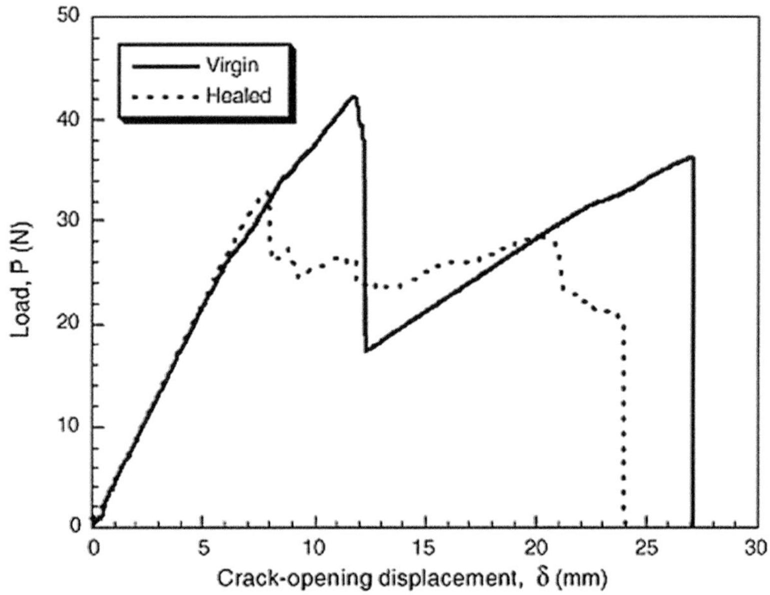

FIGURE 9.12 Normal load curves for self-healing specimens that are cleaned and healed (note: healing conditions = 48 h at 80 °C) [11] [Open access].

to the damaged area, where it reacts and initiates the repair process. Microbubbles are formed within the structure as a result of the activation of the healing process as shown in Figure 9.13. These bubbles are most likely formed as a result of the polymerization reaction that occurred during healing or as a result of interaction of healing agent with the surrounding matrix. Their development is a normal component of the internal healing mechanism of material, even though they may change the local morphology.

The overall healing efficiency of the material system is found to be exceptionally high, with recovery rates after the first injury reaching 70–80% of the previous mechanical performance. This suggests that the healing agent is effectively dispersed and chemically reacted to fill the microcracks, essentially restoring the integrity of the material [12,20,21]. However, after repeated damage and subsequent repair, the system showed a reduced but significant healing capacity, recovering about 40% of its initial damage. This reduction is expected, as the structural network can deteriorate or lose its effectiveness with repeated use, and some of the healing chemicals are consumed during the first cycle. The ability of such self-healing systems to extend the life of materials in challenging applications such as aerospace structures, where routine maintenance is difficult and material failure can be catastrophic, is nevertheless demonstrated by their ability to maintain 40% recovery after repeated damage. To create next-generation smart materials that can self-heal, this research highlights the importance of bioinspired design and autonomous repair systems that will improve reliability and reduce lifecycle costs [20,21].

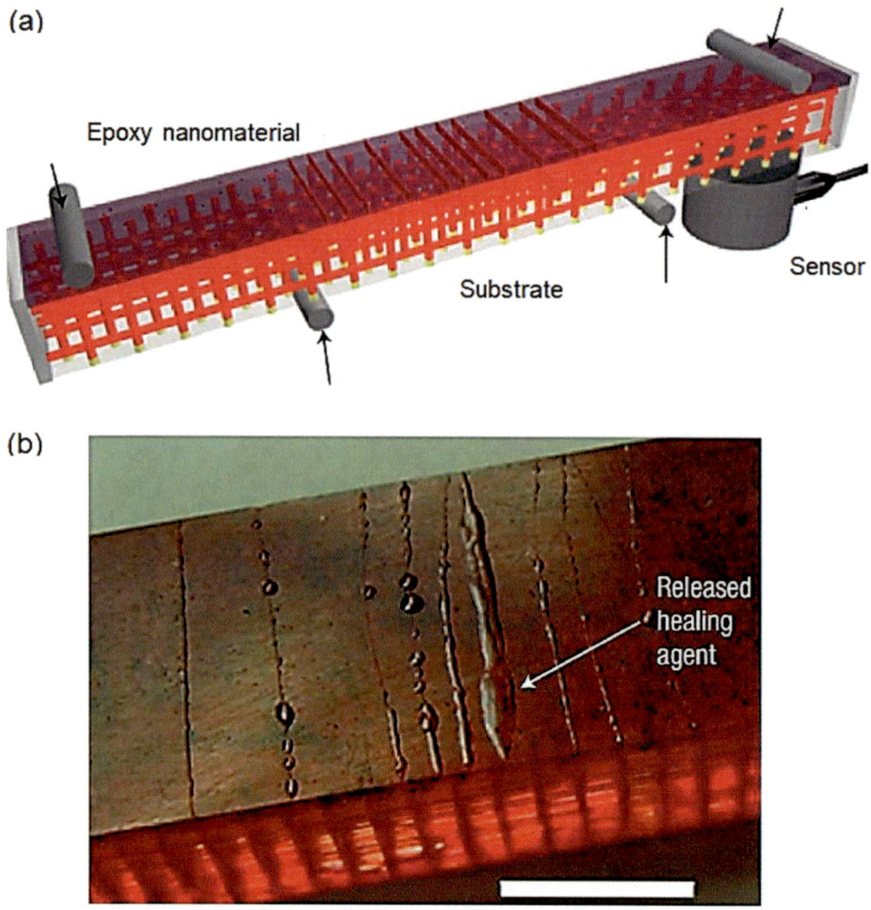

FIGURE 9.13 (a) Schematic diagram of the microvascular coating-substrate structure in a four-point bending configuration monitored with an acoustic-emission sensor. (b) Optical image of self-healing structure after cracks are formed in the coating (scale bar = 5 mm) [6] [Open access].

9.8.3 Aerospace Fuselage

The main framework containing the avionics, passengers, cargo, and vital flight systems is the aerospace fuselage, which is the central structure of an aircraft or spacecraft. One of the most important parts of aircraft engineering requires a precise balance between strength, lightweight construction, and aerodynamic efficiency. Numerous stressors, such as pressurization cycles, aerodynamic loads, vibrations, and possible impacts during flight and landing, must be supported by the fuselage. Fuselage design also includes aerodynamic considerations, ensuring minimal drag while maximizing internal volume. Cylindrical shapes with rounded nose and tail sections are the most common due to their effectiveness in managing pressurization and aerodynamics. In spacecraft, the fuselage (often called the pressure vessel or crew module) must also protect the occupants from radiation, extreme temperatures, and micrometeoroid impacts. Advanced materials, including titanium, CFRP, and aluminum–lithium alloys, are often used in the construction of modern aerospace fuselages. As composite materials offer better strength-to-weight ratios, corrosion resistance, and structural flexibility, the transition to them is especially important for next-generation aircraft. For example, more than half of the fuselage structure of commercial aircraft such as the Airbus A350 and Boeing 787 Dreamliner is made of composite materials. The results of deformation analysis are shown in Figure 9.14, which illustrates that under the applied loading conditions, the highest measured vertical displacement at the fuselage tip did not exceed 57 mm.

Even under severe mechanical stress, the overall structural integrity is maintained as evidenced by this relatively small degree of displacement [12–15,20]. However, the research also revealed the appearance of localized wrinkling in specific areas of the fuselage when stresses exceeding 60% of the critical limit are applied. This wrinkling phenomenon is typically associated with compressive instability in lightweight panels and underscores the importance of utilizing state-of-the-art materials and design techniques to mitigate these risks [20,21].

One of the most important findings of the study is that the integrity and buckling resistance of the fuselage are significantly enhanced by incorporating self-healing nanoparticles into the composite material layers. These smart materials are capable of repairing small cracks and other damage caused by mechanical loads, thereby maintaining structural integrity. The addition of these materials improved the lifespan and durability of the fuselage while increasing its load-bearing capacity. The study demonstrated improved performance in terms of mechanical stability and damage tolerance, using an innovative structural design technique that combines smart materials and optimized engineering. This development is particularly useful in aircraft applications, where maintaining airframe reliability under extreme loading and climatic conditions is critical to ensuring efficiency and safety [12–14,22–25].

Fusels now have intelligent components in addition to their traditional structural functions. For example, fatigue damage and microcracks can be automatically repaired using self-healing nanocomposites. In addition, sensors embedded in the fuselage can monitor the structural strength [14]. Modular fuselage design, which allows for greater flexibility in customization and manufacturing, is another emerging trend. As mission-specific modules can be quickly added or removed, this is especially useful for military and cargo aircraft.

The complex multipurpose structure at the center of aircraft performance and safety is the fuselage, which is more than just the body of the vehicle. The fuselage is changing to meet the needs of future space travel and aviation because of ongoing advances in materials science, structural engineering, and intelligent technologies [26–31].

FIGURE 9.14 Map of displacements of the integration structure and the fuselage nacelle under load [6] [Open access].

9.9 MICROCAPSULES' SYNTHESIS PROCESS

Polymer systems can incorporate a variety of repair agents capable of self-healing internal damage and enhancing material performance. Maintaining the stability and responsiveness of these agents during the self-healing process is crucial. Capsule shells require high thermal stability, chemical compatibility, and sufficient mechanical strength to withstand the mixing and curing phases without premature rupture, thus enhancing curing efficiency. To ensure long-term effectiveness, the catalyst used in the curing reaction must possess excellent solubility in the repair agent and maintain stability at elevated temperatures [11,32,33]. Several critical factors, such as capsule size, shell composition, core-to-shell ratio, and dispersion homogeneity within the polymer matrix, influence the design of microcapsule-based self-healing systems. Optimizing these elements directly impacts the curing mechanism's effectiveness, responsiveness, and longevity. This review also addresses recent developments in self-healing systems, which utilize supramolecular interactions or reversible chemical bonds to achieve self-healing without the need for external capsules. The research focuses specifically on their integration into nanopolymer composites, where the nanofillers enhance interfacial adhesion, stress transfer, and catalytic effects to promote the healing process. The study also examines various repair methods, mechanical performance evaluations, and emerging trends, providing statistical insights into the development and prospects of self-healing composite materials [34].

9.9.1 Micro- and Nanocapsules

Capsule-based smart composites represent a pioneering and revolutionary field in materials science and engineering. These composites consist of micro- or nanocapsules containing functional molecules that, when activated by external stimuli such as stress, temperature, or chemical exposure, can trigger specific reactions. By performing a variety of functions, including self-healing, sensing, corrosion prevention, and controlled release, these encapsulation agents can significantly enhance the functionality of the host material and extend its lifespan. This innovative discovery allows composites to react spontaneously to damage, preventing structural deterioration and reducing maintenance costs in civil engineering, automotive, and aerospace applications [35]. The self-healing mechanisms of polymers can be broadly categorized as intrinsic and extrinsic. An extrinsic phase, typically a vascular network or microcapsule, initiates the healing function in extrinsic systems by releasing a healing agent upon a fracture. The damaged area is then effectively sealed by polymerizing the depleted chemical or reacting with implanted catalysts. On the other hand, core–shell systems rely on the polymer's molecular structure, where hydrogen bonds, reverse covalent bonds, or supramolecular interactions enable the material to repair itself repeatedly without the need for external additives. Effective healing in both methods depends on the presence of a mobile phase capable of filling and dispersing the cleft region. To create customized structural shapes, such as mononuclear, multinuclear, and irregular, micro- and nanocapsules are fabricated using a variety of processes, including in situ polymerization, sol–gel processing, and emulsion methods [3,36–38,39].

In polymer matrices, these closed systems offer improved dispersion homogeneity, chemical resistance, and thermal stability. While the shell material protects the core from premature reactions or degradation, its structure between the core and shell allows for precise control of the release of active ingredients. Core-shell systems are ideal for advanced engineering and biomedical applications due to their unique geometric and surface properties, which also contribute to their exceptional mechanical performance, durability, and functionality. These systems are used in engineering environments to improve thermal management, corrosion resistance, and wear resistance. Nanoparticles between the core and shell are used in the biomedical sector [40–44].

9.9.2 Encapsulation Technologies for Micro/Nanocapsule Fabrication

The manufacturing materials and formulation techniques used significantly influence the shape and structure of capsules. By varying the processing parameters and material compositions, a wide range of capsule shapes can be produced, including spherical, multi-core, irregular, and porous. The formulation method is greatly affected by size uniformity, shell thickness, mechanical strength, and the release behavior of the encapsulating agents within the capsule [45]. Capsule construction techniques are divided into three main categories: chemical, physical, and polyphysical. Polymerization reactions, such as in situ polymerization, surface polymerization, and suspension polymerization, are the cornerstone of chemical methods. These processes utilize monomers or primary polymers to create polymer shells around a liquid or solid core. These techniques offer precise control over capsule wall structure and encapsulation efficiency, making them ideal for drug delivery, catalysis, and self-healing materials [46]. Physical methods, on the other hand, do not rely on polymerization. Alternatively, capsules are manufactured through mechanical or physical modifications such as emulsification, spray drying, or co-condensation. Using this method, capsules can be produced using only physical processes, offering simplicity and scalability, but sometimes reducing the control over consistency. Using controlled polymerization and mechanical dispersion, the physicochemical method integrates physical and chemical processes to produce more complex structures while improving stability and utility. Although common, it is difficult to precisely control the shape, size distribution, shell porosity, and compositional homogeneity of conventional microencapsulation techniques. Microfluidic methods have emerged as an advanced encapsulation technique to overcome these limitations. By precisely controlling fluid flow in microchannels, this method enables the production of highly ordered and reproducible capsules at the microscopic and nanoscale. The development of the next generation of self-healing, biomedical, and multifunctional nanopolymer compounds requires the use of microfluidic synthesis, which produces capsules with narrow volume distributions, well-defined structures, and tunable release properties [47–49].

9.10 POTENTIAL USES OF COMPOUNDS MADE FROM SELF-HEALING POLYMERS

The ability of self-healing polymers to automatically repair damage and restore their functionality after exposure to mechanical, thermal, or chemical stress is a groundbreaking development in materials research. In a variety of engineering and industrial applications, these materials aim to extend service life, enhance safety, and reduce maintenance costs by mimicking biological healing processes. Their adaptability enables integration across a wide range of industries, including electronics, medicine, automotive, and aerospace. Self-healing nanopolymers are used in fuel systems, coatings, and structural elements in the aerospace industry to prevent crack propagation and maintain aircraft safety under harsh environmental conditions. Their ability to automatically seal microcracks significantly improves safety and reduces the need for costly repairs or periodic inspections. These materials are used in tire compounds, protective coatings, and lightweight structural components in the automotive industry. By minimizing material degradation, their self-healing behavior helps maintain mechanical performance, resist fatigue, and improve fuel efficiency [50–52].

Self-healing polymers are used in electronic circuit boards, sensors, and flexible displays in electronic and flexible devices to maintain electrical conductivity and repair faults after physical damage. This property is particularly beneficial for soft robots and wearable devices, where mechanical stress is unavoidable. Self-healing polymer compounds are used in concrete admixtures, paints, and sealants in civil and structural engineering to improve waterproofing and durability. They also contribute to extending the lifespan of critical infrastructure and preventing corrosion. Furthermore, because of their superior biocompatibility and adaptive healing response, self-healing polymers are used in biomedical applications, including tissue engineering scaffolds, drug delivery systems, and biomedical coatings. Overall, materials derived from self-healing polymers have emerged as important facilitators for the production of sustainable and smart materials, leading to safer, more efficient, and more durable products across various modern technological fields. This marks the beginning of a new technological era, significantly improving the vital performance elements of many products. Structural, electronic, medical, and aerospace products are among the many applications where self-healing technologies have been integrated to enhance their overall performance (Figure 9.15).

Through localized thermal activation, self-healing nanocomposites can spontaneously restore their mechanical and electrical properties, representing a significant advance in smart materials design. The Joule effect (resistive heating) generates microscopic heat when an electric current is passed through a conductive nanonetwork composed of interconnected CNTs, metallic nanoparticles, graphite, or graphene. This localized heat stimulates the thermal reactivation of polymer chains in the surrounding matrix, enabling damaged areas to reflow, bridging cracks, and restoring continuity. In applications requiring flexible electronics, sensors, and energy storage systems, where maintaining mechanical robustness and electrical conductivity is critical, these technologies are highly beneficial [51]. Microcapsules, or microvascular networks containing healing factors such as monomers, resins, and catalysts, form the basis of external self-healing systems. These materials are released when a crack

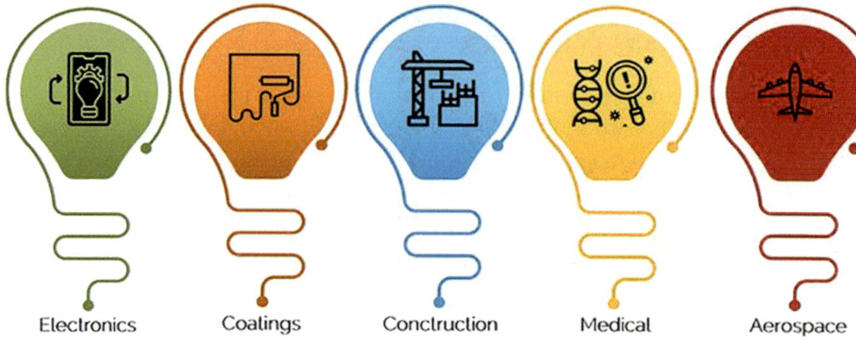

FIGURE 9.15 Integrating self-healing technology into a variety of applications [60] [Open access].

appears and undergo chemical bonding or polymerization, enabling them to fill and successfully seal the damaged area. These systems are limited by the availability of the encapsulating material; once it is depleted, the material loses its self-healing ability, although they are effective for one-time repairs.

Conversely, reversible chemical reactions directly integrated into the polymer structure form the basis of intrinsic self-healing systems.

When exposed to external stimuli such as heat, light, or pH changes, these materials possess the ability to repair themselves repeatedly. The healing process is driven by either a reversible or non-covalent bond, enabling the polymer to independently restore its structure and properties on multiple occasions. Non-covalent and covalent reactions represent the two main categories into which the bonding mechanisms underlying self-healing can be divided. Hydrogen bonding, van der Waals forces, π–π accumulation, ionic contacts, dipole–dipole forces, metal–ligand coordination, and host–guest complex formation are examples of non-covalent interactions.

These low-energy dynamic bonds are ideal for soft materials and coatings, enabling flexible and reversible repair [53–55]. In contrast, dynamic covalent bonds, including Diels–Alder (DA) processes, esterification, disulfide exchange, imine production, boron–oxygen bonds, and alkoxymine bonds, are stronger and more energy-intensive. Due to their thermally reversible behavior, which allows for efficient repair under controlled heating without compromising structural integrity, the Diels–Alder interaction stands out among these interactions. The development of next-generation self-healing polymer nanocomposites, which combine mechanical strength, electrical conductivity, and self-repair capabilities, supports these mechanisms.

These multifunctional materials enable advanced applications in aerospace structures, electronics, biomedical implants, and sustainable engineering systems, where long-term reliability and self-healing capability are key design requirements. This review focuses on Joule heating as described in the scientific literature and briefly discusses general topics related to self-healing nanocomposites (Figure 9.16). The Joule heating technique covers smart conductive materials using functionalized polymer matrices integrated with conductive nanofillers. We determine whether the

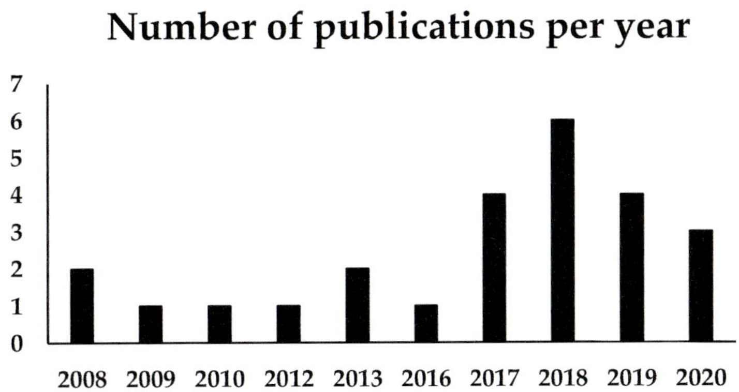

FIGURE 9.16 The number of research papers on self-healing materials based on the Joule effect [62] [Open access].

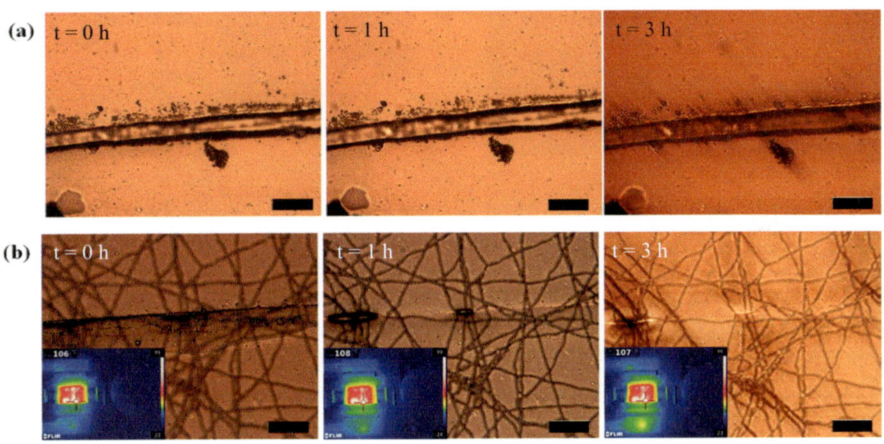

FIGURE 9.17 Optical images of the second fracture on the CuNF-heated substrate (a). 100 mm scale bars were used. (b) The pristine flat substrate [62] [Open access].

thermally stimulated chemical pathways involve dynamic covalent bonds, including Au–S bonds, alkoxyamine bonds, and Diels–Alder interactions [56].

The copper nanofiber sheets embedded in polyethylene provide a heat source for the bromobutyl polymers (BIIR rubber) through the Joule effect. Applying a voltage of 1.3 V and a current of 1.18 A stimulated heat production within them, resulting in a temperature of 100 °C, which enabled the polymer chains to move during the physical uncoupling process, resulting in the material's ability to repair itself [41] (Figure 9.17).

By applying a potential difference of 9 V, an electric current of 1.06 A was generated, raising the temperature to 126.8 °C and reducing the curing time from 24

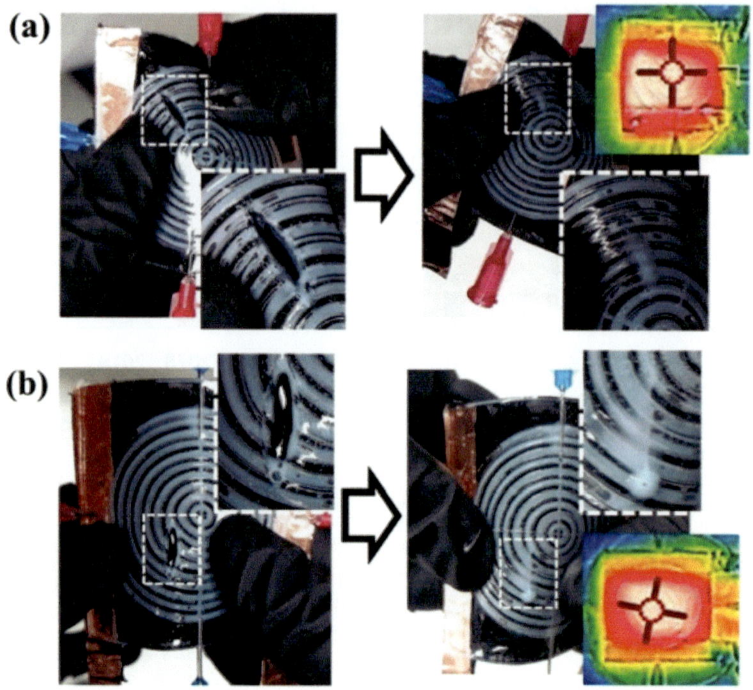

FIGURE 9.18 Various fissures self-heal when oriented at (a) 45° and (b) 90° [62] [Open access].

h to just 10 min. Up to 90% of the material's mechanical properties were restored after low temperatures (1 °C) [57] (Figure 9.18). The use of these materials may be possible in harsh climates, as the researchers examined their curative properties at.

9.11 APPROACHES

Self-healing in polymer nanocomposites is a complex interplay between material composition, nanostructure, and molecular dynamics, offering a means to enhance the durability and reliability of modern engineering materials. These systems are designed to self-heal from microscopic damage while maintaining structural integrity and functionality over time, unlike conventional polymers that degrade irreversibly under mechanical stress. The Joule effect, where an applied electric current generates localized heat within a conductive network in the polymer matrix, is one of the best methods for inducing self-healing. The nanocomposite must possess thermally activated healing capabilities and be both electrically and thermally conductive to achieve this process.

Furthermore, the shape, dispersion, and aspect ratio of the filler influence both the formation of the conductive networks and the localized heat distribution, thus affecting its homogeneity and healing rate. Well-dispersed fillers with a high aspect ratio exhibit lower leakage thresholds, faster Joule heating, and more efficient restoration of mechanical properties. Future developments should focus on designing matrix–filler interactions, optimizing filler alignment, and enhancing thermal and electrical conductivity to produce the next generation of self-healing nanocomposites with reproducible, rapid, and reversible healing cycles. Conductive fillers, such as graphene, metallic nanoparticles, CNTs, or hybrid nanostructures, are added to the polymer to produce these versatile properties. By creating permeable pathways, these fillers facilitate the efficient transfer of thermal and electrical energy to the damaged areas. The synergy between the polymer matrix and the fillers is crucial; the fillers ensure efficient energy delivery to initiate these reactions, while the matrix provides self-healing capabilities through reverse bonding interactions.

The molecular properties of polymers, such as molecular weight, elastic viscosity, and the availability and reactivity of functional groups on crack surfaces, are fundamental to intrinsic self-healing systems. In this case, the interactions between the polymer surface and the filler are critical. In addition to improving mechanical performance and load transfer, strong adhesion or chemical fixation can inhibit chain movement, potentially limiting the polymer's ability to penetrate cracks during the healing process. Optimal self-healing efficiency, on the other hand, requires a balance between surface bonding and polymer chain movement. This discussion highlights the need for a synergistic integration of matrix, filler, and interface engineering to achieve a robust and multifunctional self-healing system, rather than relying solely on the polymer or filler. In demanding applications, such as aerospace, automotive, electronics, and structural engineering, enhancing these properties enables the design of self-healing materials, leading to more robust, safer, and more sustainable material solutions.

9.11.1 Joule-Activated Self-Healing

Utilizing the Joule effect for electrically induced local heating is one of the most promising approaches for inducing self-healing. This process generates microscopic heat by passing an electric current through a conductive nanonetwork, such as interconnected CNTs, graphene sheets, metallic nanoparticles, or hybrid fillers. This heat promotes fracture closure, thermally activates reversible bonds (such as Diels–Alder, alkoxyamine, or imine bonds), and enhances local polymer chain mobility. Research has shown that high-altitude CNTs provide efficient permeability pathways, reducing the activation potential while ensuring continuous heating across the affected area [56]. Due to their superior thermal conductivity, which efficiently distributes Joule-induced heat throughout the matrix, graphene-based systems also exhibit rapid and recurrent self-healing [57].

9.11.2 Internal vs. External Healing Mechanisms

The two main categories of self-healing processes are internal and external systems: Encapsulated healing agents, stored in microcapsules and microvascular networks, form the basis of external systems. These agents are released, undergo polymerization, and seal the damaged area once a crack has formed. External methods, while effective for single use, are limited by the amount of healing agent. The material loses its ability to regenerate once depleted [58]. Internal systems utilize the reversible bonds inherent in the polymer, enabling repeated healing. To restore material integrity, these systems rely on dynamic covalent bonds (Diels–Alder, disulfide, boronic ester, etc.) and non-covalent interactions (hydrogen bonding, π–π stacking, metal-ligand coordination, etc.). Because intrinsic systems are theoretically capable of indefinite self-healing under appropriate conditions, they are particularly well-suited for applications requiring multiple cycles of damage and repair [6–8].

9.11.3 Functional Properties of the Polymer Matrix

The polymer matrix itself is fundamental to self-healing efficiency. Key elements include: Molecular weight, a high molecular weight can impede the healing process by reducing chain mobility while simultaneously increasing mechanical strength. Viscoelastic behavior: A delicate balance between fluidity and elasticity is essential; excessively soft matrices may flex under stress, while excessively rigid matrices hinder fracture closure. Functional group reactivity: The repair rate and accuracy are determined by the presence of reactive molecules at fracture contact points. Fillers and polymer chains have equally important interfacial interactions. While strong adhesion enhances mechanical performance and load transfer, it can also restrict chain mobility, thus reducing healing efficiency. Maximizing self-repair requires achieving the correct balance between interfacial bonding and polymer mobility [57,58].

9.11.4 Effect of Nanofillers

CNTs, graphene, multi-walled nanotubes, metallic nanoparticles, and nanoclay are examples of nanooastomers that support both mechanical reinforcement and healing activation: Efficient electrical and thermal energy transfer is facilitated through permeable networks; high aspect ratio (H) nanooastomers lower permeability thresholds, allowing for Joule heating at low voltage; and uniform dispersion ensures even heat distribution and protects against localized overheating. According to recent research, hybrid fillers such as CNTs combined with metallic nanoparticles or graphene improve the self-healing rate and mechanical toughness of materials, enabling them to repeatedly recover from damage under cyclic loading [11–13].

9.11.5 Operational and Environmental Aspects

Temperature, humidity, mechanical stress, and exposure conditions affect healing performance: Precise activation temperatures are essential for intrinsic systems that rely on thermally reversible processes, such as the Diels–Alder process. Non-covalent systems may be susceptible to ionic or humid conditions. Depending on the filler composition and matrix viscosity elasticity, mechanical loading during healing can either enhance chain movement or hinder crack closure. One of the biggest challenges remains designing materials capable of sustained self-repair under realistic service conditions.

9.11.6 Future Prospects and Plans

New approaches focus on hybrid curing systems and multiscale design: Hybrid systems improve the availability of limited curing agents by combining external and internal mechanisms to enable repeatable curing. Before fabrication, the filler geometry, distribution, and matrix properties can be optimized through modeling and simulation of heat transfer, crack propagation, and molecular dynamics. Improving the surface functions of the nanofiller and the interface geometry can enhance adhesion without compromising chain mobility. Joule activation can be complemented by systems that respond to stimuli using light, magnetic fields, or pH, enabling multicatalytic self-curing under complex operating conditions.

9.11.7 Applications and Results

These improved self-curing nanocomposites have numerous applications: Automotive: Lightweight, fatigue-resistant structural components. Electronics: Self-healing circuits and sensors that maintain connectivity and functionality. Aerospace: Self-healing microcracks in composite aircraft structures reduce maintenance costs and enhance safety. Biomedicine: Drug delivery, tissue structures, and implant coatings benefit from biocompatible and repeatable healing cycles. Civil engineering: Coatings, sealants, and polymer additives improve durability and corrosion resistance. By combining mechanical strength, adaptive functionality, and sustainability, self-healing polymer nanocomposites offer a transformative approach to materials design and have the potential to reduce energy consumption and material waste across various sectors [59].

9.12 FUTURE PERSPECTIVES AND RECOMMENDATIONS

The field of self-healing polymer nanocomposites is experiencing rapid, revolutionary development, responding to the growing demand for sustainable, durable, and intelligent materials that can extend the lifespan of engineering structures. Even with these significant advances, several hurdles remain to be overcome through targeted research and innovation before their full industrial application can be realized. Smart, self-healing polymer nanocomposites represent a breakthrough in materials engineering. These advanced composites exhibit remarkable mechanical strength, durability, and self-repair efficiency because of the addition of functional nanofillers such as CNTs, graphene, and nanoclay. From vascular networks and microcapsule-based systems to reverse bonding and dynamic covalent bonding, a variety of self-healing mechanisms enable the independent restoration of structural integrity after damage, improving overall performance, reliability, and service life. Future research should focus on enhancing surface bonding between polymer matrices and nanofillers to boost thermal and mechanical stability and processing efficiency. Furthermore, the development of scalable and cost-effective production technologies remains crucial for industrial deployment. Through data-driven insights, combining artificial intelligence and predictive modeling can accelerate material design, improve self-healing behavior, and reduce experimental costs.

- With their enhanced durability under harsh environmental conditions, low maintenance requirements, and high energy efficiency, self-healing polymer nanocomposites hold tremendous potential in a variety of industries, particularly the electrical, automotive, and aerospace sectors. Their recyclability, lightweight nature, and environmental resilience also make them a promising prospect for the next generation of sustainable engineering technologies. Self-healing polymer nanocomposites can reduce lifecycle costs and reshape the future of smart materials. The self-healing property should be integrated with other intelligent properties, including sensing, thermal regulation, electrical conductivity, and shape memory, in future materials. This versatility will expand their applications in aerospace, flexible electronics, biomedical implants, and renewable energy systems.
- *Scalability and Manufacturing Efficiency:* To transition from laboratory research to industrial production without compromising healing performance, cost-effective and scalable manufacturing methods must be developed, such as 3D printing, electrospinning, and green synthesis. Using surface-functional nanomaterials (such as CNTs, graphene, methane, and nanoclay) to tailor nanofiller interactions with the matrix will improve load transfer, healing response, and mechanical stability while minimizing filler agglomeration.
- Using surface-functional nanomaterials (such as CNTs, graphene, methane, and nanoclay) to tailor the interactions of nanofillers with the matrix will improve load transfer, healing response, and mechanical stability while reducing filler agglomeration.
- *Sustainable and Environmentally Friendly Systems:* Biodegradability, recyclability, and non-toxicity must be given top priority in the next generation of self-healing polymers. Bioinspired polymers and natural healing materials made from renewable resources offer an environmentally friendly path forward.
- *Real-time Monitoring and Intelligent Diagnosis:* By integrating self-sensing capabilities into polymeric nanocomposites, it is possible to create "smart materials" capable of self-diagnosis and repair, early damage detection, and real-time health monitoring.
- *Predictive Modeling and Design:* The use of computational modeling and artificial intelligence will accelerate material design and performance prediction, reducing experimental costs and increasing processing efficiency through data-driven optimization.
  - *Recommendations:* To ensure long-term reliability in harsh environments and under cyclic loads, multi-cycle and repeatable processing systems should be prioritized. Standardized testing protocols should be developed to accurately assess processing efficiency, mechanical recovery, and resistance to environmental factors. Materials scientists, chemists, and engineers should collaborate more interdisciplinarily to bridge the gap between molecular design and large-scale structural applications. Industry–academy partnerships should be encouraged to accelerate the transfer and commercialization of self-healing nanocomposite technologies. Lifecycle cost–benefit analyses should also be strengthened to assess

the practical sustainability and economic viability of these materials. Self-healing polymer nanocomposites represent a radical advance in materials science and have the potential to change how modern structures are built and maintained fundamentally. Future research may reveal autonomous, adaptable, and sustainable materials that redefine performance standards across numerous engineering disciplines by leveraging emerging technologies and overcoming current limitations.

PROBLEMS

1. Self-healing polymer nanocomposites: What are they?
2. How do nanofillers improve the properties of these materials?
3. How do self-healing materials reduce maintenance costs and extend component lifespan?
4. What makes these materials essential for automotive and aerospace applications?
5. In transportation technologies, what sustainability benefits do self-healing polymer nanocomposites offer?

REFERENCES

1. Malekkhouyan, R., Neisiany, R. E., Khorasani, S. N., Das, O., Berto, F., & Ramakrishna, S. (2021). The influence of size and healing content on the performance of extrinsic self-healing coatings. *J. Appl. Polym. Sci*, 10, 49964.
2. Panahi, P., Khorasani, S. N., Koochaki, M. S., Dinari, M., Das, O., & Neisiany, R. E. (2021). Synthesis of Cloisite 30B-acrylamide/acrylic acid nanogel composite for self-healing purposes. *Appl. Clay Sci.*, 210, 106174.
3. Wang, S., & Urban, M. W. (2020). Self-healing polymers. *Nat. Rev. Mater.*, 5, 562–583.
4. Jin, H., Miller, G. M., Sottos, N. R., & White, S. R. (2011). Fracture and fatigue response of a self-healing epoxy adhesive. *Polymer*, 52, 1628–1634.
5. Pulikkalparambil, H., Siengchin, S., & Parameswaranpillai, J. (2018). Corrosion protective self-healing epoxy resin coatings based on inhibitor and polymeric healing agents encapsulated in organic and inorganic micro and nanocontainers. *Nano-Struct. Nano-Objects*, 16, 381–395.
6. Kausar, A., Ahmad, I., Maaza, M., & Bocchetta, P. (2023). Self-healing nanocomposites—advancements and aerospace applications. *Journal of Composites Science*, 7(4), 148.
7. Guadagno, L., Vertuccio, L., Naddeo, C., Barra, G., Raimondo, M., Sorrentino, A., Binder, W. H., Michael, P., Rana, S., & Calabrese, E. (2021). Functional structural nanocomposites with integrated self-healing ability. *Mater. Today Proc.*, 34, 243–249.
8. Guadagno, L., Vertuccio, L., Naddeo, C., Calabrese, E., Barra, G., Raimondo, M., Sorrentino, A., Binder, W., Michael, P., & Rana, S. (2019). Self-healing epoxy nanocomposites via reversible hydrogen bonding. *Compos. Part B Eng.*, 157, 1–13.
9. Tong, Y.-Q., Shi, Q.-S., Liu, M.-J., Li, G.-R., Li, C.-J., Yang, G.-J. (2021). Lightweight epoxy-based abradable seal coating with high bonding strength. *J. Mater. Sci. Technol.*, 69, 129–137.

10. Ding, Q., Xu, X., Yue, Y., Mei, C., Huang, C., Jiang, S., Wu, Q., & Han, J. (2018). Nanocellulose-mediated electroconductive self-healing hydrogels with high strength, plasticity, viscoelasticity, stretchability, and biocompatibility toward multifunctional applications. *ACS Appl. Mater. Interfaces*, 10, 27987–28002.
11. Kessler, M., Sottos, N., & White, S. (2003). Self-healing structural composite materials. *Compos. Part A Appl. Sci. Manuf.*, 34, 743–753.
12. Feiz, A., & Khosravi, H. (2019). Multiscale composites based on a nanoclay-enhanced matrix and E-glass chopped strand mat. *J. Reinforc. Plast. Compos.*, 38, 591–600.
13. Naresh, K., Cantwell, W. J., Khan, K. A., & Umer, R. (2021). Single and multi-layer core designs for Pseudo-Ductile failure in honeycomb sandwich structures. *Compos. Struct.*, 256, 113059.
14. Khan, T., Irfan, M. S., Cantwell, W. J., & Umer, R. (2022). Crack healing in infusible thermoplastic composite laminates. *Compos. Part A: Appl. Sci. Manuf.*, 156, 106896.
15. Chee, S. S., Jawaid, M., Sultan, M. T. H., Alothman, O. Y., & Abdullah, L. C. (2019). Thermomechanical and dynamic mechanical properties of bamboo/woven kenaf mat reinforced epoxy hybrid composites. *Compos. Part B Eng.*, 163, 165–174.
16. Wang, X., Liang, D., & Cheng, B. (2020). Preparation and research of intrinsic self-healing elastomers based on hydrogen and ionic bond. *Compos. Sci. Technol.*, 193, 108127.
17. Rahmani, H., Eslami-Farsani, R., & Ebrahimnezhad-Khaljiri, H. (2020). High velocity impact response of aluminum- carbon fibres-epoxy laminated composites toughened by nano silica and zirconia. *Fibres Polym.*, 21, 170–178.
18. Han, J., Ding, Q., Mei, C., Wu, Q., Yue, Y., & Xu, X. (2019). An intrinsically self-healing and biocompatible electroconductive hydrogel based on nanostructured nanocellulose-polyaniline complexes embedded in a viscoelastic polymer network towards flexible conductors and electrodes. *Electrochim. Acta.*, 318, 660–672.
19. Ghosh, S. K. (Ed.). (2009). *Self-Healing Materials: Fundamentals, Design Strategies, and Applications*. Wiley-vch: Weinheim,.
20. Zhu, D. Y., Chen, F., Rong, M. Z., & Zhang, M. Q. (2022). Capsules-based self-healing polymers and polymer composites. In *Recent Advances in Smart Self-Healing Polymers and Composites* (pp. 113–140. https://doi.org/10.1016/B978-0-12-823472-3.00007-2.
21. Wang, Y., Li, P., Truong-Dinh Tran, T., Zhang, J., & Kong, L. (2016). Manufacturing techniques and surface engineering of polymer based nanoparticles for targeted drug delivery to cancer. *Nanomaterials*, 6, 26.
22. Tao, F., Qin, L., Wang, Z., & Pan, Q. (2017). Self-healable and cold-resistant supercapacitor based on a multifunctional hydrogel electrolyte. *ACS Appl. Mater. Interfaces*, 9, 15541–15548.
23. Yang, G., Feng, X., Wang, W., OuYang, Q., & Liu, L. (2021). Effective interlaminar reinforcing and delamination monitoring of carbon fibrous composites using a novel nano-carbon woven grid. *Compos. Sci. Technol.*, 213, 108959.
24. Han, J., Wang, H., Yue, Y., Mei, C., Chen, J., Huang, C., Wu, Q., & Xu, X. (2019). A self-healable and highly flexible supercapacitor integrated by dynamically cross-linked electro-conductive hydrogels based on nanocellulose-templated carbon nanotubes embedded in a viscoelastic polymer network. *Carbon*, 149, 1–18.
25. Guo, P., Su, A., Wei, Y., Liu, X., Li, Y., Guo, F., Li, J., Hu, Z., & Sun, J. (2019). Healable, highly conductive, flexible, and nonflammable supramolecular ionogel electrolytes for lithium-ion batteries. *ACS Appl. Mater. Interfaces*, 11, 19413–19420.
26. Tessler, A. (2007). Structural analysis methods for structural health management of future aerospace vehicles. *Key Eng. Mater.*, 347, 57–66. https://www.scientific.net/KEM.347.57 ;.

27. Fujii, K. (2005). Progress and future prospects of CFD in aerospace—Wind tunnel and beyond. *Prog. Aerosp. Sci.*, 41, 455–470.
28. Li, H., Cui, Y., Wang, H., Zhu, Y., & Wang, B. (2017). Preparation and application of polysulfone microcapsules containing tung oil in self-healing and self-lubricating epoxy coating. *Colloids Surf. A Physicochem. Eng. Asp.*, 518, 181–187.
29. Lorwanishpaisarn, N., Srikhao, N., Jetsrisuparb, K., Knijnenburg, J. T., Theerakulpisut, S., Okhawilai, M., & Kasemsiri, P. (2022). Self-healing ability of epoxy vitrimer nanocomposites containing bio-based curing agents and carbon nanotubes for corrosion protection. *J. Polym. Environ.*, 30, 472–482.
30. Jing, T., Heng, X., Guifeng, X., Li, L., Li, P., & Guo, X. (2022). Rapid self-healing and tough polyurethane based on the synergy of multi-level hydrogen and disulfide bonds for healing propellant microcracks. *Mater. Chem. Front.*, 6, 1161–1171.
31. Yu, Y., Shi, Y., Kurita, H., Jia, Y., Wang, Z., & Narita, F. (2022). Carbon fiber-reinforced piezoelectric nanocomposites: Design, fabrication and evaluation for damage detection and energy harvesting. *Compos. Part A Appl. Sci. Manuf.*, 172, 107587.
32. Mohd Jani, J., Leary, M., Subic, A., & Gibson, M. A. (2014). A review of shape memory alloy research, applications and opportunities. *Mater. Des.*, 56, 1078–1113.
33. Kuilla, T., Bhadra, S., Yao, D., Kim, N. H., Bose, S., & Lee, J. H. (2010). Recent advances in graphene based polymer composites. *Prog. Polym. Sci.*, 35, 1350–1375.
34. Kim, S., Jeon, H., Koo, J. M., Oh, D. X., & Park, J. (2024). Practical applications of self-healing polymers beyond mechanical and electrical recovery. *Adv. Sci.*, 11, e2302463.
35. Zhang, S., van Dijk, N., & van der Zwaag, S. (2020). A review of self-healing metals: Fundamentals, design principles and performance. *Acta Metall. Sin.*, 33, 1167–1179.
36. Sekine, N., & Nakao, W. (2023). Advanced self-healing ceramics with controlled degradation and repair by chemical reaction. *Materials*, 16, 6368.
37. Parihar, S., & Gaur, B. (2023). Self healing approaches in polymeric materials-an overview. *J. Polym. Res.*, 30, 217.
38. Venkata Chalapathi, K., Prabhakar, M. N., Lee, D.-W., & Song, J.-I. (2023). Development of thermoplastic self-healing panels by 3D printing technology and study extrinsic healing system under low-velocity impact analysis. *Polym. Test.*, 119, 107923.
39. Kontiza, A., & Kartsonakis, I. A. (2024). Smart composite materials with self-healing properties: A review on design and applications. *Polymers*, 16(15), 2115.
40. Abbasi, N. M., Yu, H., Wang, L., Amer, W. A., Akram, M., Khalid, H., ... & Shan, J. (2015). Preparation of silver nanowires and their application in conducting polymer nanocomposites. *Mater. Chem. Phys.*, 166, 1–15.
41. Orellana, J., Moreno-Villoslada, I., Bose, R. K., Picchioni, F., Flores, M. E., & Araya-Hermosilla, R. (2021). Self-healing polymer nanocomposite materials by Joule effect. *Polymers*, 13(4), 649.
42. Lee, M. W., Jo, H. S., Yoon, S. S., & Yarin, A. L. (2017). Thermally driven self-healing using copper nanofiber heater. *Appl. Phys. Lett.*, 111, 011902.
43. Shah, K. W., & Xiong, T. (2019). Multifunctional metallic nanowires in advanced building applications. *Materials*, 12, 1731.
44. Wang, L., & Aslani, F. (2019). A review on material design, performance, and practical application of electrically conductive cementitious composites. *Constr. Build. Mater.*, 229, 116892.
45. Papanastasiou, D. T., Schultheiss, A., Muñoz-Rojas, D., Celle, C., Carella, A., Simonato, J., & Bellet, D. (2020). Transparent heaters: A review. *Adv. Funct. Mater.*, 30, 1910225.

46. Liu, H., Li, Q., Zhang, S., Yin, R., Liu, X., He, Y., Dai, K., Shan, C., Guo, J., Liu, C., et al. (2018). Electrically conductive polymer composites for smart flexible strain sensors: A critical review. *J. Mater. Chem. C*, 6, 12121–12141.
47. Kaur, G., Adhikari, R., Cass, P., Bown, M., & Gunatillake, P. (2015). Electrically conductive polymers and composites for biomedical applications. *RSC Adv.*, 5, 37553–37567.
48. Aradhana, R., Mohanty, S., & Nayak, S. K. (2020). A review on epoxy-based electrically conductive adhesives. *Int. J. Adhes. Adhes.*, 99, 102596.
49. Marsden, A. J., Papageorgiou, D. G., Vallés, C., Liscio, A., Palermo, V., Bissett, M. A., Young, R. J., & Kinloch, I. A. (2018). Electrical percolation in graphene–polymer composites. *2D Mater.*, 5, 032003.
50. Park, M., Park, J., & Jeong, U. (2014). Design of conductive composite elastomers for stretchable electronics. *Nano Today*, 9, 244–260.
51. Deng, H., Lin, L., Ji, M., Zhang, S., Yang, M., & Fu, Q. (2014). Progress on the morphological control of conductive network in conductive polymer composites and the use as electroactive multifunctional materials. *Prog. Polym. Sci.*, 39, 627–655.
52. Chen, H., Ginzburg, V. V., Yang, J., Yang, Y., Liu, W., Huang, Y., Du, L., & Chen, B. (2016). Thermal conductivity of polymer-based composites: Fundamentals and applications. *Prog. Polym. Sci.*, 59, 41–85.
53. Li, A., Zhang, C., & Zhang, Y. F. (2017). Thermal conductivity of graphene-polymer composites: Mechanisms, properties, and applications. *Polymers*, 9, 437.
54. Huang, X., Zhi, C., Lin, Y., Bao, H., Wu, G., Jiang, P., & Mai, Y.-W. (2020). Thermal conductivity of graphene-based polymer nanocomposites. *Mater. Sci. Eng. R Rep.*, 142, 100577.
55. Chen, C., Huang, Z., Jiao, Y., Shi, L.-A., Zhang, Y., Li, J., Li, C., Lv, X., Wu, S., Hu, Y. et al. (2019). In situ reversible control between sliding and pinning for diverse liquids under ultra-low voltage. *ACS Nano*, 13, 5742–5752.
56. Jiménez-Suárez, A., Martín-González, J., Sánchez-Romate, X. F., & Prolongo, S. G. (2020). Carbon nanotubes to enable autonomous and volumetric self-heating in epoxy/polycaprolactone blends. *Compos. Sci. Technol.*, 199, 108321.
57. Wang, K., Zhou, Z., Zhang, J., Tang, J., Wu, P., Wang, Y., Zhao, Y., & Leng, Y. (2020). Electrical and thermal and self-healing properties of graphene-thermopolyurethane flexible conductive films. *Nanomaterials*, 10, 753.
58. Billiet, S., Hillewaere, X. K. D., Teixeira, R. F. A., & Du Prez, F. E. (2013). Chemistry of crosslinking processes for self-healing polymers. *Macromol. Rapid Commun.*, 34, 290–309.
59. Zhu, D. Y., Rong, M. Z., & Zhang, M. Q. (2015). Self-healing polymeric materials based on microencapsulated healing agents: From design to preparation. *Prog. Polym. Sci.*, 49–50, 175–220.

10 Electronics and Energy Storage Applications of Self-Healing Polymer Nanocomposites

10.1 INTRODUCTION

Rapid advances in electronics and energy technologies over the past few decades have increased the demand for smarter, more reliable, and more efficient materials. Self-healing polymer nanocomposites, a group of materials that combine the enhanced functional properties of nanoparticles with the self-healing capabilities of smart polymers, are among the most promising technologies to address these changing needs. These advanced materials exhibit a remarkable ability to "heal" or regenerate after damage. Furthermore, the addition of nano-reinforcements enhances their mechanical, electrical, and thermal performance [1–3].

These self-healing properties, which mimic biological systems that spontaneously repair themselves after damage, represent a paradigm shift in materials design. These materials can self-heal even small cracks or scratches caused by stress, corrosion, or environmental exposure, extending the life of the device or structure in which they are used and improving its functional integrity. This eliminates the need for costly maintenance or complete replacement of damaged components.

These materials are critical in flexible electronics, portable electronics, and energy storage devices such as supercapacitors and batteries. These systems are often subject to thermal cycling or mechanical deformation, where even the slightest damage can affect performance or lead to system failure. In this context, self-healing polymer nanocomposites offer significant benefits by improving long-term cost-effectiveness, performance stability, and device durability.

Functional properties are significantly improved when polymers are combined with nanomaterials, including metal oxides, graphene, and carbon nanotubes (CNTs). For example, these nanofillers improve mechanical strength and structural stability while significantly increasing electrical and thermal conductivity. These properties are essential for modern high-performance electronic applications.

In addition, the self-healing methods used in these materials vary depending on the application and design needs. Some systems utilize chemical mechanisms, including dynamic covalent bonding or microencapsulated healing agents, which can rearrange molecular structures or release chemicals to repair damage. Other

DOI: 10.1201/9781003758556-10

systems rely on thermal activation, where exposure to heat stimulates the healing process. With limited external assistance, these diverse approaches can restore mechanical and electrical properties, offering useful advantages for everyday applications [4].

The potential applications of these self-healing nanocomposites have expanded significantly because of ongoing advances in materials science, computational modeling, and nanotechnology. Recent studies have demonstrated their usefulness not only in conventional electronics and batteries but also in advanced industries where flexibility, durability, and safety are key design factors, such as wearable electronics, electronic coatings, biomedical devices, and others [5].

This article aims to provide a comprehensive review of self-healing polymer nanocomposites, focusing on their use in electronics and energy storage in light of recent encouraging developments. This article will address the fundamental ideas behind self-healing behavior, how nanomaterials can enhance performance, and the opportunities and challenges of integrating this technology into advanced systems. [6].

10.2 EXTRINSIC SELF-HEALING

Extrinsic self-healing is one of the key processes for self-healing materials. It is based on the addition of external healing agents into the structure of the material during manufacturing. When damage occurs, such as a crack or fracture, these agents are released from their microscopic reservoirs or networks. The release and interaction of these chemicals at the site of damage initiate the healing process, resulting in the restoration of the material [4–6].

The basic idea of extrinsic self-healing is the release of healing agents from microcapsules or microvascular networks embedded in the host material. These reservoirs rupture as the crack propagates through the material, allowing the healing agents to be released. Upon exposure to air or other external stimuli, the healing agent either heals or interacts with a pre-embedded catalyst, forming a strong bond. To seal the cracks, the released monomers react with catalysts scattered throughout the matrix, causing polymerization. The capsule-type self-healing method is shown in Figure 10.1.

Microcapsule Systems: These systems use small capsules loaded with adhesives or resins that act as healing agents. The capsules are randomly distributed within the matrix, bursting and releasing their contents when they break [5–7].

Microvascular Networks: Modeled after biological systems such as capillaries, these systems consist of interconnected channels capable of continuously delivering therapeutic agents. They are ideal because they allow for multiple treatment cycles.

Multi-Chamber Systems: These systems contain separate compartments for the catalyst and the therapeutic agent. To repair the material after damage, the components combine and initiate a chemical process.

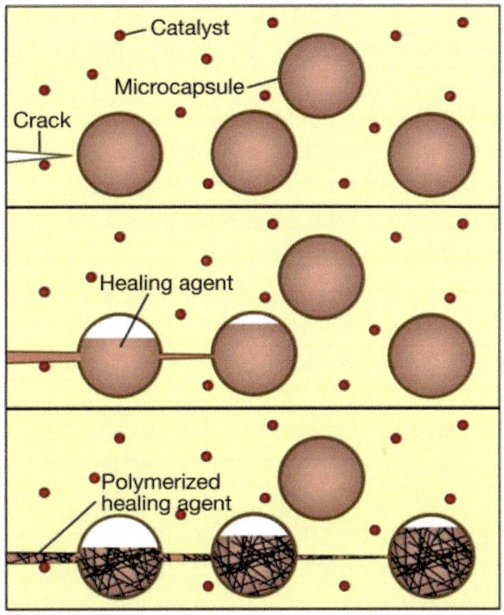

FIGURE 10.1 Diagrammatic representation of therapeutic drugs microencapsulated in a composite matrix that incorporates catalysts to polymerize the agents. (a) There are cracks and damage to the matrix. (b) Capillary action releases the healing ingredients into the fissures once they reach the microcapsules. (c) The healing agents polymerize when they come into touch with the catalyst, closing the crack [2] [Open access].

Electronic systems, protective coatings, and spacecraft make extensive use of external self-healing technology. Self-healing is essential for performance and safety in critical situations where even minor cracks can lead to major failures, such as in spacecraft structures or battery housings.

The main advantage of external self-healing is that it responds to damage quickly, locally, and repeatedly without the need for external assistance. A significant drawback is that many capsule-based systems only allow one recovery event per capsule. Another major technical challenge is ensuring uniform distribution and effective capsule activation in complex situations [1,5].

10.2.1 Capsular Type

One of the oldest and most studied methods in the field of self-healing materials is the extrinsic capsule approach to self-healing. This approach uses microcapsules, small spherical containers implanted in the host polymer matrix during manufacturing, containing the healing components. These capsules are stable. The microcapsules rupture along the crack at its inception and disperse throughout the material. The healing agent is released at the crack level upon rupture, where it fills the damaged area and flows out. The healing agent undergoes a polymerization reaction at room temperature or higher temperatures, sealing the crack surfaces and restoring mechanical integrity if the system is constructed with an integrated catalyst or hardener dispersed in the matrix (such as dicyclopentadiene diacetyl (DCPD) with a Grubbs catalyst).

- *Self-Response:* Healing occurs spontaneously without external assistance.
- *Ease of Fabrication:* The methods of implanting and implanting capsules have proven their effectiveness over time. Localized repair minimizes material waste by focusing only on the damaged area.
- *Broad Compatibility:* Adapts to various polymer matrices.
- *Single-Healing:* The system can only perform one healing cycle at a time, as a ruptured capsule cannot be reused.
- *Capsule Distribution:* Adequate capsule concentration and uniform distribution are essential for effective healing.
- *Balance in Mechanical Properties:* The use of capsules may result in a decrease in the overall strength or stiffness of the host material.

Capsule systems have been investigated in structural polymers, coatings, aerospace vehicles, and microelectronics, where early and localized damage management is required. They are particularly useful in protecting components from damage caused by fatigue, corrosion, and microcracking.

10.2.2 Vascular Type

Inspired by the biological circulatory networks found in living organisms, the vascularized form of external self-healing represents a significant advancement over capsular systems. This approach involves embedding interconnected vascular networks or microchannels into the material, which serve as delivery routes for therapeutic chemicals to affected areas. Unlike single-use microcapsules, vascularized systems are capable of undergoing multiple healing cycles, making them highly attractive for applications requiring recurrent damage mitigation and long-term durability. Figure 10.2 illustrates how the basic self-healing mechanism of vascular-type and capsular-type materials compares.

Vascularized self-healing systems are typically designed using fiber-reinforced composites with embedded vascular channels, hollow microtubes, or three-dimensional (3D) channel networks. Healing agents such as reactive polymers, stiffeners, or monomers are already present in these networks. The healing agent is released spontaneously when the crack crosses a channel and enters the damaged area through capillary action. One of the key advantages of vascularized systems is their ability to refill and maintain themselves. Healing fluids can be continuously supplied and recharged in some advanced systems by connecting external reservoirs and pumps to the internal network. This property significantly increases the lifetime and functionality of the self-healing system, making it particularly suitable for high-performance electronics, civil infrastructure, automotive parts, and aerospace structures.

However, significant challenges exist in creating and constructing such networks due to their complexity. It is essential to carefully develop elements such as channel size, direction, distribution, and fluid flow dynamics to ensure effective healing

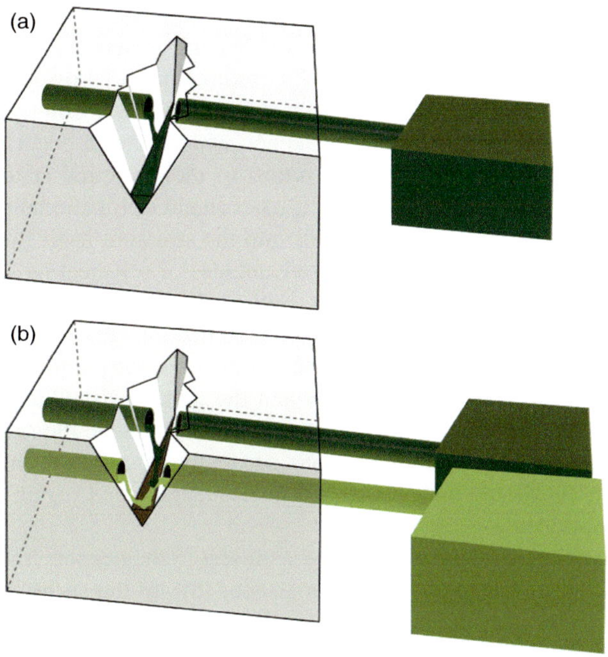

FIGURE 10.2 Schematic illustration of a vascular-type self-healing material. The reservoir releases healing factors in response to damage via hydrostatic pressure, capillary forces, and gravity. Single-channel (A) and multi-channel (B) designs are used to illustrate vascular self-healing systems [2] [Open access].

without compromising the structural integrity of the material. Maintaining mechanical strength and thermal stability also depends on the compatibility of the healing agents with the host matrix [7–9].

Despite these obstacles, self-healing vascular systems represent a practical avenue for developing next-generation smart materials. The growing need for robust, low-maintenance, and sustainable engineering solutions is met by their ability to independently heal repeated damage in a controlled and sustainable manner.

10.3 INTRINSIC SELF-HEALING

The term "intrinsic self-healing" describes the self-healing ability of a material that derives from its natural molecular structure rather than from external factors such as vascular networks or capsules. Typically, reversible chemical bonds, physical interactions, or dynamic covalent mechanisms that can be broken and reformed under specific conditions—often triggered by heat, light, pH changes, or mechanical force—enable this type of self-healing. Intrinsic systems offer a distinct advantage over extrinsic methods, as the healing process is typically long-lasting and reproducible.

The dynamic nature of certain molecular interactions within the polymer network is the fundamental idea behind intrinsic self-healing materials. These interactions can include π–π stacking, metal–ligand coordination, hydrogen bonding, ionic interactions, and reversible covalent bonds such as disulfide bonds or Diels–Alder (DA) processes. These responses are weakened when damage occurs, but when the appropriate trigger is present, they can return to their original state, allowing the material to "close" the crack and restore its mechanical or functional properties.

Because the healing mechanism is built into the structure itself rather than being consumed in the process, one of the key advantages of self-healing systems is their ability to undergo multiple healing cycles. Self-healing systems also avoid problems such as limited healing volume, complex channel design, or interference with mechanical properties due to filler content, as they do not require external healing chemicals. These dynamic bonds break when the material is subjected to mechanical damage, such as a crack or microfracture. However, due to their reversibility, the interactions begin to reorganize and connect across the fractured surfaces. Over time, this mechanism allows the material to self-heal the crack and regain its mechanical integrity and usefulness.

As the molecular chains or fragments are closest to the deepest fracture sites, this healing process often begins there and progresses toward the surface. Similar to the effect of lightning, this behavior gradually realigns the fractured surfaces and restores their integrity through a series of bonding processes that occur along the fracture path (as shown in Figure 10.3).

To produce materials capable of undergoing multiple self-healing cycles without the need for external agents, intrinsic self-healing systems have received significant attention as a promising approach. The healing ability of these systems stems from dynamic intermolecular interactions or reversible molecular mechanisms inherent in the polymer matrix. Examples of these mechanisms include reversible covalent bonds, hydrogen bonds, ionic contacts, and supramolecular assemblies that can dissociate and reassemble in response to specific stimuli such as heat, light, or mechanical stress. As the healing activity is inherent in the polymer network itself and is not limited by the presence of an external factor, one of the most significant advantages of self-healing systems is their ability to repeat the healing cycle multiple times. Therefore, self-healing materials are excellent options for long-term applications in dynamic environments, such as flexible electronics, wearable technologies, soft robotics, and structural coatings, where sustained mechanical stress can lead to recurring damage.

In areas where microcracks can occur as a result of repeated deformation or environmental stress, such as flexible electronics, coatings, structural polymers, wearable technologies, and biomedical applications, self-healing materials have shown great potential. Their ability to spontaneously repair damage can lead to increased product life, improved reliability, and reduced maintenance costs.

Unlike some exogenous systems, other endogenous systems may have slower healing rates and rely on external stimuli (such as high temperatures) to initiate the healing process. Research is ongoing to develop materials that heal quickly at room temperature without sacrificing mechanical strength.

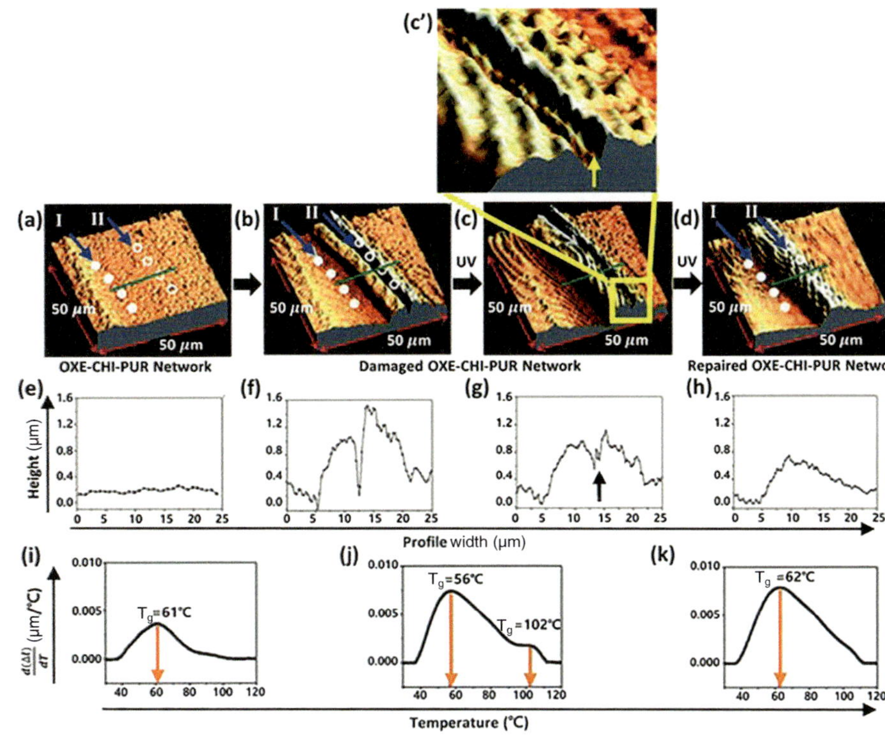

FIGURE 10.3 OXE-CHI-PUR meshes were imaged using an atomic force microscope (AFM) under the following conditions: intact (a), damaged (b), UV-exposed (c, c'), and cured (d). The HDI formulation ratio was 1.0:1.33:1.17×10–4:2×10–5. The symbols I and II in (a) indicate the thermal expansion measurement locations before and after self-healing, respectively, while the green line in (a) indicates the depth mapping direction. Included in these data are height versus width graphs for intact (e), damaged (f), UV-exposed (g), and cured (h) meshes, as well as d(Δl)/dT versus temperature (i–k) patterns for each mesh [2] [Open access].

Despite these obstacles, self-healing polymers offer a smart and sustainable approach to materials development, opening the door to next-generation applications that require self-sufficiency, flexibility, and adaptability.

10.3.1 Chemical Reaction-Based Self-Healing

A class of materials known as "chemically based self-healing" relies on specific chemical processes to restore their structure and function after injury. These systems often involve intrinsic or extrinsic healing processes, where chemical changes at the molecular level trigger and maintain the healing process. Carefully engineered reactions, which can occur under mild or normal conditions, are essential to the success of these systems, ensuring their suitability for a range of applications [8–10].

Reactive materials stored in microcapsules or vascular networks implanted within the material are often released to achieve chemical healing in extrinsic systems. These carriers rupture upon a crack or fracture, releasing monomers or crosslinkers that interact with healing agents or catalysts dispersed throughout the matrix [11]. A common example is DCPD, which reacts with a Grubbs catalyst to form a robust polymer that fills and seals the crack. These reactions must be localized, rapid, and capable of restoring mechanical strength with minimal external assistance. In intrinsic systems, the polymer structure incorporates a chemical healing mechanism. These often involve dynamic covalent bonds that break and reversibly reform in response to heat or light, such as imine bonds, DA processes, and disulfide exchanges. By allowing the material to "rearrange" itself at the molecular level, these processes enable self-healing without compromising the properties of the original material. One of the key advantages of these mechanisms is their ability to facilitate multiple healing cycles, ensuring long-term functional performance and sustainability. The healing mechanism can be explained as follows, based on the above information, and Figure 10.4 shows a schematic representation of the method.

In a variety of industries where maintaining mechanical integrity and extending service life are critical, such as coatings, structural composites, electronics, and biomedical equipment, self-healing based on chemical reactions has shown promising results. However, challenges remain in improving environmental stability, healing efficiency, and reaction time. To ensure self-healing even under the harshest conditions, research is ongoing to develop faster, more selective, and more stimuli-sensitive responses [10–12].

10.3.1.1 DA Reactions

A well-known thermoreversible chemical process required for self-healing polymer systems is the DA reaction. This process forms a hexagonal ring via cycloaddition between a diene and a [4 + 2] dienophile. The reversible nature of the DA reaction makes it very useful for self-healing applications. Heat can initiate the reverse

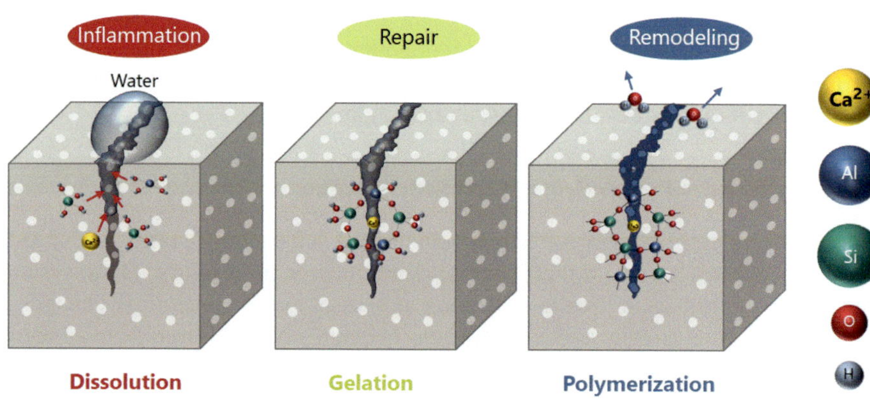

FIGURE 10.4 This material's healing mechanism [3] [Open access].

reaction, known as the reverse Diels–Alder (rDA) reaction, which breaks down the cycloaddition product and returns it to the original diene and dienophile.

When heated, materials can rearrange and reform through this reversible bonding mechanism, and when cooled, they can return to their original structure. When exposed to high temperatures, typically between 80 and 150 °C, diene-based self-healing polymers are produced, allowing the diene bonds to dissociate and reform while remaining stable under operating conditions. Multiple cure cycles are possible due to temperature-controlled reversibility, eliminating the need for complex delivery systems or external curing chemicals.

Due to their permanently crosslinked networks, thermoset polymers are typically not recyclable or reusable, making the DA reaction highly attractive. These materials can be made recyclable and reusable by adding chemical DA, which increases the sustainability and durability of the materials. Furan as a diene, and maleimide as a dienophile are often used in DA-based systems due to their high reactivity and compatibility with a wide range of polymer substrates. By using near-infrared (NIR) radiation to indirectly generate heat in polydopamine, this device achieved remarkably high self-healing efficiency without the need for direct heating. Synthesis of aliphatic or aromatic cyclic products, physiologically active substances, monomers, and multifunctional structures is among the common application pathways. Oxanobornenes, structural analogs of the small biologically active molecule cantharidin (a natural terpenoid isolated from Spanish blighted beetles), and their synthetic analogs norcantharidin and norcantharimides, which also exhibit strong biological activity, are produced by the reduction of the double bond in furan-derived oxanobornenes (Figure 10.5).

Coatings, adhesives, aircraft components, and flexible electronics are among the applications for DA-based self-healing materials, where mechanical performance, thermal stability, and repairability are key. Additionally, the reaction is attractive for high-performance, eco-friendly materials due to the absence of solvent and precise molecular design.

Despite their advantages, DA-based systems may require external heating to initiate the curing process, which can be difficult to achieve.

10.3.1.2 Disulfide Reactions

In self-healing polymer systems, disulfide reactions, a series of important reversible chemical reactions, are fundamental. The dynamic exchange of disulfide (S–S)

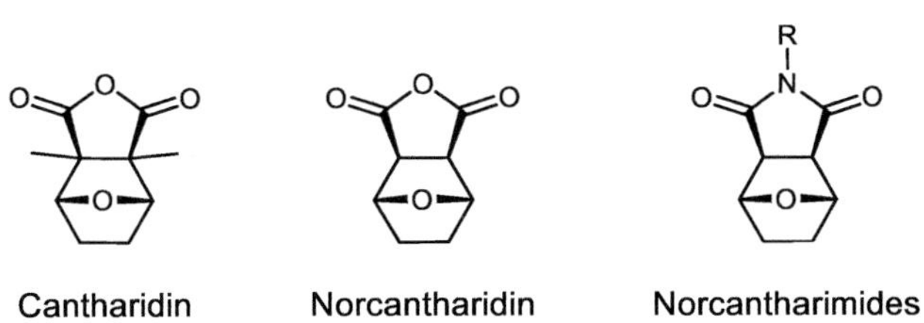

FIGURE 10.5 The chemical structure of cantharidine, norcantharidin, and norcantharidin [32] [Open access].

bonds, which can be reversibly broken and reformed under appropriate conditions such as heat, light, or redox catalysts, is a key element of this mechanism. This feature makes disulfide-based systems very attractive for applications requiring efficient, independent, and reproducible repair [12–14].

The covalent bonds between two sulfur atoms are known as disulfide bonds and are often formed by thiol (–SH) groups. These bonds are intentionally included in the crosslinking network or backbone of the polymer in self-healing materials. The disulfide bonds at the fracture surface can undergo bond exchange or rearrangement reactions in response to mechanical damage such as crack formation. This allows the polymer chains to reconnect and successfully seal the damaged area.

One of the most significant advantages of disulfide chemistry is its catalytic susceptibility. For example, the dynamic nature of disulfide bonds upon thermal activation (typically 60 to 120 °C) promotes chain migration and exchange, leading to self-healing. Alternatively, cleavage and remodeling of disulfide bonds can be induced by ultraviolet light or mild redox agents, allowing for recovery at lower temperatures.

Self-healing materials based on disulfides are widely used in biomedical devices, flexible electronics, wearable technologies, coatings, and soft robotics, where robustness, mechanical recovery, and environmental responsiveness are critical. In addition, the durability and lifetime of the material are increased by repeated recovery cycles enabled by the reversible nature of disulfide bonds [15–18].

However, under ambient conditions, disulfide systems may face challenges such as oxidation sensitivity and slow recovery rates. Current research focuses on improving performance, stability, and healing rates using stimuli-responsive hybrid systems or integrating disulfide chemistry with other dynamic bonds.

In the world of smart self-healing polymers, disulfide bond exchange is a powerful and adaptive, versatile mechanism that paves the way for robust, reconfigurable, and resilient material systems [1–3]. The atomic force microscopy (AFM) images shown in Figure 10.6 illustrate homodimers with a slightly different relative arrangement of their disulfide domains compared to the main protein core.. However, all images of GvFFTR and GaFFTR2 clearly show FR configurations, which are consistent with the evaluated oxidized states.

10.3.1.3 Exchange Reactions of Boronic Ester

Borate ester exchange reactions are an interesting dynamic covalent chemistry used in self-healing polymer systems. The reversible reaction of boronic acids with disalcohols produces boron esters, covalent bonds that can be broken and reformed in mild environments. Boron ester chemistry is particularly attractive for the production of reproducible self-healing materials because of its remarkable reversibility.

A key feature of boron ester exchange is its responsiveness to stimuli. Through esterification processes triggered by variables such as temperature, humidity, and pH changes, these bonds can be dynamically exchanged. In particular, the exchange occurs under mildly acidic or basic conditions, allowing the bonds to rearrange and reform at room temperature. This versatility allows the development of self-healing materials with tunable reaction rates and recovery conditions.

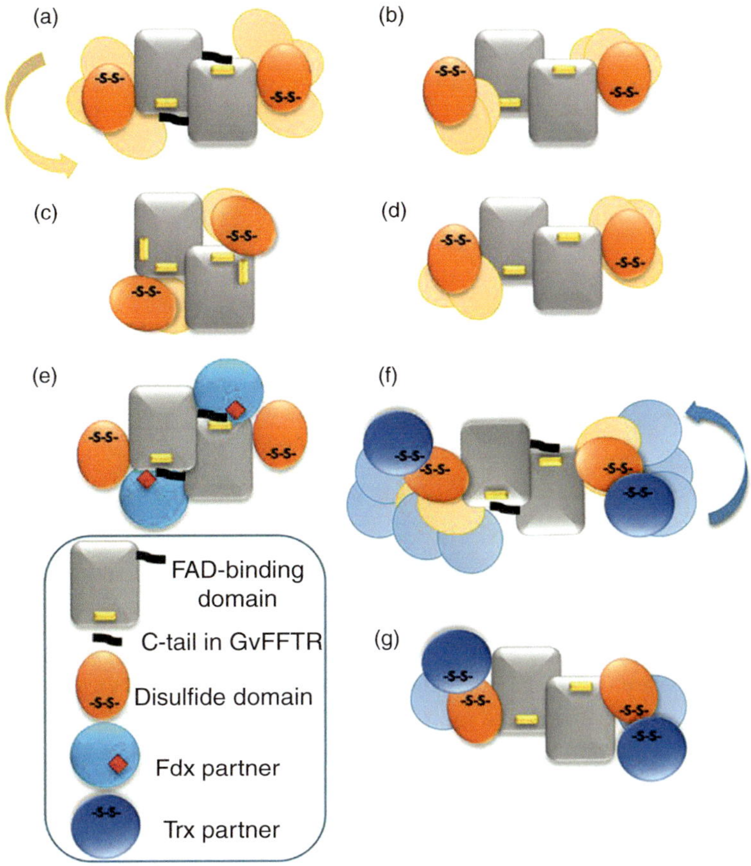

FIGURE 10.6 The variability of most representative AFM features is shown schematically for (a) GvFFTR, (b) CaFFTR2, (c) GvDDOR, and (d) GvFFTR_Δtail homodimers, free and in GvFFTR_Δtail and GvFdx1 mixtures, as well as for (e) transient complexes formed in GvFFTR and GvFdx1 mixtures, (f) transient complexes formed in GvFFTR and GvTrxm mixtures and covalent complexes of GvFFTR: GvTrxm and CaFFTR: CaTrx2 heterotetramers, and (g) transient complexes formed in GvFFTR_Δtail and GvTrxm mixtures and some covalent complexes of CaFFTR: CaTrx2 heterotetramers. Protein and domain codes are shown at the bottom left. Orders based on crystal structures are presented as solid forms. Faint colors indicate expected AFM imaging variability of disulfide domains or partner proteins relative to the major FAD binding regions of the two protomers [7] [Open access].

Boronic acid esters are often added to polymer chains or crosslinking points in self-healing contexts due to their ability to rearrange, which helps the polymer network recover from damage. This material can rebuild its network and restore its mechanical integrity when cracked or broken, as the broken boronic ester bonds near the damaged area undergo exchange reactions with neighboring functional groups [19].

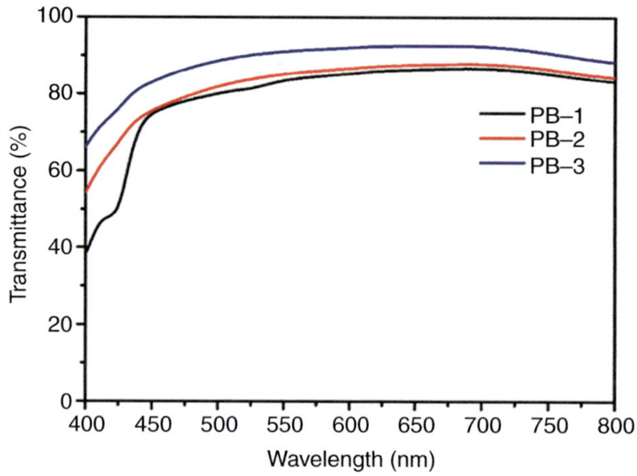

FIGURE 10.7 Optical transmission PB-1, PB-2, and PB-3 [32] [Open access].

In applications where materials are often subjected to mechanical stress or environmental changes, such as flexible electronics, soft robotics, wearable sensors, and biomedical devices, boronic ester-based self-healing systems have been the subject of extensive research. They are also ideal for smart hydrogels and are biocompatible due to their reversible nature and moisture sensitivity.

Despite their advantages, boronic esters can become unstable in aqueous conditions or when exposed to high humidity levels, which can lead to deterioration or weakening of the polymer structure. To overcome these limitations, scientists are exploring hybrid systems that improve overall performance by combining boronic esters with other dynamic covalent bonds or supramolecular interactions [20–24].

Boronic ester exchange reactions combine chemical reactivity, structural adaptability, and environmental sensitivity, providing a flexible and controllable self-healing method ideal for creating next-generation smart and functional materials. The transmittance of three materials was tested, as shown in Figure 10.7. The three resulting polymers showed a visible light transmittance of over 80%, indicating good transmittance of visible light.

10.3.2 Supramolecular Interactions-Based Self-Healing

A type of material known as supramolecular self-healing utilizes reversible non-covalent interactions between molecular units to effect healing. Supramolecular interactions are generally weaker and more dynamic than covalent bonds, allowing materials to self-heal after physical damage without the need for chemical additives or external catalysts. The movement and reassembly of polymer chains after fracture largely depend on these interactions, which include ionic interactions, host–guest inclusion complexes, π–π stacking, hydrogen bonding, and metal–ligand coordination. Hydrogen bonding is one of the most common processes in supramolecular self-healing. Reversible association and dissociation between polymer chains is possible due to the directional nature of hydrogen bonds and their moderate strength. Mechanical integrity can be gradually restored after damage because broken hydrogen bonds near the fracture surface can reform through molecular rearrangement. At room temperature, materials with multiple hydrogen bonding sites exhibit excellent healing efficiency, making them suitable for use in wearables, coatings, and soft electronics [25–28].

Another effective mechanism in supramolecular self-healing systems is the metal–ligand coordination mechanism. When certain ligands are introduced, transition metal ions such as Zn^{2+}, Fe^{3+}, and Cu^{2+} can form dynamic coordination complexes. Healing at desired temperatures is possible through these interactions, which are not only reversible but also adjustable in strength depending on the metal and ligand composition. In addition, metal–ligand systems often include stimuli-responsiveness, including conductivity or color change, adding further versatility to the material. To compare the experimental mechanical properties with the computational analysis, the self-healing of the shape-memory polymer was modeled using constitutive equations. The authors found that the models and experiments were in good agreement [28]. Regarding the self-healing of the polymer, the key variables shown in Figure 10.8 determine the rate of self-healing

Other non-covalent healing methods include π–π stacking and host–guest interactions (e.g., using cucurbits or cyclodextrins). Under normal conditions,

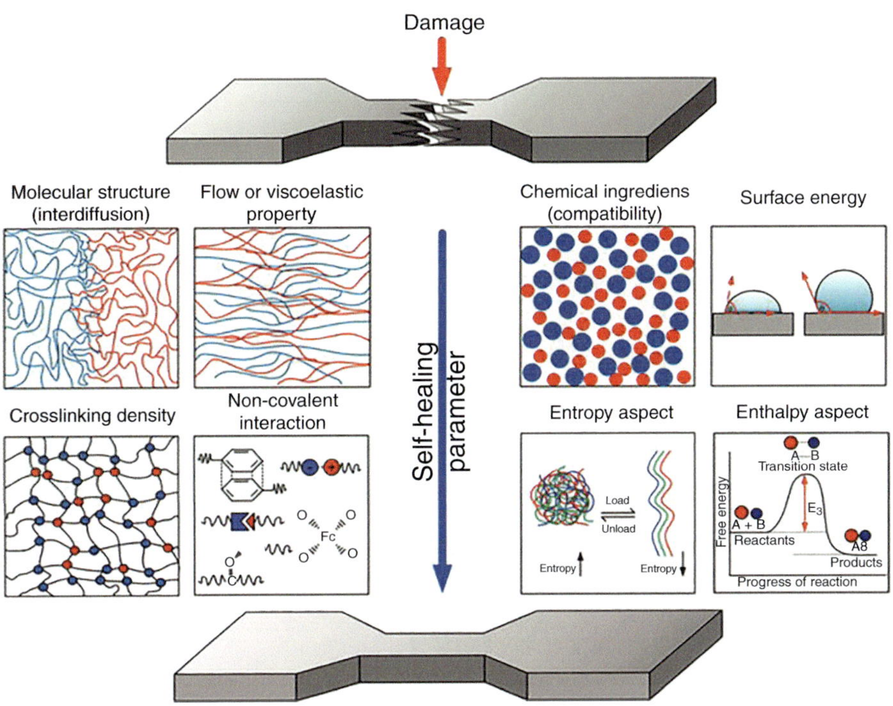

FIGURE 10.8 The self-healing polymer's model of key parameters [23] [Open access].

these interactions can facilitate multiple healing cycles and are reversible and non-destructive. However, the challenge is to achieve a balance between the strength of the supramolecular bond and the flexibility of the polymer chains, which is necessary for effective healing without compromising mechanical properties.

Overall, supramolecular self-healing materials offer reversible, lightweight, and energy-efficient alternatives to covalently bonded systems. They are ideal for sustainable design in advanced technologies, including flexible electronics, sensors, biomedical devices, and smart coatings, due to their reversible nature, which allows for the production of adaptable and recyclable materials. According to the data (Figure 10.9), multiple hydrogen bonds are formed between the carbonyl and imino groups in addition to the resulting hydroxyl groups. When comparing the mechanical parameters of the restored sample with the original sample, the tensile strength and elongation at break after 24 hours can be achieved at a level of about 90%. The mechanical properties of the original highly branched silicone rubber sample are comparable to or superior to those of traditional vulcanized silicone rubber.

Using a solution mixing method, carboxymethyl chitosan (CMCS) was successfully added to oxidized natural rubber (ENR) to create a non-covalent supramolecular network. This strategy aimed to improve the mechanical performance and self-healing efficiency of the resulting composite by leveraging its hydrogen bonding capabilities. CMCS is an excellent choice for creating multiple hydrogen bonds with the epoxy groups present in ENR polymer chains due to its high concentration of hydrophilic functional groups, such as hydroxyl (–OH) and amine ($–NH_2$) groups.

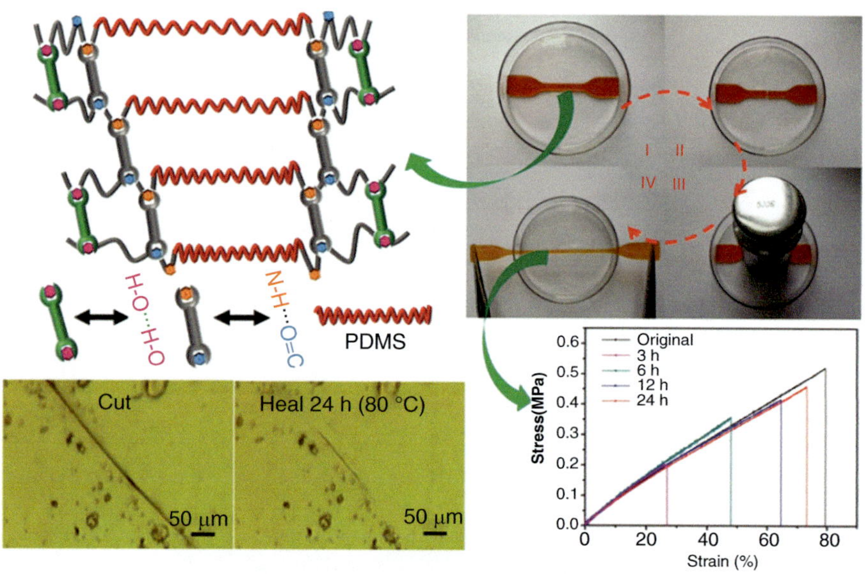

FIGURE 10.9 The self-healing properties of silicone rubber (PDMS) are due to the network of hydrogen bonds [23] [Open access].

The interactions between CMCS and ENR during mixing resulted in the creation of extensive hydrogen bond networks that function as multi-linked supramolecular structures at the molecular level. As dynamic physical crosslinks, these non-covalent bonds allow for reversible bond formation and dissociation in response to external stimuli or mechanical damage. This network structure facilitates self-healing by supporting stress displacement and enabling microscopic reorganization of the material, as illustrated in Figure 10.10.

Furthermore, the addition of CMCS to the ENR matrix was found to significantly improve the mechanical properties of the composite, such as its elasticity and tensile strength. In addition to strengthening the polymer structure, the extensive hydrogen bonding network increases the structure's resistance to crack propagation.

Nevertheless, the incorporation of CMCS into ENR offers a practical approach to producing smart self-healing elastomers with enhanced elasticity, flexibility, and

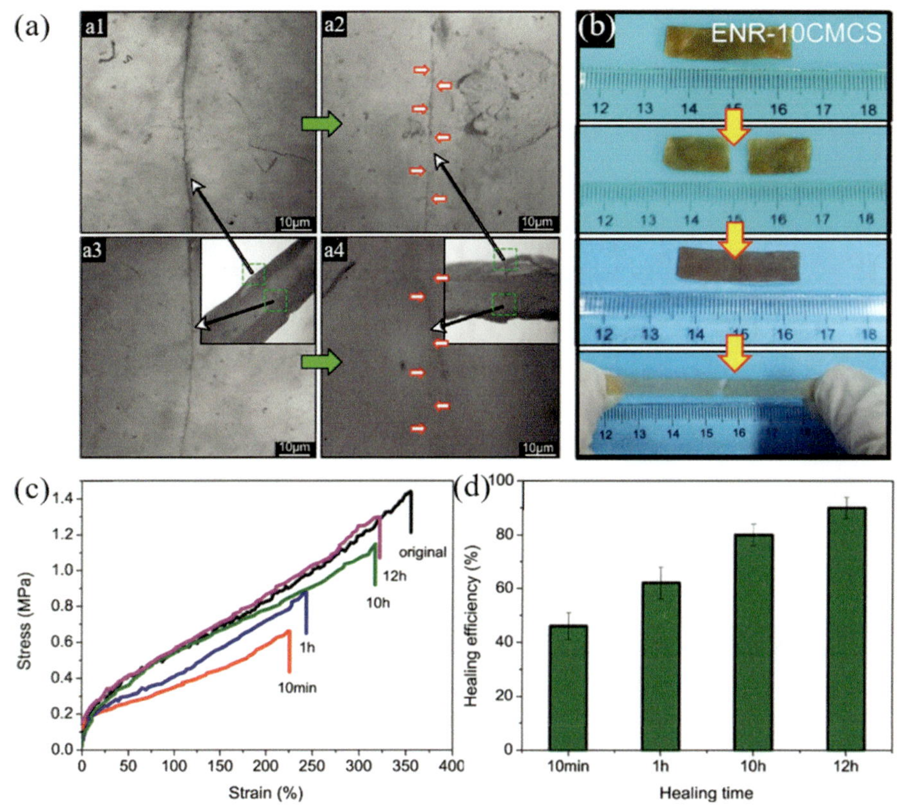

FIGURE 10.10 Self-curing microscope stress–strain curves and images (a, b) under different curing times (c) and curing efficiency (d) of natural rubber oxidized with CMCS composites ((A1) notch line on the surface before self-curing; (A2) notch line on the surface after self-curing; (A3) inner part of notch position before self-curing; (A4) inner part of notch position after self-curing for 12 h at room temperature) [23] [Open access].

versatility. These materials may find applications in biomedical devices, flexible electronics, environmentally friendly packaging, and automotive components, all of which rely on durability and long service life.

The self-healing mechanisms are shown in Figure 10.11. While Al^{3+} was distributed close to the fracture surface to promote the mobility of Alg-DA chains, Al^{3+} at the fracture interface continued to be reactivated to Alginate-dopamine or Alg-DA chains through the coordination contact. Consequently, the coordination and electrostatic interactions were rearranged [23].

The results presented in Figure 10.12 demonstrate that during the 12-hour healing process at 50 °C, the material's mechanical properties are completely restored to their original state. The molecular structure of the material contains hydrophobic ionic–dipole interactions, which are largely responsible for this remarkable recovery.

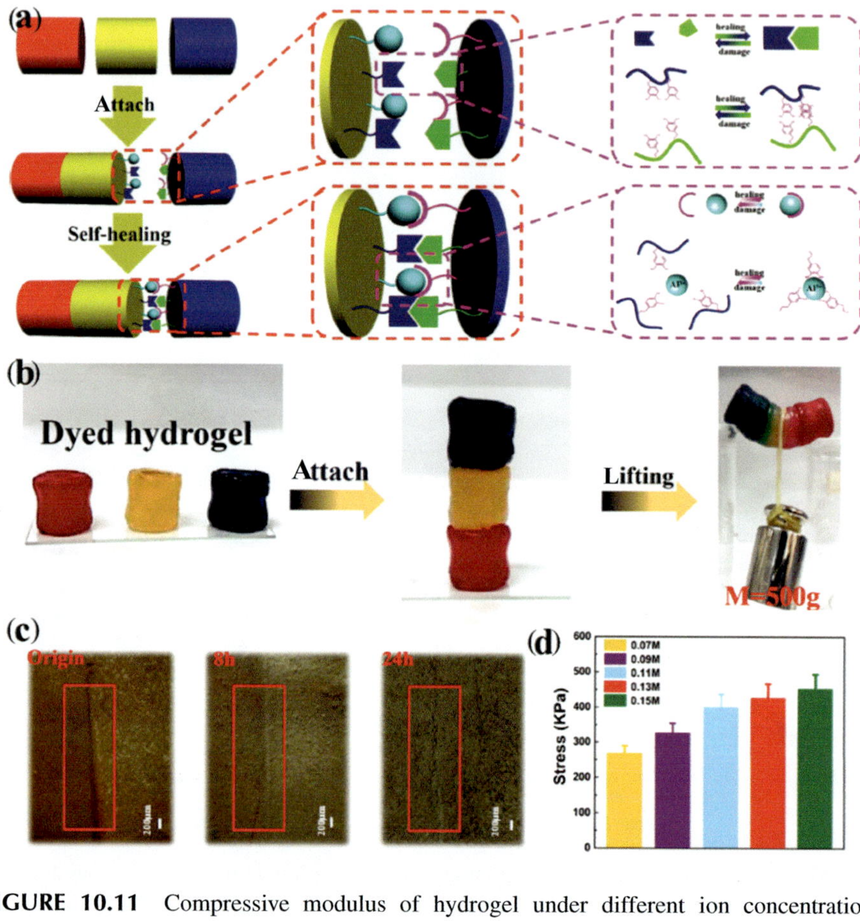

FIGURE 10.11 Compressive modulus of hydrogel under different ion concentrations, (a) mechanism of self-healing process, (b) healing process and its weightlifting experiment, (c) ultra-deep microscopic images of damaged and repaired samples, and (d) self-healing in mussel-inspired hydrogel system [23] [Open access].

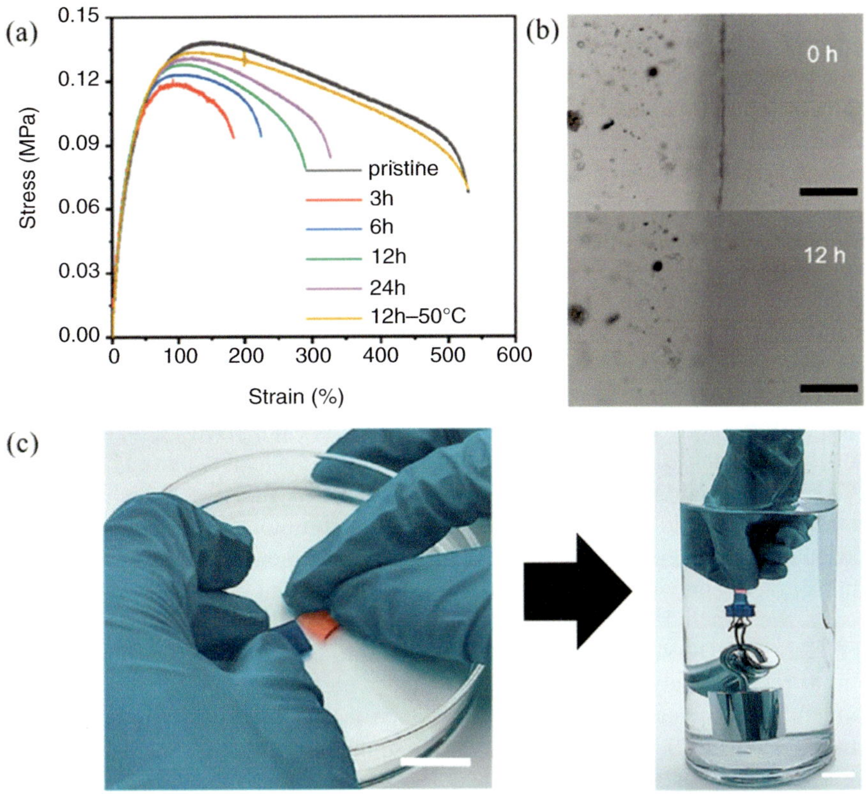

FIGURE 10.12 The stress–strain curves for the original and healed specimens with different healing times, optical images of the healed specimens after a 12-hour healing period at 25 °C under water, and the experimental part of the healed specimen using underwater weightlifting represent the experimental part of the self-healing process of fluorinated elastomer [23] [Open access].

These interactions are essential for promoting the reorganization and reconstruction of broken bonds at the damaged surface. Because of the improved molecular mobility and alignment resulting from these interactions, the material can self-heal its structural defects and return to its original mechanical integrity.

Furthermore, visual inspection of the damaged surface shows that the crack completely disappears after the 12-hour healing period, indicating that the self-healing process provides complete morphological recovery and restores the material's strength. The system does not require chemical additives or external adhesives to maintain surface regeneration and structural cohesion.

The experimental results also indicate that temperature and healing time significantly influence the efficiency of self-healing. High temperatures enhance molecular mobility, facilitating more efficient reconnection and accelerating the healing of the damaged area. On the other hand, if the temperature is too low or the healing period is too short, the molecular components may not have enough energy or mobility to reestablish the required contact points, which can lead to partial healing [23].

To achieve optimal healing efficiency in such systems, it is important to optimize healing conditions, particularly temperature and time. These findings highlight the need for precise tuning of kinetic and thermodynamic parameters in self-healing materials based on ion–dipole interactions, particularly for applications that require reliable mechanical stability under operating pressure, high toughness, and repeatability of properties [29,30].

The experimental results also indicate that temperature and healing time have a significant impact on the efficiency of self-healing. High temperatures enhance molecular mobility, facilitating more efficient reconnection and accelerating the healing of the damaged area. However, if the temperature is too low or the healing period is too short, the molecular components may not have enough energy or mobility to reestablish the necessary contacts, which can lead to partial healing [31].

The very low protein concentrations used in AFM imaging, which are fundamentally different from the crowding conditions common in real cells, may have contributed to the detection of a limited proportion of monomers in natural environments. Based on previously published data on the biological functional profile of these proteins, we concluded (Figure 10.13) that the main features observed in GvFFTR, CaFFTR2, GvDDOR, and GvFFTR_Δtail in natural environments are homodimers.

In a recent analysis, non-covalent gels were divided into supermolecular polymers, low-molecular-weight gels, and colloidal and macromolecular solid-particle systems [32]. This research highlighted the molecular design techniques and

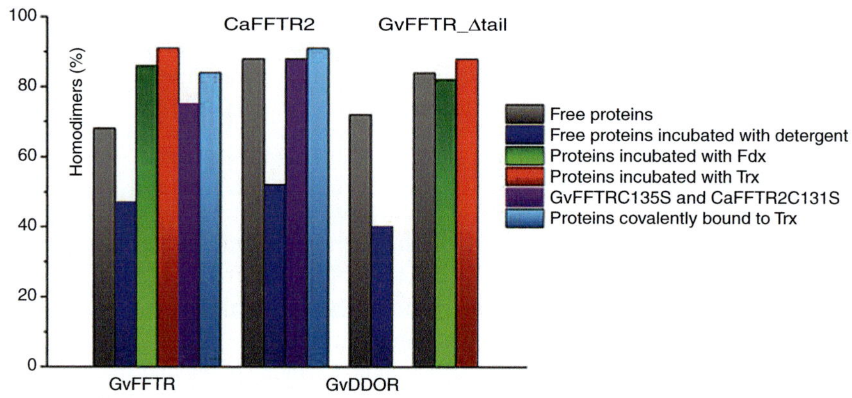

FIGURE 10.13 The bar chart shows the percentages of homodimers of GvFFTR, CaFFTR2, GvDDOR, and GvFFTR_Δtail proteins, as well as their mixtures and covalent complexes with protein variations, as detected by atomic force microscopy (AFM). The right panel shows the color codes for each selected condition. PBS pH 7.0 was used to measure the samples. The total number of protomers identified as flavoprotein homodimers (n = 100) is expressed as a percentage. In the presence of detergent, this percentage is 0.1% SDS/0.2% Tween 20 [7] [Open access].

requirements for 3D-printable hydrogel inks. Because of their improved biocompatibility and ability to enhance cell survival and differentiation in culture, natural and slightly modified biopolymers have thus far shown great potential for biofabrication. A team of researchers successfully printed mesenchymal stem cells (MSCs) and 3T3 fibroblasts, loaded within a biocompatible network of biopolymers including adamantane, β-CD, and hyaluronic acid (HA), into their own cavities using self-healing gels based on host–guest complexes to create engineered multicellular structures (Figure 10.14).

Tensile strength tests were performed on several hydrogel samples before and after the reconstitution of the fractured surfaces of the torn gel to evaluate its self-healing property. The results showed that water-soluble surfactants [33], and possibly a charged co-monomer to trap oppositely charged mesylates, such as AAc to trap mixed mesylates of C18 and CTAB (Figure 10.15), are essential for the self-healing ability of the modified hydrophobic PAAm gel.

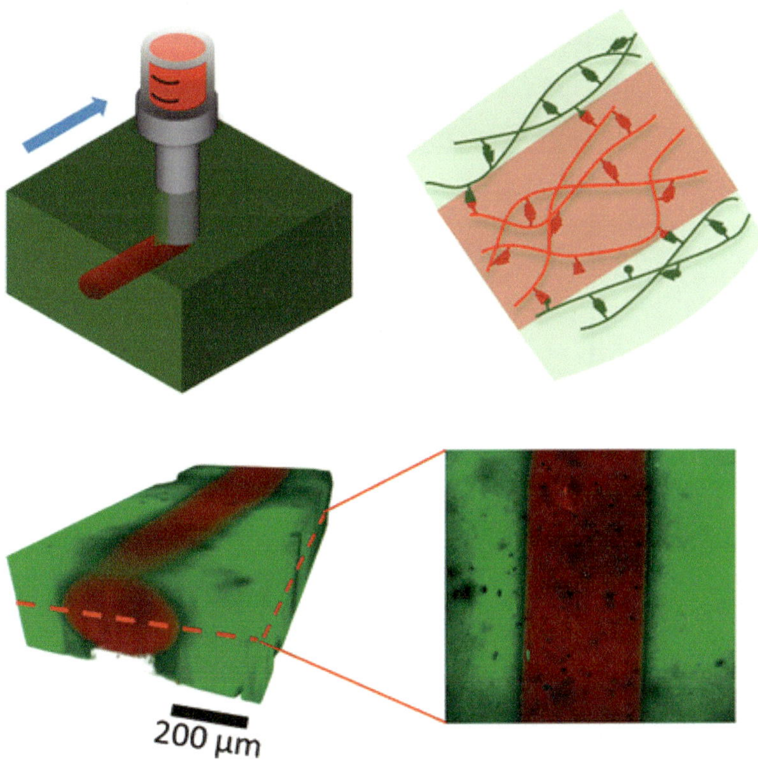

FIGURE 10.14 A graphical representation of the 3D printing of a beta-cyclodextrin (β-CD, host) supermolecule coupled to HA and adamantane (Ad, guest). (Below) 3D reconstruction of a confocal Z-stack image of an ink filament (labeled with rhodamine, red) printed on a support gel (labeled with fluorescein, green); green indicates the supermolecule support gel, and red means the printed ink. 200 μm is the scale bar [33] [Open access].

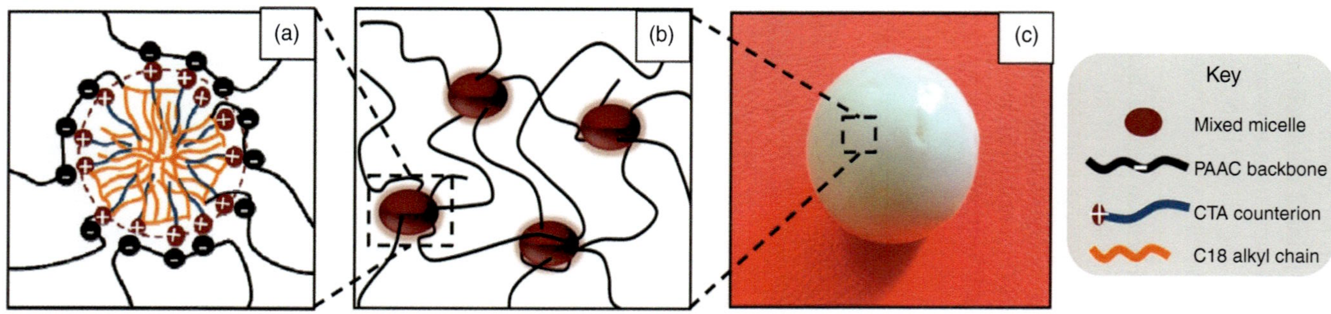

FIGURE 10.15 (a, b) Cartoon illustrating the bonding process in self-healing, hydrophobic poly(acrylic) gels (PAAc); (c) Photograph of a spherical poly(acrylic) gel sample in hydrostatic equilibrium. The mixed meselles consist of stearyl methacrylate (C18) and CTAB in an aqueous NaBr solution. (Reprinted from [80] with permission. American Chemical Society, 2016 [33] [Open access]).

PROBLEMS

1. How do self-healing polymer nanocomposites enhance the durability of electronic equipment?
2. How can nanomaterials improve conductivity and self-healing efficiency?
3. What role do these composites play in wearable or flexible electronics?
4. How do self-healing polymers enhance the functionality of energy storage devices such as supercapacitors and batteries?
5. What are the barriers to the widespread use of self-healing nanocomposites in the energy industry?

REFERENCES

1. Main, K., Provan, J. I., Haynes, P. J., Wells, G., Hartley, J. A., & Pyne, A. Atomic Force Microscopy-A Tool for Structural and Translational DNA Research. APL Bioeng. 2021, 5, 031504.
2. Choi, K., Noh, A., Kim, J., Hong, P. H., Ko, M. J., & Hong, S. W. (2023). Properties and applications of self-healing polymeric materials: A review. *Polymers*, *15*(22), 4408.
3. Sekine, N., & Nakao, W. (2023). Advanced self-healing ceramics with controlled degradation and repair by chemical reaction. *Materials*, *16*(19), 6368.
4. Ekinovic, S., & Ekinovic, E. (2003). Some advantages of high speed machining in aspect of machined surface quality. In *3rd Research/Expert Conference with International Participations "Quality"* (pp. 453–460). B&H:Zenica.
5. Hon, K., & Baharudin, B. H. T. (2006). The impact of high speed machining on computing and automation. *Int. J. Autom. Comput.*, 3, 63–68. https://doi.org/10.1007/s11633-006-0063-3
6. Gari, R. R. S., Frey, N. C., Mao, C., Randall, L. L., & King, G. M. (2013). Dynamic structure of the translocon SecYEG in membrane: Direct single molecule observations. *J. Biol. Chem.*, 288, 16848–16854.
7. Marcuello, C., Frempong, G. A., Balsera, M., Medina, M., & Lostao, A. (2021). Atomic force microscopy to elicit conformational transitions of ferredoxin-dependent flavin thioredoxin reductases. *Antioxidants*, 10(9), 1437.
8. Lira-Navarrete, E., Pallares, M. C., Castello, F., Ruedas-Rama, M. J., Orte, A., Lostao, A., & Hurtado-Guerrero, R. (2021). Protein O-Fucosyltransferase 1 undergoes interdomain flexibility in solution. *Molecules*, 26, 2105.
9. Osada, T., Kamoda, K., Mitome, M., Hara, T., Abe, T., Tamagawa, Y., Nakao, W., & Ohmura, T. (2017). A novel design approach for self-crack-healing structural ceramics with 3D networks of healing activator. *Sci. Rep.*, 7, 17853.
10. Hu, D., Fu, Q., Zhou, L., Liu, B., Cheng, C., Li, X., & Sun, J. Self-healing improvement strategy of thermally sprayed MoSi2 coating at 1773 K: From calculation to experiment. *Corros. Sci.*, 189, 109599.
11. Monteverde, F., Saraga, F., Reimer, T., & Sciti, D. (2021). Thermally stimulated self-healing capabilities of Zrb2-Sic ceramics. *J. Eur. Ceram. Soc.*, 41, 7423–7433.
12. Smeacetto, F., Fabiana, D., Valentina, C., Peter, T., & Milena, S. (2021). Ytterbium disilicate-based glass-ceramic as joining material for ceramic matrix composites. *J. Eur. Ceram. Soc.*, 41, 1099–1106.
13. Song, Q., & Zhang, Z. (2022). Microstructure and self-healing mechanism of B4c–Tib2–Sic composite ceramic after pre-oxidation behaviour. *Ceram. Int.*, 48, 25458–25464.

14. Yuan, M., Liu, L., Wang, J., Hu, Q., Zhang, H., Zhang, S., & Zhou, X. (2022). Crack-healing behaviour of MoSi2 dispersed Yb2Si2O7 environmental barrier coatings. *Ceram. Int.*, 48, 29919–29928.
15. Cai, H., Wei, Y., Wang, B., & Jiang, K. (2023). Oxidation induced crack-healing behavior of Sic-Al2O3-Tib2 composites at 600–800 °C. *Int. J. Appl. Ceram. Technol.*, 20, 3635–3646.
16. Cui, X., Huang, C., Shi, Z., Liu, H., Li, S., Han, Y., Ji, L., Wang, Z., Xu, L., & Huang, S. (2023). The crack-healing and strength recovery mechanisms for the new developed Ti (C, N)-TiSi2-Wc composite ceramic tool materials with crack-healing ability. *Ceram. Int.*, 49, 35382–35391.
17. Liu, Y., Liu, H., Huang, C., Ji, L., Wang, L., Yuan, Y.;,Liu, Q., & Han, Q. (2023). Study on crack healing performance of Al2O3/Sicw/TiSi2 new ceramic tool material. Ceram. Int., 49, 13790–13798.
18. Maruoka, D., Itaya, T., Misaki, T., & Nanko, M. (2012). Recovery of mechanical property on nano-Co particles dispersed Al2O3 via high-temperature oxidation. *Mater. Trans.*, 53, 1816–1821.
19. Maruoka, D., MurakaMi, T., & Kasai, E. (2020). Influence of heat treatment temperature on self-healing effect of fe particle/Mullite ceramic composites. *Tetsu-Hagane/J. Iron Steel Inst. Jpn.*, 106, 844–850.
20. Jiang, X., Zeng, F., Yang, X., Jian, C., Zhang, L., Yu, A., & Lu, A. (2022). Injectable self-healing cellulose hydrogel based on host-guest interactions and acylhydrazone bonds for sustained cancer therapy. *Acta Biomater.*, 141, 102–113.
21. Jin, Q., Du, R., Tang, H., Zhao, Y., Peng, W., Li, Y., ... & Jia, X. (2023). Quadruple H-Bonding and polyrotaxanes dual cross-linking supramolecular elastomer for high toughness and self-healing conductors. *Angew. Chem. Int. Ed.*, 62(26), e202305282. .
22. Zou, X., Wang, X., Gou, M., Yue, O., Bai, Z., Zhang, H., & Liu, X. (2022). Ultra-strong adhesive, self-healing and electroactive bio-based hydrogels for the on-demand fabrication of sandwich-inspired smart electronic sensing floors. *J. Mater. Chem. A*, 10, 14555–14567.
23. Buaksuntear, K., Limarun, P., Suethao, S., & Smitthipong, W. (2022). Non-covalent interaction on the self-healing of mechanical properties in supramolecular polymers. *Int. J. Mol. Sci.*, *23*(13), 6902.
24. Mei, J. F., Jia, X. Y., Lai, J. C., Sun, Y., Li, C. H., Wu, J. H., Cao, Y., You, X. Z., & Bao, Z. (2016). A highly stretchable and autonomous self-healing polymer based on combination of pt··· pt and π–π interactions. *Macromol. Rapid Commun.*, 37, 1667–1675.
25. Burattini, S., Greenland, B. W., Merino, D. H., Weng, W., Seppala, J., Colquhoun, H. M., Hayes, W., Mackay, M. E., Hamley, I. W., & Rowan, S. J. (2013). A healable supramolecular polymer blend based on aromatic π– π stacking and hydrogen-bonding interactions. *J. Am. Chem. Soc.*, 132, 12051–12058.
26. Dai, J., Wang, Z., Wu, Z., Fang, Z., Heliu, S., Yang, W. T., Bai, Y., & Zhang, X. (2023). Shape memory polymer constructed by π–π stacking with ultrafast photoresponse and self-healing performance. *ACS Appl. Polym. Mater.*, 5, 2575–2582.
27. Li, Y., Yang, D., Wu, Z., Gao, F.-L., Gao, X.-Z., Zhao, H.-Y., Li, X., Yu, Z.-Z. (2023). Self-adhesive, self-healing, biocompatible and conductive polyacrylamide nanocomposite hydrogels for reliable strain and pressure sensors. *Nano Energy*, 109, 108324.
28. Zhang, L., Wang, D., Xu, L., Zhang, X., Zhang, A., & Xu, Y. (2020). A highly stretchable, transparent, notch-insensitive self-healing elastomer for coating. *J. Mater. Chem. C*, 8, 2043–2053.

29. Son, D. H., Bae, H. E., Bae, M. J., Lee, S.-H., Cheong, I. W., Park, Y. I., Jeong, J.-E., & Kim, J. C. (2022). Fast, localized, and low-energy consumption self-healing of automotive clearcoats using a photothermal effect triggered by NIR radiation. *ACS Appl. Polym. Mater.*, 4, 3802–3810.
30. Zhu, Y., Cao, K., Chen, M., & Wu, L. (2019). Synthesis of UV-responsive self-healing microcapsules and their potential application in aerospace coatings. *ACS Appl. Mater. Interfaces*, 11, 33314–33322.
31. Yuan, Z., Riaz, A., & Chohan, B. S. (2023). Precision machining by dicing blades: a systematic review. *Machines*, *11*(2), 259.
32. Wang, L., Wang, X., Feng, S., & Li, L. (2024). Dynamic boronic ester cross-linked polymers with tunable properties via side-group engineering. *Polymers*, *16*(24), 3567.
33. Strandman, S., & Zhu, X. X. (2016). Self-healing supramolecular hydrogels based on reversible physical interactions. *Gels*, *2*(2), 16.

Index

M

N

O

P

For Product Safety Concerns and Information please contact our EU representative GPSR@taylorandfrancis.com Taylor & Francis Verlag GmbH, Kaufingerstraße 24, 80331 München, Germany